Building Information Modelling, Building Performance, Design and Smart Construction

Mohammad Dastbaz • Chris Gorse
Alice Moncaster
Editors

Building Information Modelling, Building Performance, Design and Smart Construction

Springer

Editors
Mohammad Dastbaz
Deputy Vice Chancellor
University of Suffolk
UK

Chris Gorse
Leeds Sustainability Institute
Leeds Beckett University
Leeds, UK

Alice Moncaster
Fellow, Newnham College
University of Cambridge
Cambridge, UK

ISBN 978-3-319-50345-5 ISBN 978-3-319-50346-2 (eBook)
DOI 10.1007/978-3-319-50346-2

Library of Congress Control Number: 2017931959

Printed on acid-free paper

This Springer imprint is published by Springer Nature
The registered company is Springer International Publishing AG
The registered company address is: Gewerbestrasse 11, 6330 Cham, Switzerland

Preface

Introduction

In a resolution adopted by the UN General Assembly, 25th September 2015, titled "Transforming our World: The 2030 Agenda for Sustainable Development," the UN identifies significant challenges to our future sustainable development over the next 15 years that includes extreme poverty as one of the greatest nemeses humanity faces in the twenty-first century.

Seventeen Sustainable Development Goals (SDGs) and 169 targets are identified by the UN General Assembly, which indicate the scale of the task that we face.

One of the key debates around sustainable development, in recent decades, has been around the impact of technology and whether technology is "a solution or a problem." United Nations, 2016, Global Sustainable Development Report 2016 suggests that: "*Technology has greatly shaped society, economy and environment. Indeed, technology is a double edged tool, while technology progress has been a solution to many ills and problems, it has also added ever new challenges.*"

Clearly the emergence of the technology has had an immense positive and negative impact on our environment. The carbon footprint of our technological usage and requirements (running over two billion smart devices and over two billion computers, laptops, tablets, etc.) as well as the energy required to keep our "connected world" running 24/7 365 days a year is enormous and while it will be difficult to measure all this while it is rapidly expanding, some research have indicated that these requirements are fast getting out of control.

In an interesting report by Mark Mills (CEO of Digital Power and sponsored by the National Mining Association American Coalition for Clean Coal Electricity) produced in August 2013, Mills states that: "*The information economy is a blue whale economy with its energy uses mostly out of sight. Based on a mid-range estimate, the world's Information Communications Technologies (ICT) ecosystem uses about 1,500 TWh of electricity annually, equal to all the electric generation of Japan and Germany combined as much electricity as was used for global illumination in 1985. The ICT ecosystem now approaches 10% of world electricity* *generation. Or in other energy terms the zettabyte [1000^7] era already uses about 50% more energy than global aviation.*"

The impact of technology on built environment has also been significant. From one side we can see that technological development and research in the area of built environment has been used as enablers providing the bases for new and more environmentally friendly design, smart materials and smart construction techniques, and smarter way of generating and using energy.

Our Focus

The main focus for this book, in its broadest remit, is the "Built Environment and Environmental Sustainability" with particular attention to Building Information Modelling (BIM), building performance and sustainable design, and smart construction.

One of the challenges identified in the literature dealing with "sustainable design and built environment" is the different viewpoints and approaches between industry, business and environmental campaigners, and researchers and academia and how to bridge the gap between the differences and more importantly how to tackle the issues facing our environment.

This edited volume is divided into four parts and includes interesting collaborative research work between the Industry and Academia challenging some of the current perspectives and norms and offering interesting perspectives.

Part I of this volume is dedicated to presenting some of the key conceptual discussions around what sustainability agenda is all about.

Peter Young and Patricia A. Aloise-Young in their chapter "*The Problem Is Also the Solution: The Sustainability Paradox*" point out that sustainability shares the same word root as sustenance. It isn't a coincidence. Food, water, air, and energy—sustainability is at the very heart of our long-term survival. Furthermore, they argue that people and technology are at the centre of our climate crisis. Technological advances, particularly since the industrial revolution, have contributed to the accumulation of GHG. On the other side of the coin, technological advances such as renewable energy hold promise for ameliorating our environmental woes.

Barbara Colledge in her chapter "*Appreciating the Wicked Problem: A Systems Approach to Sustainable Cities*" argues that sustainable city place making is a complex process and can deliver systemic unintended or undesirable development paths such as poverty, health inequality, or environmental degradation over generations. The chapter goes on to suggest a new conceptual model and alternative reference frames to understand and influence transformative action necessary to realise sustainable cities.

Francesco Pomponi and Alice Moncaster in their chapter "*A Theoretical Framework for Circular Economy Research in the Built Environment*" discuss the new and emerging research area of *Circular Economy* and state that the founding principles of circular economies lie in a different perspective on, and management of, resources under the idea that an ever-growing economic development and profitability can happen without an ever-growing pressure on the environment. They go on to propose a framework to formulate building research from within a circular economy perspective.

Part II is dedicated to BIM and some key research questions associated with BIM. Farzad Khosrowshahi, in his chapter "*Building Information Modelling (BIM): A Paradigm Shift in Construction*," argues that BIM has been hailed as a catalyst for a fundamental change in the way the industry conducts its business in a data-intensive and complex environment that significantly relies on effective collaboration of a diverse range of disciplines. He then goes to point out that there are numerous ways by which BIM can contribute to the sustainability agenda. Energy modelling, building orientation (saving energy) lifecycle evaluation, building massing (optimise the building envelope), daylighting analysis, water harvesting, and sustainable materials (to reduce material needs and to use recycled materials) are only a few examples where all three sustainability parameters come together. In the second chapter of this part titled "*Using Agile Project Management and BIM for Improved Building Performance*" by Mohammad Sakikhales and Spyros Stravoravdis, the authors argue that the early design stage is the most crucial stage to achieve sustainability targets because this is when major design decisions that affect sustainability performance are taken. They further emphasise that their work will be discussing the advantages of agile project management through an extended literature review and analyse the potential benefits from the adoption of this methodology in the construction industry and sustainable design process. It introduces an iterative design framework for the design phase of construction projects, using agile principles. The chapter further explores how BIM can facilitate the implementation of this framework to achieve improved building performance.

Muhammad Khalid, Muhammad Bashir, and Darryl Newport in their chapter "*Development of a Building Information Modelling (BIM) Based Real-Time Data Integration System Using a Building Management System (BMS)*" point out that the aim of BIM is to provide a complete solution for the life cycle of the built environment from the design stage to construction and then operation. Their interesting research work investigates the integration of real-time data from the BMS system into a BIM model, which would potentially aid facility managers to interact with the real world environment inside the BIM model.

Part III is dedicated to building performance and design. The first chapter in the part is an interesting collaborative work between Academia and Saint-Gobain Recherche. Johann Meulemans, Florent Alzetto, David Farmer, and Christopher Gorse in their chapter titled "*Qub/E—A Novel Transient Experimental Method for In Situ Measurements of the Thermal Performance of Building Fabrics*" present a novel transient experimental method developed in order to perform in situ measurements of the thermal performance of building fabrics: the QUB/e method.

Al kanani, Dawood, and Vukovic, in their chapter titled "*Energy Efficiency in Residential Buildings in the Kingdom Of Saudi Arabia*," present an interesting case study related to challenges in providing energy efficient buildings in Saudi Arabia. They emphasise that due to a rapidly escalating population and a high level of economic growth, the Kingdom of Saudi Arabia is experiencing a vigorous infrastructure expansion, especially with respect to residential buildings. As a result, energy demand for residential buildings is of a very high level whereby approximately 70% of electricity is consumed by air conditioning systems alone for interior cooling throughout the year due to the hot and humid Saudi climate. They go on to suggest that adding a thermal insulation of polyurethane to external walls and adopting an appropriate construction type could reduce energy consumption by over 30%.

Rajat Gupta and Matt Gregg in their chapter "*Local Energy Mapping Using Publicly Available Data for Urban Energy Retrofit*" make an important case for the urgent need to improve the energy performance of the built environment, so as to help alleviate fuel poverty, meet national carbon targets, and improve the local economy. They go on to point out how publicly available datasets on housing and energy can be used to plan mass retrofit and provide targeted low carbon measures across a city, in order to address the challenges of having incomplete data on which homes could benefit from which retrofit measures and the inability to aggregate private sector housing retrofit activities to minimise installation costs.

Part IV: The final part of this edited volume is dedicated to issues around "smart construction." Alison Pooley in her chapter titled "*Things Change: Exploring Transformational Experiences Within the UK Construction Industry*" states that the built environment has a significant impact on energy consumption, resource depletion, and ecological degradation—reducing this impact is imperative. Existing policies and research are dominated by the assumption that increased regulation, and an improvement in professional skills and knowledge, will address these issues. She goes on to explain that her work is looking beyond a technical or regulatory fix, by exploring the potential opportunities for change that lie within the relationships between experience, learning, and the transformation of individual and professional perspectives.

Cormac Flood, Lloyd Scott, and William Gleeson in their chapter titled "*Comparison of Transient Hygrothermal Modelling Against In Situ Measurement for Thermal Transmittance*"—a joint work between academia and a firm of architects—point out that their work provides the context, research process, and analysis of four case studies situated in Dublin, Ireland. The case studies offer an account of the in situ thermal transmittance of exterior walls and link these to hygrothermally simulated comparisons along with more traditional design U-values. They further point out that their work can form the basis for further research on retrofit of the Irish housing stock.

Craig White and Oliver Styles in their chapter titled "*Decarbonising Construction Using Renewable Photosynthetic Materials*" discuss an important issue: the need to reduce CO_2 emissions from the operational energy use in buildings is more pressing as we seek to mitigate the effects of climate change. They go on to point out that the use of bio-based materials in construction might allow us to tackle both operational and embodied CO_2 emissions. According to them the ModCell Straw Technology system achieves this by using the renewable materials timber and straw.

A Final Note

The UK's Sustainable Development Commission (www.sd-commission.org.uk) states that: "Sustainable development is a development that meets the needs of the present, without compromising the ability of future generations to meet their own needs."

Technological advances over the past four decades have brought significant changes to our lives. The technological revolution has opened new possibilities to develop new innovative solutions in health, education, and in planning our future. But the IT revolution has not been without a cost unless we take responsible steps to use our technological advances wisely and for the benefit of the society rather than for the short-term financial gains of the large conglomerates that control and own them.

A sustainable future requires new ways of urban living, new ways of production and consumption. In small, but significant ways, the issues discussed by the authors in this book have in many ways responded to that call and, more importantly, offered both socially informed and technically literate responses to the global and local challenge of working to make the place and spaces we inhabit more sustainable.

Mohammad Dastbaz

Acknowledgements

We would like to thank all the contributing authors for their tireless work and for providing us with their valuable research work which has made this edited volume possible. We would also like to thank the Springer editorial and production team, Amanda Quinn, Brian Halm, and Brinda Megasyamalan for their patience and valuable advice and support.

Special thanks go to Ellen Glover, whose help and support was critical in organising the SEEDS conference and creating the links with all our contributors and Fiona Scarth, Laura Messer who worked tirelessly in organising our schedule, getting all the necessary forms done and sending numerous e-mails and gentle reminders when necessary.

Finally, our thanks go to all our colleagues at Leeds Sustainability Institute, and Department of the Engineering University of Cambridge whose work has made a significant contribution to our sustainable development agenda and has informed some of the ideas and core discussions, which are presented in this edited volume.

Mohammad Dastbaz
Chris Gorse
Alice Moncaster

Contents

About the Editors and Contributors

Editors

Mohammad Dastbaz, B.Sc., P.Dip., Ph.D., F.B.C.S. is a Professor of Informatics and his main research work over the recent years has been focused on the use and impact of emerging technologies in society, particularly learning, training, and the development of "government." Mohammad has led European Union and UK-based funded research projects and has been the Symposium Chair of Multimedia Systems in IEEE's Information Visualisation (IV) conference since 2002. He has over 50 refereed publications, including numerous journal paper articles, conference papers, book chapters, and books on sustainable development, Green ICT, e-learning, eGovernment, and design and development of multimedia systems. Professor Dastbaz is a Fellow of the Royal Society of Arts, Fellow of British Computer Society and UK's Higher Education Academy as well as the professional member of ACM and IEEE's computer society.

Christopher Gorse, B.Sc., M.Sc., Ph.D., M.C.I.O.B. is Director of Leeds Sustainability Institute, Head of the Centre for the Built Environment and Professor of Construction and Project Management at Leeds Beckett University. He is an Engineering Professors Council Member, Chartered Builder, Vice Chair for the Association of Researchers in Construction Management, and leads research projects in the energy, sustainability, and building performance. Chris holds research positions within International Energy Agency projects, has reported on government-led research, and provided evidence to all party enquiries held at the Houses of Parliament. He is an established author with publications in management, law, energy, construction, and refurbishment. On site, he operated as an engineer and project manager working on both building and civil engineering projects and continues to undertake legal and technical consultancy within the built environment.

Alice Moncaster, Ph.D., C.Eng., M.I.C.E. is Lecturer at the University of Cambridge Engineering Department, Deputy Director of the master's course on Interdisciplinary Design for the Built Environment (IDBE), and Fellow of Newnham College. Her research into sustainable construction crosses the boundaries between academia and practice and between engineering, architecture, and the social sciences. Interests include embodied carbon and energy of buildings, performance of vernacular building materials and typologies, and transdisciplinary research approaches for the built environment.

Contributors

Salam Al bizri has more than 24 years of experience in higher education as well as the construction industry in the UK. His research interest is around exploring integration of design and construction processes and how, through education, a better understanding of issues can be achieved and best practice delivered. By acquiring and exploiting the knowledge and skills of experts in the field, students can be educated in the cause and effect relationships, which result from decision-making. This is demonstrated as best practice as well as illustrating the weaknesses in the way activities are currently carried out. With input from leading organisations in the field such as Mace, Lend Lease and Rise, I have developed several courses and lectures and sourced and developed site-based case studies, which at the same time have enabled these organisations to question their management approaches. These courses at postgraduate level, as well as CPD modules, were delivered at the University of Reading, School of Construction Management and Engineering. I also developed courses that I delivered at the Mace Business School to potential site teams and their trade contractors.

A. Al kanani is currently a Ph.D. research student at the Technology Futures Institute at Teesside University, United Kingdom. His research interest lies in the areas of BIM technologies, with a focus on the development of processes and policies regarding energy efficiency of construction industry of Saudi Arabia.

Ahmed is an architect by profession. After completing his MSc in Project Management at Teesside University in 2011, he continued his research studies as it led to a sparked interest into the use of innovative new technologies within the construction industry of Saudi Arabia to improve energy efficiency in residential buildings of Saudi Arabia which have a high energy consumption rate.

Ahmed is working under the supervisor of Professor Nashwan Dawood who is a leading specialist in construction. Currently, Ahmed is in the process of writing an academic journal paper in conjunction with his research supervisory team as part of his on-going research.

Mohammad Alhusban is a doctoral student in construction management at the University of Portsmouth School of Civil Engineering and Surveying. Before that he obtained an M.Sc. in quantity surveying from the University of Portsmouth (2013). Before coming to Portsmouth, he received his undergraduate degree in civil engineering from Hashemite University in Jordan (2012). The working title for his Ph.D. thesis is "A model of a procurement approach for effective implementation of Building Information Modeling (BIM) to achieve sustainable high performance buildings over their whole life cycle." His research is focused on investigating whether construction procurement approaches influence the ability to use Building Information Modeling techniques to deliver high performance buildings. Currently, he is a Graduate Teaching Assistant in the department and has spoken at a number of conferences on construction management and sustainability research areas. He can be contacted at: mohammad.alhusban@port.ac.uk.

Patricia A. Aloise-Young, Ph.D. is Associate Professor of Psychology and co-Director of the Center for Energy and Behavior at Colorado State University. She is also a Visiting Professor and Fellow of the Leeds Sustainability Institute at Leeds Beckett University. In both her research and consulting she focuses on applying the principles of behaviour change from social psychology to the domains of health and energy. Her research includes consumers' responses to utility programmes such as home energy reports and smart metres and energy conservation programmes for low-income housing. She has published 30 journal articles and 2 book chapters, and given 37 conference presentations. She has also written a curriculum for communities to use to help homeowners reduce their energy usage.

Florent Alzetto is R&D Project Leader and Thematic Domain Coordinator for Building Thermal Performance in Saint-Gobain Recherche. He got a Ph.D. in quantum physics and is expert in building in situ measurements such as coheating or tracer gas experiments. He is author of three patents in the field of short methodologies for in situ measurements of heat loss coefficient, thermal transmittance and air renewal rate. He is involved in FP7 European Project PERFORMER and IEA Annex 71 related to building energy performance assessment based on in situ measurements. He is author of several publications and conferences in condensed matter, radiative transfers and building physics.

Alolote Amadi is a lecturer at the Rivers State University of Science and Technology, Port Harcourt. With a Bachelor's Degree in Quantity Surveying, Masters/M.Phil. in Environmental Management, and currently concluding a Ph.D. in Construction Management at the University of Salford, UK. The researchers academic interest has been largely defined by cross-disciplinary research via the interface between the financial and technical aspects of construction and the natural environment. Over 15 peer-reviewed publications have thus far been published along this line of research interest. This was further enabled by a background in Meteorology, Environmental Geotechnics, and Hydro-geology courses, which were undertaken as part of the requirements of the M.Phil. degree.

Muhammad Khalid Bashir is the Regional Director of Engineering for Starwood Hotels and Resorts in the Kingdom of Saudi Arabia. With a professional industry experience of 29 years working with the world's leading hotel chains, he has been awarded President's Award on two occasions by Starwood Hotels and Resorts. His area of expertise includes hotel maintenance, operations, renovation, construction, and take over of acquired hotels. He has extensive knowledge of Building Management System (BMS) and computerised preventive maintenance programme. His expertise lies within devising energy conversation plans, implementation of Green House programmes as well as the preventive maintenance programme with the ultimate purpose to ensure proper measures are taken to protect the environment of the hotels, including the physical building structure, assets, and related equipment.

Quintin Bradley is a Senior Lecturer in Planning and Housing at Leeds Beckett University and leads post-graduate study in planning, housing, and regeneration at the School of Built Environment & Engineering. He heads research in the fields of community planning, housing studies, and social policy. His latest book on neighbourhood planning will be published by Policy Press in January next year and a previous book on the UK tenants' movement was published by Routledge in 2015. He has a Ph.D. in Housing Studies from Leeds Metropolitan University and his research has been published extensively in peer-reviewed international journals. He has previous experience as a practitioner in community involvement and has worked for resident-led organisations, as well as local housing authorities and housing associations. He is active in campaigns and social movements and has a background as an investigative journalist.

Barbara Colledge is Dean of Quality at Leeds Beckett University with responsibility for institutional academic standards and quality and leads the University Quality Assurance services. With an early career and commercial background in the construction industry as a Chartered Surveyor on major building and civil engineering projects, Barbara's academic career spans 28 years in a range of senior academic roles and disciplines including Built Environment, Information and Technology, and Business and Law. With a specialism in construction law, partnerships, city-region development, and sustainable communities, Barbara's contribution to professional education was recognised through a Teaching Fellowship award made by the Royal Institution of Chartered Surveyors and she is a Principal Fellow of the Higher Education Academy. Barbara is an experienced Quality Assurance Agency Higher Education Reviewer/Auditor, is experienced in professional body accreditations, and was recognised by the Higher Education Academy for work as a teaching grant reviewer. She has contributed to city-regional development through city-regional groups and through the development of business and educational partnerships. She has served as a Director on the former Board of the Yorkshire Humber Metropolitan Area Network (YHMAN Ltd.), which delivered infrastructure services for the region, and is a Fellow of the Chartered Institute of Directors and The Royal Society for the Encouragement of Arts, Manufactures and Commerce.

Mark Danso-Amoako is a senior lecturer with the School of Civil Engineering and Surveying, University of Portsmouth. Mark's areas of interest are in computer information systems, automation in construction, construction software development for mobile devices, stand-alone desktop computers and also web-based application, Building Information Modelling (BIM), data exchange and interoperability between construction applications, estimating and scheduling and application of Radio Frequency Identification (RFID) in the Architectural, Engineering and Construction industry. Mark has worked for construction and consulting companies in Ghana and in the USA in various areas such as quantity surveying, estimating, CAD, planning and scheduling, project management and construction software development. One of my notable projects in Ghana is the $400 million Takoradi Thermal Power Project (Phase 1) where I worked as a QS. I have developed several IT solutions including custom web-based, stand-alone and mobile construction applications for construction companies in the USA. Prior to joining University of Portsmouth, Mark was an Assistant Professor at Texas A&M University (Department of Construction Science) where I taught both undergraduate and postgraduate courses.

N. Dawood is a specialist in project construction management and the application of IT in the construction process. This has ranged across a number of research topics including BIM technologies and processes, sustainability, Information Technologies and Systems (5D, VR, Integrated databases), the planning and management of off-site production, risk management, intelligent decision support systems, cost forecasting and control and business processes.

Professor Dawood is currently Director of the Centre for Construction Research & Innovation (CCIR) and Professor of Construction Management and IT at the University of Teesside, UK. He is also Director of the Technology Futures Institute, through which the engineering and technology research is structured and supported. This role includes responsibility for developing and promoting research polices throughout the institution.

Susan Dawson, B.Sc., A.C.I.A.T., P.G. H.E.P. is Senior Lecturer at Northumbria University, where she teaches on a range of programmes related to Architecture and Built Environment. Her work emphasises the value in maximising the tripartite relationship between employer, student, and academia. This is reflected in her research interests which span education and training, professional development, and multi-disciplinary collaboration. Susan is an experienced Built Environment practitioner, an associate of the Chartered Institute of Architectural Technologists (ACIAT), and a Fellow of the Higher Education Academy (FHEA).

David Farmer studied construction management at Leeds Metropolitan University, specialising in energy efficient refurbishment; his dissertation was highly commended in CIOB International Research and Innovation Awards. Following graduation David joined the Centre for the Built Environment research group. David's research activity is primarily involved with assessing the energy performance of buildings, combining in situ fabric and services test methodologies with data analysis techniques. His work has included building performance evaluation on new and existing dwellings, as well as in-use monitoring projects.

Eric R.P. Farr is a critic, educator, researcher, architect, and urban designer, with a track record of promoting the role of women and ethnical groups in architecture and design. He is founding principal of NONAMES design research foundation and acts as the coordinator of architecture and urban theory and history at New School of Architecture and Design where he runs comprehensive design studios and a super studio on verticality. His research focuses are intramural built environment, objects, methods and processes, digital design and fabrication, integrated systems design, fuzzy logic, (Mass-) customization and personalization, and facade and envelope systems.

Martin Fletcher is a Research Assistant in the Leeds Sustainability Institute. Prior to joining Leeds Beckett, Martin gained his Master's in Renewable Energy from Newcastle University before working with the Centre for the Green Knowledge Economy at Bournemouth University. His current research focuses on thermal comfort in the built environment, primarily the domestic housing sector.

Cormac Flood is an Architectural Technologist and Ph.D. Researcher with Coady Architects and the School of Surveying and Construction Management at Dublin Institute of Technology (DIT) respectively. Cormac joined the Dublin School of Architecture in 2006 completing a B.Sc.(Hons) in Architectural Technology in June 2011 specialising in external insulation technology. Following a year in practice in London, Cormac was selected for the M.Phil. Research Programme in the Dublin School of Architecture, transferring to Ph.D. level in 2015 with the School of Surveying and Construction Management. Cormac has a strong focus on existing and retrofitted building performance exemplified in his presentations and publications on retrofit-related research topics including existing housing figures, thermal upgrade, thermal transmittance, hygrothermal performance, and in situ building evaluation and analysis.

Abdulhakeem Garba, M.Sc., M.N.I.Q.S. is a senior lecturer with the department of Quality Surveying, Kaduna Polytechnic, Nigeria, and is at the final stage of his Ph.D. study at the Robert Gordon University, Aberdeen. He has significant industrial experience of approximately two decade in the practice of Quantity Surveying and Project Management and is currently a Partner with AGF Partnership. His role in this position includes project conceptualisation and development, bills of quantities development, project cost management, resource management, financial statement, and final account preparation. Abdulhakeem research interests include construction economics, sustainability in construction, and recently sustainable energy provision to rural areas. He has published more than ten papers.

Mark Gaterell is Professor of Sustainable Construction and Associate Dean for Research in the Faculty of Technology at the University of Portsmouth. He has been involved with different aspects of the field of sustainable built environments for over twenty years, working for companies such as Thames Water, Scott Wilson and the Building Research Establishment as well as receiving research degrees from the University of Cambridge and Imperial College. Current research activities consider a broad range of issues relating to sustainable buildings, from the analysis of the relationship between buildings and open spaces at a redevelopment scale and the implications of different urban futures, to the consideration of elements of both new build and the existing building stock at an individual building scale. This work is characterised by multidisciplinary approaches, working together with ecologists, social scientists, urban designers, architects, economists and engineers.

Barry J. Gledson, B.Sc.(Hons)., M.C.I.O.B. is Programme Leader and Senior Lecturer at Northumbria University, where he teaches on a range of Construction Engineering and Project Management courses. Barry engages in peer-review and publication activity for a range of academic journal, such as *Automation in Construction*, *Construction Innovation*, *Construction Management and Economics*, *Engineering Construction and Architectural Management*, and *The Journal of Information Technology in Construction* (ITcon). He is an experienced construction project management practitioner, a Chartered Construction Manager (MCIOB), and a Fellow of the Higher Education Academy (FHEA) whose research interests revolve around innovation in the Architectural Engineering and Construction (AEC) industries and across wider Project Management practice.

William Gleeson joined the practice on graduation from University College Dublin in 2005 and became an associate in 2013. He is a qualified accessibility auditor and leads the firm's drive for inclusive design. William is a skilled designer with excellent project management and delivery skills. His completed work includes large residential, leisure, workspace, and education buildings. William has a strong focus on low energy building solutions and has completed Colaiste Choilm, Tullamore, the first A2 rated post-primary school in Ireland. He is currently completing large residential schemes in both private and public sectors and primary schools in Ballymun and Haddington Road, Dublin.

Matt Gregg is a Research Fellow in Architecture and Climate Change, based in the Low Carbon Building Group of the Oxford Institute for Sustainable Development at the School of Architecture, Oxford Brookes University. Matt specialises in modelling and simulation of low carbon design and retrofit in both current and future climate conditions evaluating the impact of architecture on climate change and vice versa. He also has ample experience in building performance evaluation having studied non-domestic, new low carbon housing and retrofitted homes. Matt has a Master of Science in Sustainable Building: Performance and Design from Oxford Brookes University and has accreditation in Leadership in Energy and Environmental Design.

Rajat Gupta B.Arch., M.Sc., Ph.D., F.R.S.A. is Professor of Sustainable Architecture and Climate Change, Director of the Oxford Institute for Sustainable

Development and Director of the Low Carbon Building Research Group at Oxford Brookes University. His research interests lie in advanced low carbon refurbishment, building performance evaluation, and climate change adaptation of buildings. As Principal Investigator, he has won research grants of over £7 million and has produced over 100 publications.

Anthony Higham M.R.I.C.S., M.C.I.O.B., F.H.E.A. is a Senior Lecturer, Programme Director and Chair of the Science and Technology Ethics Committee. His research focuses on business ethics, value led procurement and social sustainability with a focus on the measurement and evaluation of sustainable benefit delivery at project and organisational levels.

Yahya Ibraheem is a practising architect and a lecturer at the Department of Architectural Engineering, University of Technology, in Baghdad, Iraq, where he acted as a module leader for Building Construction and Freehand Drawing, and also served as a member of architectural design studio team for more than 12 years. His research, in Architectural Technology, focuses on improving building energy performance using integrated solutions. He is currently a Ph.D. student at the School of Environment and Technology, University of Brighton, United Kingdom.

David Johnston B.Eng. (Hons.), M.Sc., Ph.D. is a Professor of Building Performance Evaluation within the Centre of the Built Environment Group, Leeds Sustainability Institute. David has over 20 years experience of applied and theoretical research and consultancy in low carbon housing and is a leading expert in coheating testing and building performance evaluation.

Muhammad Umar Khalid is a doctoral student at the University of East London before completing his M.Sc. in Civil Engineering, achieving a distinction. He has won a number of different academic competitions representing UEL. His areas of interest include sustainability, Building Information Modelling (BIM), and construction automation. He also lecturers on BIM and Autodesk Revit software to final year architectural engineering students at the University of East London. His Ph.D. is an industry sponsored research project to develop a mathematical model of the pan pelletisation process using empirical mathematical modelling techniques such as machine learning and artificial intelligence.

Farzad Khosrowshahi is a Professor of Construction IT and has expertise in financial forecasting, modelling, and management. He is well published and has a broad track record in multidisciplinary research at both national and international levels. Farzad served as the Director of Construct IT from 2003 to 2012. He founded BAF (BIM Academic Forum) and served as its first chair. He is an ambassador and Fellow member of the Chartered Institute of Building. His national and international standing is reinforced by his membership of several scientific and industry committees and as assessor/evaluator of many international funding bodies. His committee memberships include ECCREDI (European Council for Construction Research, Development & Innovation); AEEBC (Association d'Experts Européens du Bâtiment et de la Construction); CIOB Innovation & Research; CIOB Discipline Network; Innovation Focus group; IAI; buildingSMART; and British Standards BS555. Farzad has been involved in several UK and EU research projects including leadership of two TEMPUS projects. He served as the chair of ARCOM (Association of Researchers in Construction management) and founded Information Visualisation Society (IVS), serving as its chair.

Mohammed Kishk is the Group Lead and Postgraduate programme leader of operations and project management within Aberdeen Business School, Robert Gordon University. He has extensive industrial experience as a Civil Engineer, Software Developer, Project Manager, Senior Structural Engineer, Senior Consultant, Manager, and a Director for an international engineering consulting firm. Duties within these roles included, among others, conceptual and detailed design, resource management, quality assurance, cost management, staff line management, staff development, feasibility studies, customer care, planning, and strategic management. Dr. Kishk has published more than 80 academic and professional papers in the areas of whole life costing, asset management, risk management, project management, operations management, sustainability, and structural analysis.

Johann Meulemans is Research Engineer at Saint-Gobain Recherche. He holds master's degrees in both mechanical engineering and scientific computing. He is expert in the field of in situ building energy characterisation (e.g. coheating, ISO 9869-1, infrared thermography, QUB/e) and is involved in international collaborative projects (e.g. IEA EBC Annex 67). Johann's research activities are also related to the characterisation of the thermal and radiative properties of materials. Johann authored several journal articles and conference papers in thermal science, radiative heat transfer and building physics. He also authored one patent related to the measurement of the thermal transmittance of building elements.

David R. Moore, Ph.D., B.Sc., M.C.I.O.B. is a reader in project management at Robert Gordon University, Aberdeen. David has been involved in funded research examining the behaviours of superior performing project managers, which was rated as tending to international significance by the EPSRC. His research activity covers areas ranging from buildability, through sustainable design and the use of solar technologies, to perception and cognition in a construction industry context. David has authored, co-authored, and contributed to seven books and over 80 papers. His published work reflects his research interests, including sustainable development (in terms of encouraging design acceptance of ingenious materials and methods), use of collar technologies for material production and wealth generation, low carbon construction skills generally, factors in the visual perception of information embedded in complex charts, culture and conflict in design and build teams, and project management competences and performance level.

Darryl Newport is Professor of Sustainable Materials Engineering and an Environment and Sustainable technologies specialist. He is Director of the Sustainability Research Institute at the University of East London and has over 20 years Resource Efficiency and Built Environment research experience. He has collaborated on a number of international, national, and regional research programmes. The Sustainability Research Institute contributes and leads on a number of high-quality research projects funded through the TSB/EPSRC, ERDF, ESF, FP7, and private commercial. He has published over 30 peer-reviewed journal papers and commercial reports.

James Parker is a Senior Research Fellow in the Leeds Sustainability Institute at Leeds Beckett University. James specialises in building performance simulation and his work has entailed analysis of domestic and non-domestic buildings, including Passivhaus dwellings, Supermarkets, Educational Facilities, Office buildings, and Airport Terminals. His current work includes domestic projects that focus on deep retrofit, low energy new-build housing, and overheating, plus non-domestic projects investigating process-driven internal heat exchanges and the potential for energy flexibility in buildings whilst maintaining the thermal comfort of occupants.

Poorang Piroozfar is an architect, a principal lecturer, and the Academic Programme Leader for Architectural Technology at the University of Brighton, where he also serves on School Research Strategy Committee and acts as the director of Built Environment Research Group (BERG). The interaction between theory and practice has been in the centre of his attention where he has examined manifestation of green urbanisation, all-inclusivity in urban design, and, most recently, values and frames in design for sustainability as well as the judgment processes in architecture.

Francesco Pomponi, B.Eng., M.Sc., Ph.D., M.I.E.T. is an academic and engineer with over 10 years' experience in the construction industry. He is a researcher at the Centre for Sustainable Development, University of Cambridge. His interdisciplinary research focuses on sustainable design and embodied carbon in buildings, and the measurement, management, and mitigation of the impact of the built environment on the natural environment. Interests include life cycle thinking, global goals for sustainable development, and the integration of physical, social, and management sciences.

Alison Pooley joined Anglia Ruskin University in 2012 and is course leader for B.A. (Hons) Architecture. Prior to joining ARU, Alison was programme leader for the Professional Diploma in Architecture at the Centre for Alternative Technology, Wales, where she taught for nearly 10 years. Before turning to teaching in higher education, Alison worked for several years in architectural practice in London and was a housing officer in East London prior to that. Alison's research examines learning in the built environment professions, with particular emphasis on the transformative nature of learning and the impact on one's own practice, profession, and the wider community.

Mohammad H. Sakikhales is a doctoral researcher in the built environment department at the University of Greenwich. He is interested in Building Information Modelling (BIM) and sustainability. Mohammad received his B.A. in Architecture from Art University of Tehran and his MSc in Project Management from the University of Greenwich. His research interests include digital construction, BIM processes and technologies building performance and sustainable design. More specifically, his work examines how BIM can facilitate building performance at the early stage design. Additionally, he serves as a part-time lecturer and teaches BIM, Design and Construction Management (DCM) professional skills, and management of the built environment.

Paul Sapsford is a Chartered Accountant with more than 30 years industrial experience. Paul's interest in continuous improvements stems from working initially in the defence sector for the aerospace specialist Flight Refuelling Ltd. before becoming Finance Director for the small BAE subsidiary Nanoquest. Paul joined Lynwood Products as Finance Director in 2005 and with more than 10 years' experience in plastic recycling and manufacturing, Paul has recently resumed his business relationship with Howard Waghorn at Hahn Plastics Ltd.

Lloyd Scott is Academic Advisor and Partnership Co-Ordinator in the School of Surveying and Construction Management at Dublin Institute of Technology (DIT). He joined the DIT as a lecturer in Construction Management and Technology in 2000. Apart from his lecturing, supervision, research, and academic administrative duties, Lloyd has completed a Ph.D. in the field of Built Environment Education and has developed a framework for assessment-led learning strategies for Built Environment education. In 2014 he accepted the position of "Professor of Practice" at the Haskell and Irene Lemon Construction Science Division in the College of Architecture at the University of Oklahoma. Along with this he has produced many peer-reviewed conference and journal papers. He serves as the Associate Editor of the *International Journal of Construction Education and Research*. His research interests include modern approaches to thermal performance in domestic construction, development of sustainable energy sources and their practical application in Ireland, and project delivery methods for a sustainable environment in Ireland. He also serves on the editorial board of *Structural Survey*, an academic journal that publishes contemporary and original research in building pathology and building forensics, refurbishment and adaptation. Recently, Lloyd accepted the position of Research Fellow at the Sustainability Institute at Leeds Beckett University.

Spyros Stravoravdis is a senior lecturer of sustainable design and construction at the University of Greenwich. His interests lie in the fields of environmental and sustainable design, building performance modelling, tool development, BIM, daylighting and comfort. He has extensive research and consulting experience having worked in these fields for the past 14 years from small- to large-scale international projects while consulting governments on future developments and their legislation.

Oliver Styles is a Passivhaus consultant and co-founder of Progetic environmental engineers based in Barcelona, Spain. Oliver pursues a fabric-first approach to design using natural materials. He has certified PassivHaus straw bale projects and, working with ModCell, Passivhaus Component certified a whole house straw bale building system.

V. Vukovic [M] is a Deputy Director of Doctoral Training Alliance in Energy established between 12 UK universities and a Senior Research Lecturer in BIM and Energy Reduction in Built Environment at Teesside University, UK. His research interests and 10-year experience include sustainable (energy efficient, socially acceptable, environmentally friendly) building design and operation: Building Information Modelling, cyber physical system modelling and optimisation, predictive controls, indoor environmental quality, energy efficiency implications on occupants' well-being and productivity and energy-aware social networking enabled via smart metering. He had a leading role in more than 10 funded EU and international research projects and consulting experience for UNDP, Portuguese Foundation for Science and Technology, Siemens Building Technologies and EUREKA Secretariat. He is author or co-author of over 20 peer-reviewed publications and recipient of numerous awards (e.g. ASHRAE, A&WMA).

Howard Waghorn is the UK director of Hahn Plastics Ltd., which together with an associated company in Germany, Hahn Kunststoffe, form Europe's largest manufacturer of thick section plastic profiles made from 100% recycled plastics, producing over 30,000 tonnes of finished products every year. Howard has been involved in plastics recycling and the manufacture of products from recycled waste plastic since 2001, when he became Managing Director of Lynwood Products/i-plas Ltd. Howard's commitment to product quality and excellent customer service was first demonstrated in an earlier role when Howard was the winner of the Management Today Best Small Factory award whilst Operations Director at Hamilton Acorn.

Craig White is founding Director of White Design, ModCell, and Coobio Circular Materials, Chair of Wood for Good, Director at TRADA, and senior lecturer in Planning and Architecture, University of the West of England. Craig is a renewable material entrepreneur and combines these roles to help deliver sustainable outcomes in the built environment.

Peter M. Young is a Professor of Electrical & Computer Engineering at Colorado State University, and a Visiting Professor and Fellow of the Leeds Sustainability Institute at Leeds Beckett University. His recent research interests include the development of analysis and design techniques for large-scale uncertain systems and robust learning controllers. He has applied these tools to a number of specific areas, including building energy management systems, disk drive control, human machine interfaces, and smart power distribution grids incorporating renewable energy sources. He is a Senior Member of the IEEE, with 35 journal publications, 7 book chapters, and 58 conference papers. He holds four US patents, with two more currently pending. He has won numerous awards for his achievements in both research and education, including most recently the George T. Abell Outstanding Teaching & Service Faculty Award (2015) from the College of Engineering at Colorado State University.

Part I
Concepts in Sustainability

Chapter 1
The Problem Is Also the Solution: The Sustainability Paradox

Patricia A. Aloise-Young and Peter M. Young

Abstract People and technology are at the centre of our climate crisis. Technological advances have contributed to the accumulation of GHG. On the other side of the coin, technological advances such as renewable energy hold promise for ameliorating our environmental woes. Similarly, rapid population growth has the potential to exacerbate our environmental problems, whilst behaviour change holds the key to conserving our natural resources. This is the sustainability paradox: the problem is also the solution.

Sustainability shares the same word root as sustenance. It isn't a coincidence. Food, water, air, and energy—sustainability is at the very heart of our long-term survival. According to the U.S. Environmental Protection Agency (EPA), current greenhouse gas (GHG) emission levels are already projected to lead to warming temperatures, changes in precipitation, increases in the frequency or intensity of some extreme weather events, and rising sea levels. This climate change could be accompanied by other serious consequences such as species extinction and major health issues (EPA 2016). Energy use is one of the primary sources of these GHG emissions in the U.S.A. (EPA 2016).

People and technology are at the centre of our climate crisis. Technological advances, particularly since the industrial revolution, have contributed to the accumulation of GHG (http://www.acs.org/content/acs/en/climatescience/greenhousegases/industrialrevolution.html). On the other side of the coin, technological advances such as renewable energy hold promise for ameliorating our environmental woes. Similarly, rapid population growth has the potential to exacerbate our environmental problems (http://pai.org/wp-content/uploads/2012/02/PAI-1293-Climate-Change_compressed.pdf), whilst behaviour change holds the key to conserving our natural resources. This is the sustainability paradox: the problem is also the solution. However, the answer lies in viewing technology and people not as independent entities but rather as parts of a sociotechnical system.

P.A. Aloise-Young • P.M. Young (✉)
Colorado State University, Fort Collins, CO, USA
e-mail: patricia.aloise-young@colostate.edu; peter.young@colostate.edu

M. Dastbaz et al. (eds.), *Building Information Modelling, Building Performance, Design and Smart Construction,* DOI 10.1007/978-3-319-50346-2_1

For example, looking at energy use in buildings and focusing in on the U.S.A., it can be seen that electricity demand in the U.S. is expected to grow by 1163 billion kWh from 2009 to 2035 compared to a 260 billion kWh increase in renewable electricity generation (excluding hydropower (DOE 2011)). Thus, slowing the growth in demand is essential to the U.S. meeting its sustainable energy goals. This can be achieved without adversely affecting economic output by increasing efficiency and conservation. Per capita growth in energy consumption has been slowed by the introduction of improved standards for designing efficient buildings (e.g. BREEAM, LEED, and most recently, IgCC). However, "Even as standards for building shells and energy efficiency are being tightened in the commercial sector, the growth rate for commercial energy use … is the fastest rate among the end-use sectors" (DOE 2011). This is likely because high efficiency building standards and control approaches focus heavily on the physical elements of the building, whereas the vast majority of a building's energy usage is related to human occupancy of the building (e.g. temperature control, lighting, and plug loads). In fact, a recent sensitivity analysis concluded that "a significant percentage of building energy use is driven directly by operational and occupant habits", but that these "are currently outside the scope of energy codes, policy initiatives, and general perceptions in the building industry" (Heller et al. 2011). Thus, current approaches fall far short of addressing the fundamental problem at hand, namely that *a building and its occupants form an integrated system of behavioural and physical components which interact to determine energy use*.

Currently, engineers and architects consider occupant behaviour in their designs by incorporating features such as occupancy sensors and motion detectors. However, behavioural research shows that the potential energy savings of a technological advance are often offset by less efficient behavioural choices (Sorrell 2007). Many design solutions fail to realise their energy efficiency potential because models of energy-saving technologies typically do not consider the social, psychological, or organisational patterns that influence energy use. While engineers emphasise technological solutions to energy conservation and sustainability, behavioural scientists have focussed on highlighting pro-conservation norms, engaging consumers, and providing feedback (Ehrhardt-Martinez et al. 2010). There has been almost no true integration of these two philosophies.

It is estimated that energy consumption in America's commercial sector (representing nearly one-fifth of America's energy-related GHG emissions (EIA 2015)) can be reduced by 5–30% by operational changes (Moezzi et al. 2014) in addition to the savings made possible by occupant behaviour (Hong and Lin 2013) and automated energy management systems. Current approaches in such systems have attempted to get such behaviour change via informational feedback. However, one of the most pervasive problems with web portals and other energy feedback delivery systems that have been developed by technology experts to date is that they have failed to leverage the social science of behaviour change (Aloise-Young et al. 2016). As a result, they have often resulted in disappointing energy savings. To date, energy conservation programmes have largely operated independently from building energy management systems (BEMS). However, conceptualising the building as an integrated sociotechnical system recognises that conservation messages change people's behaviour and impact the performance of the building in the process.

It is true that there are some industrial products that have considered occupant behaviour in the design process (e.g. Honeywell Lyric and NEST thermostats), and some researchers have considered building occupants as part of their control and energy efficiency approach (Hassanabadi et al. 2012; Page et al. 2008; Mamidi et al. 2012; Klein et al. 2011). However, these studies have rarely included rigorous behavioural science or taken a systems engineering view of the complete integrated system.

Ultimately, we envisage that the way forward will be via integrated system-level approaches for building energy management that are able to adapt to physical conditions and occupant behaviour to optimise the performance of the *overall* socio-technical building energy system. This observation is paramount—most of us have observed buildings where people use space heaters in summer because the heating–ventilating–air-conditioning (HVAC) system has their space too cold for them. The *overall* system is performing poorly, and indeed it is not useful to command a temperature drop via the building HVAC system at a time and place that has consistently prompted users to plug in additional local heating, *unless* a corresponding informational feedback has been identified that effectively prevents this behaviour. A system-level building energy management solution that accounted for the socio-technical nature of the problem would be aware of this, and always able to take the appropriate action that optimised the occupant/technology interactions to deliver true energy savings, even during circumstances where traditional energy management approaches might fail.

We believe that sociotechnical systems are pervasive in sustainability problems (such as building energy management). In order to truly optimise the energy usage of such systems, it is necessary to devise approaches which also implement, model, measure, and actuate these different components in an integrated fashion. This system-level approach will require multidisciplinary research teams with expertise in engineering (sensing, power systems, and building systems), computational modelling and controls, and occupant behaviour. In order to maximise the energy efficiency potential of such systems, each component of the system must be state of the art, and this can only be achieved when each component of the system is designed, measured, and implemented with a high degree of precision. At the same time component-level optimisation is not enough. The interactions between system components (e.g. people and technology) must also be carefully considered, and hence the need for multidisciplinary research teams and system-level thinking is of paramount importance.

The same issues can be observed in energy-efficient residential buildings. For example, the most sophisticated building fabric fails to meet energy efficiency standards when it is installed incorrectly (Gorse et al. 2015). Fluctuations in funding for government programmes make it next to impossible to maintain a skilled and experienced workforce capable of correctly retrofitting homes with insulation

(Harak 2012). Moreover, without proper engagement strategies technological solutions designed to help occupants save energy fall short as well. For example, a couple who sleep with a window open every night will save energy when a mechanical-ventilation-with-heat-recovery (MVHR) is installed in the home, but only if the system is explained to the couple and they are encouraged to change their behaviour. Similarly, studies in the U.S. have shown that programmable thermostats that automatically reduce heating temperatures at night are capable of achieving substantial energy and financial savings. However, when these thermostats are installed without educating the occupants and taking into account their comfort preferences, the majority of occupants override their settings and fail to derive those benefits (Urban and Gomez 2012).

Smart home design has been criticised on the basis that smart home features have been designed without asking users what they wanted, and that almost no usability research was being conducted (Aloise-Young et al. 2016). However, a recent project in the UK was a departure from that model. Developer Hill built a concept home and allowed a family of five to live in the home free of charge for a full year (http://be-zero.co.uk/). They obtained feedback from the family continuously throughout the year to inform a revision of the design of the home. One example of how the design changed was the drying cupboard. Together with a team member from Leeds Beckett University, the Hill team was able to spec out a dehumidifier for the drying cupboard that enabled the drying cupboard to meet the occupants' needs. In the words of the occupants, 'When we moved into the concept house the drying cupboard wasn't really working that well, but then they put in a dehumidifier and then it worked a lot better'. Without this revision, future occupants would likely have reverted to installing conventional clothes dryers and would not have reaped the benefits of the more efficient design.

At the community level the sociotechnical nature of the system again plays a major role in the degree of success of proposed solutions. For example, in 1999 Nogee reviewed the barriers to the adoption of renewable energy technologies, grouping them into four categories (with example subcategories) as follows (Nogee et al. 1999):

1. Commercialisation barriers not faced by mature technologies
    a. Prospecting for appropriate sites with access to transmission lines
    b. New permitting issues (e.g. ecosystem impacts)
    c. Workforce development
2. Price distortions from existing subsidies and unequal tax burdens between renewables and other energy sources
3. Failure of the market to value the public benefits (a through e) of renewables
    a. Decreased pollution
    b. Reduced GHG emissions
    c. Green jobs
    d. Energy independence
    e. Price stability

4. Market barriers
   a. Inadequate information (consumers and utility decision-makers)
   b. Lack of access to capital/high financing costs
   c. Discriminatory regulatory and market practices
   d. Market deregulation

It is noteworthy that these are barriers to **adoption** after major technological barriers to the creation of the photovoltaic (PV) and wind generation had been overcome. As we see, the majority of these barriers are social, whether the source of the obstacle is a policy-maker, end-user, or financier. Sovacool (2009a, b) identified many of these same barriers in his analysis and interviews with industry leaders. Interestingly, stakeholder interviews identified an additional barrier. Several of his interviewees noted that the renewable energy industry had shown promise in the 1970s and then failed to deliver (in part, due to a changing political landscape). As a consequence, consumers, policy-makers, and, likely, financial institutions were now more sceptical and cautious in throwing their support behind the technology. Similarly, in 2010 a group of panellists well versed in financing solar projects, including William Lee, Vice President of Project Finance & Corporate Development for SunEdison, discussed issues with financing utility-scale solar projects. They agreed that the financing climate was improving but that challenges still existed with inaccuracy in banks' perceptions of project risk (Greentechmedia 2010).

It is also important to remember that sustainability is not restricted to environmental concerns. When communities are designed to maximise social sustainability they incorporate policies that enhance the physical, mental, and social well-being of their residents. Our view of social sustainability is rooted in the rich tradition of Healthy Cities. Healthy Cities initiatives have been underway in Europe and the U.S. for nearly three decades. One of the founders of the Healthy Cities movement, Trevor Hancock in 1993 identified characteristics of social sustainability including (Hancock 1993; Hancock n.d.):

1. Preserves the community's cultural and ecological heritage, thus strengthening citizens' sense of connectedness to their history and environment;
2. Supports education and the arts so that everyone within the community will reach their potential;
3. Promotes a spirit of neighbourhood, with people living together in mutual support of each other and in harmony with each other and the environment;
4. Meets basic needs for food, shelter, work, and income so that both food and housing are affordable;
5. Provides for safe living and working conditions;
6. Is democratic, promoting citizen participation and involvement through transparency and empowerment;
7. Ensures that the city's environmental and economic resources can be enjoyed by all its citizens;
8. Enhances the physical well-being of its citizens through an active lifestyle.

Thus, we see that sustainability is crucial to our long-term survival and well-being in many fundamental ways. We also see that at the heart of sustainability problems are sociotechnical systems—a complex interaction of technology and people. It is apparent therefore that tackling sustainability issues will require socio-technical solutions, driving the need for innovative multidisciplinary research. We envisage new types of solutions being created that combine ideas from previously disparate disciplines using an integrated system-level approach, and we firmly believe that this is the way forward to unlock the sustainability paradox.

References

Aloise-Young, P. A., Cross, J. E., & Sandoval, P. K. (2016). If you build it, will they come? Getting consumers on board with the future of the smart grid. In S. Suryanarayanan, R. Roche, & T. Hansen (Eds.), *Cyber-physical-social systems and constructs in electric power engineering* (pp. 319–344). Institution of Engineering & Technology: London, UK.

DOE (2011). 2010 Buildings energy data book. Retrieved from http://buildingsdatabook.eren.doe.gov/

Ehrhardt-Martinez, K., Donnelly, K.A., & Laitner, J.A. (2010). *Advanced metering initiatives and residential feedback programs: A meta-review for household electricity saving opportunities.* American Council for an Energy Efficient Economy (Report #E105).

EIA (2015). *Monthly Energy Review*, 12.3 Carbon dioxide emissions from energy consumption: Commercial sector.

EPA (2016). Retrieved from https://www3.epa.gov/climatechange/

Gorse, C., Smith, M., Glew, D., Thomas, F., Shenton, D. M., & Farmer, D. (2015). Surveying and measuring the thermal properties of buildings. In M. Dastbaz, I. Strange, & S. Selkowitz (Eds.), *Building sustainable futures design and the built environment* (pp. 15–34). Basel, Switzerland: Springer. doi:10.1007/978-3-319-19348-9_2.

Greentechmedia (2010). Opportunities and barriers in the U.S. utility-scale solar market. Retrieved from http://www.greentechmedia.com/articles/read/opportunities-and-barriers-in-the-u.s.-utility-scale-solar-market

Hancock, T. (n.d.). Social Sustainability. Retrieved May 30, 2016, from http://newcity.ca/Pages/social_sustainability.html#FNT0

Hancock, T. (1993). The evolution, impact and significance of the health cities/healthy communities movement. *Journal of Public Health Policy, 14*(1), 5–18 Retrieved from http://www.jstor.org/stable/3342823.

Harak, C. (2012). *National Low-income weatherization: Stimulus-funded program shines but storm clouds are on the horizon.* Retrieved from https://www.nclc.org/images/pdf/energy_utility_telecom/weatherization/weatherize-report.pdf

Hassanabadi, M. S., Banihashemi, S., & Javaheri, A. R. (2012). Analysis and comparison of impacts of design optimization approaches with occupant behavior on energy consumption reduction in residential buildings. *IACSIT International Journal of Engineering and Technology, 4*(6), 680–683.

Heller, J., Heater, M., & Frankel, M. (2011). Sensitivity analysis: Comparing the impact of design, operation, and tenant behavior on building energy performance. *New Buildings Institute White Paper.*

Hong, T., & Lin, H. W. (2013). *Occupant behavior: Impact on energy use in private offices.* Berkeley, CA: Lawrence Berkeley National Lab Retrieved from http://eande.lbl.gov/sites/all/files/lbnl-6128e.pdf.

Klein, L., Kavulya, G., Jazizadeh, F., Kwak, J.-Y., Becerik-Gerber, B., Varakantham, P., et al. (2011). *Towards optimization of building energy and occupant comfort using multi-agent simulation*. International Symposium on Automation and Robotics in Construction. Retrieved from http://www.iaarc.org/publications/fulltext/S07-5.pdf

Mamidi, S., Chang, Y., & Maheswaran, R. (2012). Improving building energy efficiency with a network of sensing, learning and prediction agents. *Proceedings of the 11th International Conference on Autonomous Agents and Multiagent Systems—Innovative Applications Track*. Retrieved from http://www.aamas-conference.org/Proceedings/aamas2012/papers/1A_5.pdf

Moezzi, M., Hammer, C., Goins, J., & Meier, A. (2014). *Behavioral strategies to reduce the gap between potential and actual savings in commercial buildings* (Contract Number: 09-327). Sacramento: Air Resources Board.

Nogee, A., Clemmer, S., Paulos, B., & Haddad, B. (1999). Powerful solutions: 7 ways to switrch America to renewable electricity. Retrieved from http://www.ucsusa.org/sites/default/files/legacy/assets/documents/clean_energy/ps-all.pdf

Page, J. J., Robinson, D. D., Morel, N. N., & Scartezzini, J. L. (2008). A generalized stochastic model for the simulation of occupant presence. *Energy & Buildings, 40*(2), 83–98.

Sorrell, S. (2007). The rebound effect: An assessment of the evidence for economy-wide energy savings from improved energy efficiency. UK Energy Research Centre. Retrieved from http://www.ukerc.ac.uk/publications/the-rebound-effect-an-assessment-of-the-evidence-for-economy-wide-energy-savings-from-improved-energy-efficiency.html

Sovacool, B. K. (2009a). Rejecting renewables: The socio-technical impediments to renewable electricity in the United States. *Energy Policy, 37*, 4500–4513.

Sovacool, B. K. (2009b). The intermittency of wind, solar, and renewable electricity generators: Technical barrier or rhetorical excuse? *Utilities Policy, 17*, 288–296.

Urban, B., & Gomez, C. (2012). A case for thermostat user models. *Proceedings of BS2013: 13th Conference of International Building Performance Simulation Association, Chambéry, France, August 26–28*, 1483–1490.

Chapter 2
Appreciating the Wicked Problem: A Systems Approach to Sustainable Cities

Barbara Colledge

Abstract Sustainable city place-making is a complex process and can deliver systemic unintended or undesirable development paths such as poverty, health inequality, or environmental degradation over generations. The application of socio-technical and socio-ecological systems thinking is applied to this critical challenge of how to create sustainable cities. Creating sustainable cities demands a different process of inquiry by decision-makers, policy-makers, and practitioners to support sustainable holistic thinking and transformational outcomes (Lonsdale et al. 2015. Transformative adaptation: What is it, why it matters and what is needed. UK Climate Impacts Programme, University of Oxford, Oxford, UK). The application of complex systems theory (Santa Fe Institute Bulletin, summer fall 2(1):8–10, 1987; Daedalus 121(1):17–30, 1992; Journal of Systems Science and Complexity 19(1):1–8, 2006), and socio-technical systems thinking, such as "appreciative systems" theories (The art of judgement, London, 1965; American Behavioral Scientist 38(1):75–91, 1994) and "systemic learning cycles" from soft systems methodologies (HRDI 3(3):377–383, 2000, pp. 380–381) are explored as mechanisms to support this new dynamic of skills, behaviour and mindset to foster transformational leadership of place. A new conceptual model and alternative reference frames are proposed as a way to understand and influence transformative action necessary to realise sustainable cities.

2.1 Introduction

The prevalence of persistent 'wicked problems' (Anderson et al. 2014, pp. 147–149; Wimsatt 1976) such as poverty, health inequality or environmental degradation in urban societies highlights a critical challenge for creating sustainable cities. Place-making is a complex process and can deliver systemic unintended, inequitable or undesirable development paths over generations (Lee et al. 2014; UN-Habitat 2016, p. 169). This replication of might be viewed as unsustainable development

B. Colledge (✉)
Leeds Sustainability Institute, Leeds Beckett University, Leeds, LS1 3HE, UK
e-mail: B.Colledge@leedsbeckett.ac.uk

M. Dastbaz et al. (eds.), *Building Information Modelling, Building Performance, Design and Smart Construction*, DOI 10.1007/978-3-319-50346-2_2

paths is a feature of cities globally and is evident in the cyclical nature of economies, in the evolution of businesses and in shocks to political or social systems. Cities and societies are continuously evolving and reshaping with non-linear patterns of development and self-organisation of complex city systems resulting in multiple development paths, some unpredictable, which emerge from the multiple interactions at work.

The expansion of urban development and cities on a global scale has created arenas for economic growth, populations and communities which are vital for quality of life and wellbeing but in contrast deliver unsustainable outcomes or trajectories contrary to these essential human requirements for long-term sustainability. This evidences a tension in our desire for and in our understanding of the complexity of cities. As identified by UN-Habitat (2016): *"cities have become sites of structural transformation"* (UN-Habitat 2016, p. 161) and there is *"an urgent need to reframe the global debate and national agenda for policy and action*". (UN-Habitat 2016, p. 163).

This chapter offers a contribution to this reframing of the debate, policy and action, through a systems approach to our understanding of sustainable cities. Cities are complex evolutionary and ecological systems which are continually shaped and formed through historical and social construction and multiple interdependencies over time (Martin and Sunley 2014). Drawing on soft systems methodologies, complex systems theories, system dynamic factors and learning processes, an analysis of the sustainable development of cities is undertaken identifying factors that could be addressed for more inclusive and sustainable development pathways.

2.2 Unlocking Socio-Technical Thinking: Research Study and Methodology

The fundamental role of people and their actions (Norström et al. 2014) in the process of city place-making needs to be examined if the critical and urgent challenges of city sustainability are to be addressed. It is argued that there is a breakdown in or a fragility of the process of place-making and in societal understanding of the complexity of cities and how sustainable urban transformation might be realised.

Cities, communities and organisations are biological settlements, involving multiple actors (Emirbayer and Mishe 1998, p. 1003, cited in Davies and Msengana-Ndlela 2014, p. 6; Healey 2006; Martin and Sunley 2014) and actants (Law 1992) and rely on people to design, plan, engage in and shape city development processes. This is a complex picture of connectivity, interdependencies and interaction, with direct cause and effect outcomes not able to be fully scoped. Actors and agents within society or decision-makers within organisations and institutions, fulfil a critical role in determining the outcomes of places explicitly and implicitly through this social construction (Berger and Luckmann 1991) and historical construction of every day decision-making.

This chapter explores the way in which sustainable cities are understood, conceptualised and informed by complex systems theory and systems thinking. It is proposed that creating sustainable cities demands a different process of inquiry by decision-makers, policy-makers, and practitioners to support sustainable holistic thinking and transformational outcomes (Lonsdale et al. 2015). How city ecosystems evolve and how development paths emerge is considered through analysis of complex adaptive systems theory (Holland 1987, 1992, 2006; Martin and Sunley 2006; Martin and Sunley 2014) and the important role of people and their actions in shaping the sustainable development of cities. The prevailing reference frames and "world views" held by people, actors, and agents and are identified as critical differentiators in terms of understanding the complexity and required solutions for city sustainability complexity and judgements which inform action.

The application of soft systems methodology (Checkland 2011) and critical systems heuristics (Ulrich 1983) highlights the importance of engaging with multiple world views to inform understanding of complex systems to support a richer understanding of the problem and potential solutions. Consideration is given to how this systems-led approach could inform a different process of inquiry or policy development. Understanding the way action is shaped and influenced by reference frames (Silverman 1970; Ulrich 2005) or appreciative systems (Vickers 1965) is discussed. The potential for enhancing city sustainability and the effectiveness of judgements and actions through a richer understanding of multiple perspectives at different dimensions of stages of the place-making process and the development of pro-sustainability reference frames is explored. It is posited that the approach to sustainable cities needs reframing to address this complexity and that there is a need for pro-sustainability reference frames if progress is to be made in the future.

A social constructivism perspective informs this analysis (Berger and Luckmann 1991 p1/1991), reflecting multiple realities and perspectives, the social and relational factors involved in city-regional development (Paasi 2001; Brenner 1998, pp. 463; 467) and the systemic interactions, and multiple interdependencies in operation. This embraces realities of pragmatism, including social action theory (Joass 1996) and factors of power which exert influence in human systems (Healey 2006).

2.3 The Research Study Method

This conceptual, theoretical study utilises a qualitative desk-based approach to the analysis of selected literature relating to systems theories and sustainable cities. The study is informed by aspects of a constructivist grounded theory approach (Charmaz 2006, pp. 130–131) to enhance qualitative analysis (Charmaz 2006, p. 9). A constructed theory and contribution is developed (Charmaz 2006, p. 10) relating to "*the main concern*" (Glaser 2001, cited in Charmaz 2006, p. 133; 149): how to address the challenge of fostering sustainable cities and why the challenges appear to be replicated over time in different contexts. Through this a theoretical and conceptual contribution is offered with "*grab and fit*" (Bryant 2009, p. 78) for the reframing of the debate, policy, and action for sustainable cities (UN-Habitat 2016, p. 163). The research is grounded via iterative analysis, theoretical sampling, and abduction (Charmaz, 2006, p. 188; Bryant 2009, pp. 88–100) using relevant secondary data and extant literature considered useful (Bryant 2009, p. 106) and valid as data when using Grounded Theory (Bryant 2009, p. 64) and Glaser's "*All is data*" principle (Glaser 2002, p. 1). Whilst primary empirical data in not utilised within this study, interviews with elites undertaken in the context of separate research on the role and contribution of anchor institutions and decision-makers in the sustainable development of city-regions in England informs prior knowledge of the researcher which will have an influence on the iterative analysis.

The focus in this chapter is first on the development of the conceptual framework and the way in which the dimensions of the process of city place-making is shaped and needs to be analysed from the perspective of cities as ecosystems informed by complex adaptive systems and systems thinking. An abstract situational map advocated by Clarke (2003, pp. 558–565; 2005 cited in Charmaz 2006, pp. 118–119) is used to enhance data analysis and the construction of theory adopting a flexible, reflexive approach and "open mind" (Bryant 2009, p. 63). Theories of "Appreciative Systems" (Vickers 1965), "Action Reference Frames", (Silverman 1970) and the "Reference System" (Ulrich 2005) are applied to the challenge of the shaping of sustainable cities and how these can be used to view the city differently. The proposition is that engagement with reference frames is necessary to fully understand complex urban challenges and to realise more sustainable cities. Different ways of understanding or alternative "world views" are required to unblock hidden pathways and widen the choices available for creating different and more sustainable critical development paths for cities. The implications of this for policy and practice to transform our understanding and process of sustainable city place-making are explored through a synthesis and construction of concepts and theory.

2.3.1 A Conceptual Framework for Analysis of Sustainable Cities

Cities as complex adaptive systems (Holland 1987, 2006) are in a constant and continuous process of reshaping (Martin and Sunley 2014) with implications for the way in which policy and action can effect transformation towards sustainable city outcomes that are always in motion (Holland 1992, p. 18). The abstract situational map (Table 2.1) as advocated by Clarke (2003, pp. 558–565; 2005 cited in Charmaz 2006, pp. 118–119) developed from a messy working relationship map illustrates the complex and complicated dimensions that contribute to the process of evolution and adaptation of sustainable cities. This highlights five dimensions, people, location, temporal, resource and interaction, and interdependence that influence and shape sustainable cities. These dimensions are considered interconnected and interdependent. The reality is more complex involving the richness of diverse societies

Table 2.1 Abstract situational map of sustainable city dimensions: ordered working version

A. People dimensions	B. Location dimensions	C. Temporal dimensions	D. Resource dimensions	E. Interaction and interdependence dimensions
Diverse actors Individual; group; communities; societies; organisations; institutions and government; governance; independent actors; firms; representational; agents; people characteristics: gender, ethnicity, beliefs, religious, faith, political; diversity and inclusivity; tolerance **Action and organisation** Act and action; leadership; decision; judgement; belief; values, morals, and ethics; mindset or appreciative/reference system; custom, norm, or habit; discourses; institutions (soft and hard); idea; influence; plan, policy, strategy; process; practice; governance; organisation and structures; negotiation or discourse; situation and context; education, skills; resources **Belief** Mindset or appreciative system; belief; faith; custom, norm, or habit; idea; influence; political; values, morals, and ethics; emotional or affective; understanding and concept of sustainable city **Power dimension** Role or position; authority; agency; responsibility; influence; size and scale; symbolism; governance	**Geography** Type, e.g. urban, rural; mixed; resources; bounded or fuzzy boundaries; connected or isolated; 2D or 3D **Context or situation dependent** All dimensions A-E relevant **Scale** Micro-macro; local, sub-regional, regional, national, international, multiple, and overlapping	**Connectivity and interdependence** Historical, present, future; intergenerational; non-linear; systemic; life cycles **Context or situation** Seasonal; intermittent; conditional, e.g. crisis, economic condition or state, firm or city growth state; resource availability **Context or situation dependency** Social and historical construction; path dependency; life cycles **Change and transitions** Evolution and emergence; flows, e.g. increase; decrease; expansion; contraction; stage, e.g. early, mature, scale; changed states; pipeline; outcomes at points in space or place-time	**Physical—non-human** Built environment infrastructure; natural environment; digital and infrastructure; actants; protection, investment, and renewal; technological development **Biological and ecological** People dimensions; education and skills; innovation; non-human ecologies and biological entities; economies; cultural and social amenities; organisations; protection, investment, and renewal **Natural environment** Flora and fauna; natural landscape and resources, e.g. water; air; protection, investment, and renewal	**Interaction and interdependence** Influences; connectivity; overlapping scales; path dependencies; relationships; actants and actors; networks; all dimensions A-E relevant; holistic and integrated **Evolution** Continuous change; emergence; adaptation; self-organisation; systemic; multiple paths and possibilities depending on situation; process; transitions **Scale** Complexity; complicatedness; situational and interconnected to multiple scales; locational; 3D **Temporal** Past, present, future, intergenerational

and communities with diverse cultures, economic organisation, businesses, communities, institutions, political ideologies, and power influences, resulting in multiple overlapping ecologies, types of environment, and settlement. Dynamic co-evolution of the entity (organism or people) and the environment shape each other and lead to self-organisation of cities, economies, and societal systems. Development paths, path dependency (Martin and Sunley 2006; Martin 2010), and non-linear adaptation emerge though this complex process of interdependencies between multiple actors, organisations and institutions, and environment (Simmie and Martin 2010, Martin 2010). This can lead to institutional or city-regional form with agglomeration forces (Brenner 1998) influenced by people's actions which may or may not lead to sustainable cities over time. In this way:

> *Cities, clusters, and regional economies arise out of the myriad individual actions and interactions of economic agents (firms, workers, households, institutions) that generate outcomes (behaviours, investment and employment decisions, knowledges...) that serve to reproduce ...spatial systems.*
> (Martin and Sunley 2014, p. 11)

It is through this complex evolutionary and interactive process involving historical and social construction (Eder 1996) that cities and regions and their sustainability are continuously shaped, re-created, and transformed. Increasing globalisation of society with rapid digital and physical connectivity between people and places enhances and can strengthen local, national, and global interaction and the influences or impacts (positive and negative) that this can bring to a place. Urban challenges are outcomes of this process of place-making or place-shaping which emerge or evolve from a complex web of uniquely configured interactions and interdependencies between people, actions, and the environment at multiple and overlapping scales in time (see Fig. 1.1).

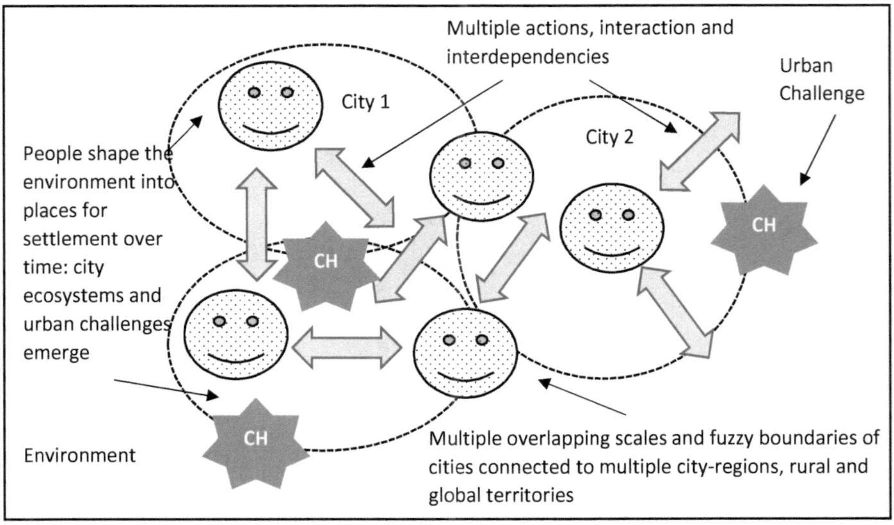

Fig. 1.1 The city ecosystem

The persistence of socio-economic and environmental challenges, inequality, or poverty is an intergenerational feature of cities globally and highlights an underlying societal problem and a potential systemic failure in the process of place-making. This can be viewed as a form of "extended inheritance" or replication of unsustainable practices with urban challenges being represented or reconstructed at each generation via institutional culture or decision-making or from continued practice or beliefs (Martin and Sunley 2014, p. 8). Cities and sustainable cities are created and shaped by people, agency, and actions (Stone 1989; Mossberger and Stoker 2001; Davies and Msengana-Ndlela 2014) within complex organisational and societal structures. This role of people and their actions is under developed in our understanding of sustainable cities (Martin and Sunley 2010, p. 16) and yet is fundamental to addressing the root causes and solutions for sustainable city transformation. Systems theories are applied to this problem to provide a different view of how this can enhance sustainable cities.

2.4 The Results

2.4.1 Systems Thinking Applied to Sustainable City Transformation

Complex systems theories as set out in the conceptual framework can explain how cities evolve, the dimensions involved, and how development paths emerge at a macro- or meso-level but do not fully explain why people's actions and agency continue to produce urban challenges and unsustainable practices. Navigating development paths through this system and influencing their trajectory is a human endeavour. How people interpret the situational context, the options, future possibilities, and resultant actions in their everyday practice and process of shaping sustainable cities is critical (World Bank Group 2015). Actions can build resilience to shocks and support the maintenance of the development path (Perrings 1998); or action, such as innovation or entrepreneurship, can support transformation to a new or different path. Replication of unsustainable development paths such as urban poverty, environmental degradation, or social inequality over generations is a factor of people's actions, interdependence, and evolutionary systemic forces.

Actions are shaped and influenced by people's beliefs developed by culture, values, institutions, experiences, professional practices, social norms, and habits which are historically and socially constructed. Practices can impact negatively on a society, economy, or a place when actions, decisions, or habitual practices are informed by reference frames that are perceived as unethical or against the perceived societal standards or regulation. The way in which collective cultures and habitual practices in the financial crisis in 2008–2010 contributed to fraudulent behaviour and rule breaking is illustrative (Bachmann 2011, p. 209). Such perceived malpractice or insufficiently regulated practice, as standards in journalism, the financial sector and sports have highlighted, show how habits, norms, or standards of practice are vital for maintaining stability of a system or to deliver the right transformation for improvement. Values and a mindset for the creation of shared more inclusive value are important elements in the application of systems thinking.

Mindset, social norms, and culture play a significant role in determining the way in which decisions and actions are framed (World Bank Group 2015, pp. 62–75). Cities and their conceptualisation have multiple meanings influenced by history, culture, and society (Hall 1984, pp. 346–348; Batty and Marshall 2009, p. 551) and are interpreted, conceptualised, or perceived in diverse ways in a similar way to sustainable development (Gibson 1991; Pezzoli 1997, p. 550; Haughton and Counsell 2004). This different way of seeing and understanding, what Vickers terms an "appreciative system" (Vickers 1965) influences and can constrain appreciation of the problem and will frame solutions and actions. Institutional practices and professional norms and habits contribute to the development of individual reference frames through this continuous interaction and reshaping informed by relevant legislation, professional, or organisational rules. Appreciative systems, termed "mental models" by the World Development Report (World Bank Group 2015, p. 11) are becoming identified as important in shaping and influencing actions but some may also "*contribute to the intergenerational transmission of poverty*" (World Bank Group 2015, p. 11). As Herepath (2014) citing Joseph (2000) highlights

> *the interplay of structure and agency is sensitized to the emergence of the contested hegemonic control that fosters advocacy for, and resistance to, strategic change, so providing the requisite insight into strategic direction and ensuing outcome.*
> (Joseph 2000 cited by Herepath 2014, p. 874)

Empirical studies such as that of Bristow and Healy (2015) focused on the study of agency in Wales highlights the systemic nature of agency within complex economic systems and how in this case during the recession, agency appeared to create a dominant think mindset of

> *'getting by'—rather than a more reflexive interrogation of ...the need for and means of pursuing longer term, more transformative change.* (Bristow and Healy 2015, p. 13)

Reference frames and path dependencies in complex city systems that are influenced by unsustainable or negative "collective patterns of behaviour", habits, and norms can sometimes act as a constraint or hidden barrier to taking the necessary course action for transformation resulting in sub-optimal outcomes (World Development Report, 2015, p. 55). Internationally, global business executives conceptualised cities primarily as places to do business and for access to customers, markets, and investors (McKinsey Global Survey 2015 in Global Cities Business Alliance 2015, p. 2). Factors such as improving cities as places to live scored low (5–6%) along with the importance for their firm of having a city-level strategy (27%) (McKinsey Global Survey 2015 in Global Cities Business Alliance 2015, pp. 14; 17–18). This has implications for the way in which sustainable cities are understood and conceptualised and for framing actions. Research undertaken by Ibrahim, El-zaart, and Adams (2015) in the Arab region highlighted a gap in knowledge relating to effecting transformation towards smart sustainable cities and a need to address challenges at city and national levels (Ibrahim, El-zaart and Adams 2015, p. 573). This illustrates a need to address the development of pro-sustainable reference frames and to reframe understanding, policy, practice, and outcomes dimensions for realising Sustainable Cities.

It is evident that "Appreciative Systems" (Vickers 1965), "Action Reference Frames' (Silverman 1970), or "World Views" (Checkland, and Scholes 1990, p. 40) influence understanding, judgements, and decision-making of a particular situation, concept, objective, problem, or decision. As Silverman (1970) identified

> *The overall set of expectations and meanings through which the members of organisations are able to act and to interpret the actions of others is a social construct...participants continually shape and re-shape the pattern of expectation by means of their actions. For, as they act, they validate, deny or create prevailing definitions of the situation. In doing so, they are influenced by the changing stock of knowledge in the wider social world, by their own particular interpretations of the situation, and by the form of their attachment to the existing system.*
> (Silverman 1970, p. 196)

Soft Systems Methodology (Checkland and Scholes 1990), Critical Systems Heuristics (Ulrich 1983), Appreciative Systems (Vickers 1965), or Silverman's Action Reference Frames (1970) offers a relevant approach to understanding decision-making and judgements made regarding action for transformation sustainable city discourse. These approaches go some way to enabling diverse world views to be integrated into the judgement or decision-making processes and in the interpretation or understanding of the complex nature of cities. Soft Systems Methodology (Checkland and Scholes 1990) and Critical Systems Heuristics (Ulrich 1983) integrate different perspectives or "world views" into the process of understanding the complex object under analysis, for example, a wicked problem, a societal challenge, or a city system. Checkland (in Checkland, and Winter 2000) sees these perceived problems as being multiple world views of people as observers and the process of problem solving as a "*learning process*" in trying to understand this complexity and address it through purposeful "*action to improve it*" which necessitates an holistic view informed by multiple perspectives (Checkland and Winter 2000, pp. 379–383). For Ulrich (1983) informed by similar systems thinking, this process of understanding is set in a world reality that involves "*social planning*" which requires an emphasis on "*the art of promoting improvement*" (Ulrich 1996 and 2014, pp. 7–9) involving understanding multiple perspectives not merely "*purposeful-rational action*" (Ulrich 1983, pp. 6–7).

In this regard, Ulrich engages with "*critical holism*" (Ulrich 1993, pp. 5–7), a way of addressing the integral challenge of "holistic thinking" necessary for sustainable development through a practical methodology (Ulrich 1993, pp. 3–5). This critical inquiry process or "*systemic triangulation*" (Ulrich 2005, p. 6) of discourse involving views on Values, Facts, and the System that inform boundary judgements (Ulrich 1993, p. 14; 1996 and 2014, pp. 15–16) requires engagement with potentially different or conflicting views on values, purpose, power, knowledge, and legitimacy to inform purposeful action (Ulrich and Reynolds 2010). This enables a dialogue on "*what is ideal*" and "*what ought to be*" (Ulrich 1996 and 2014, pp. 20;

23–42) supporting a reframing of the reference frames that inform the different views or stances. For whom is left to negotiation via discourse but for a Sustainable City Reference Frame or Lens, it is proposed that this ought to be inclusive of all people which the city supports. Soft systems methodology adopts a similar integration of different world views and perspectives to shape the conceptual model with Simonsen (1994) proposing a "*united perspective*" rather than using a single perspective which can reflect power or vested interests (Simonsen 1994, p. 17).

2.4.2 A "Sustainable City Lens": Implications for Transformation of the Sustainable City Debate, Policy, and Action

Applying a systems approach to the process of development of sustainable cities offers a different view of conceptualising and framing city place-making processes through a "Sustainable City Reference Frame" termed a "Sustainable City Lens" for practical application (see Table 2.2). A systems perspective offers a different, more integrated and holistic view of Sustainable Cities and has implications for the transformation of conceptual models, policy development, sustainable city planning and practice, and the way in which this action and emergent outcomes might be enhanced. Ellingsen and Leknes' three dimensions (2012, pp. 227–229) for understanding city-regional development, Concept, Object, and Practice are considered helpful as a foundation for developing a practical and useful explanation of the stages of Sustainable City place-making. This recognises that cities and regions are relational and institutional spaces, as well as socially and historically constructed territories (Ellingsen and Leknes 2012, pp. 227–229). This on its own is considered insufficient for addressing outcomes and the earlier five conceptual dimensions which are not fully covered by Concept, Object, and Practice dimensions of place-making.

A fourth dimension, "Outcome Emergence", is introduced to reflect the continuous evolution of cities and the associated conceptual dimensions discussed earlier in this chapter. The "Outcome Emergence" dimension reflects the nature of outcomes and indicators as progress made in the shaping of a sustainable city as at transitory moments at a point in time, over a particular geography on the time continuum. This provides a lens or transitory snapshot of sustainable city outcomes at different and overlapping scales which can be observed in "place-time" and opens up options for path dependencies to be explored and improved through a learning system and reflexive approach informed by the interaction and interdependence conceptual dimension. The term "place-time" is utilised from the work of Einstein, frequently referred to as "space-time" (Minkowski 1908). The literal translation "place-time" of Lorentz's "*Ortszeit*" is used (Lehmkuhl 2010, p. xli) to reflect the geography of cities and supports an explanation and understanding of cities in development as evolutionary processes.

Table 2.2 Reframing city development policy and action: towards a sustainable city lens

Reference frame and definition	Present approach (UK focus)	Systems approach	Sustainable city lens
Principles and value systems	• Market and economic growth orientation • Secondary attention to social welfare and protecting environment • Increasing consideration of social deprivation and access to education • Shift towards individual responsibility and reductions in state welfare with austerity • London centric and skewed investment • Northern powerhouse and devolved administrations policies transferring some powers and responsibilities to some city-regions with elected mayor model • Increasing role of private sector through local enterprise partnerships, health, education, and other privatisation models • Weakening of sustainable development commitment through re-interpretation and restatement of previous labour government policy. Criticism of progress made in Sustainable Development Goals (SDGS) • Short—medium-term timescales for return • City as a business, market, designed and planned for business, markets, retail—viewed as a bounded place • A built environment emphasis; infrastructure investment lags in some cities; • Individualistic, institutional, and organisational interests predominate within society	• Action for improvement and inclusive benefit with action informed by multiple world views • Long-term timescales • Whole system sustainability • Holistic and Integration of all factors and elements • A continuous process of historical and social construction • Shaped by interaction, evolution, and interdependences • Society interests to support all individuals • Values informed • Economic, social, environmental, and governance dimensions addressed together with conceptual dimensions relating to people, location, temporal, resource, and interaction and interdependence	• Sustainable city over time with inclusive continuous improvement • Long-term timescales • Whole system sustainability • Holistic and integration of all factors and elements • People and society centric—shared vision and value • Whole system focus • Values informed • Continuous development of sustainable practices and investment in renewal, up skilling, infrastructure, and people • SDGs enacted and prioritised • Long-term timescales for return and short-term outcomes enhanced • Economic, social, environmental, governance, cultural and education dimensions addressed together with conceptual dimensions relating to people, location, temporal, resource, and interaction and interdependence • Community and collaborative approach with responsibility

(continued)

Table 2.2 (continued)

Reference frame and definition	Present approach (UK focus)	Systems approach	Sustainable city lens
Concept How a sustainable city is defined and understood	• Different and diverse prevailing views, e.g. design, plan, and build; the green city; the smart city; the business city; the cultural city • Sustainability seen as tick box not embedded into habitual practices and norms • Sustainability considered as environmental rather than integrated • City as bounded within administrative boundaries for policy and investment	• Complex adaptive system with evolution and emergence • Historically and socially constructed with path dependency via multiple interactions and interdependencies • Appreciative systems and reference frames shape action • Fuzzy boundaries and overlapping multiple scales and territories	• Sustainable city shared vision—inclusive, integrated, and holistic • Complex adaptive system with evolution and emergence • Focus on continuous improvement for city sustainability • Historically and socially constructed with path dependency—learning from history and projective action for future • Multiple interactions and interdependencies with pro-sustainable cities appreciative systems and reference frames shape action
Object How a sustainable city is designed, planned, and interpreted into intent as objectives, goals, and action	• Individual or stakeholder interest led • Design, plan, build, review • Pro-development institutional planning framework • Object is diverse—individual, organisational, economy, place, people not unified vision • Some collaboration but generally not joined up nor shared and agreed approach	• Shared or community/societal objective enables inclusive stakeholder interests • Whole system approach balancing economic, social, environmental, governance and cultural aspects for quality of life and sustainable society	• Unified vision of sustainable city and priorities and processes to support continuous improvement • Process of development understood as a continuous and as a complex system • Social learning processes support action for improvement and reflexive practice • Collaborative structures and engagement mechanisms to enable multiple views to inform policy and action—boundary judgements debated • Agreed progress stages, long-term vision and outcomes to evaluate progress across all dimensions

Practice The continual process of forming sustainable cities in practice, through action and use	• Unsustainable practices prevalent • Lack of integration leads to tensions • Foresight and long-term planning not habitually used • Piece meal and disjointed approach • Strategy and plan led often on specific single dimensions	• Inclusive, integrated, diversity embraced • Collaborative and reflective practice • Multi-stakeholder and systems leadership • Long-term focus and techniques used for forecasting and big data for monitoring trends and patterns • Sustainability approaches more habitual and expected	• Reflexive practice and learning system approach • Continuous improvement focus • Inclusive, integrated, diversity, and values embraced • Collaborative and reflective practice • Multi-stakeholder governance, decision-making, and systems leadership • Long-term focus and techniques used for forecasting and big data for monitoring trends and patterns • Transition management focus and transformational adaption for system change and continuous improvement • Institutional and societal norms and values foster pro-sustainability practices across all dimensions
Outcome Emergence The actual representation and evidence of a sustainable city in place-time	• Success led—primarily economic • Economic Growth, Productivity, and GDP indicators used for comparisons at different scales • Social welfare, environmental priorities, sustainable development lag • Indicators and metrics and goals measured and inform action • Reducing influence of European institutions over time	• System led—sustainable system over time • Integrated and balanced system, e.g. economic, social, and environmental all prioritised to achieve sustainability • Holistic focus • Outcomes continuously reviewed and inform reflexive practice and continuous improvement	• Society and people led • Shared vision of sustainable city and of ideal outcomes and priorities • Continuous monitoring of progress, flows, pipelines, transitions, and path dependencies to achieve equilibrium for sustainable cities • Careful management of transitions • Outcomes inform future action for improvement

2.4.2.1 The Conceptual Dimension

Systems techniques and approaches have the potential to inform conceptual understanding, development of inclusive shared visions of sustainable cities, and better connected policy and practices. The development of a systems informed "sustainable city reference" frame is explored to enhance understanding and practical usefulness (Table 2.2). Systems methodologies and complex adaptive systems understanding foster and enable a more inclusive people-centred approach to concepts of sustainable cities and the system changes needed to secure and maintain them. Transformation necessitates systems change which is likely to involve multiple stakeholders and a wider perspective than that which is within the purview and responsibility of any single decision-maker or organisation. This engagement with the concerns of others and the way in which more inclusive, collaborative, and shared understandings of the real issue or solutions capable of addressing the root cause of the problem has the potential to enable transformation of the system beyond line management or organisational responsibilities or roles. This moves the judgement boundaries towards the whole system, societal learning, and engagement collaboratively with others who are capable of taking action or implementing the solution across professional or policy concepts and fields.

2.4.2.2 Process Dimension

Processes for inclusive and more collaborative models of decision-making and governance will require changes to mindset and practices. This will necessitate a focus on shared community interests and value (i.e. the Sustainable City Frame) as distinct from individual or partisan interests and organisational short-term benefit. As identified by the OECD (2012, p. 10), case studies of under-performing (against national GDP) and successful regions highlights that policy-making, governance, and policy coordination/integration can act as inhibitors or contributors to success. This is relevant for a sustainable city reference frame with the complexity of decision-making necessitating discourse on diverse and potentially conflicting views of priorities and policies. Using techniques, structural and organisational models that are effective in enabling critical inquiry and reflective practice are needed as well as use of monitoring and forecasting over the long term of flows and data indicators to maintain equilibrium and inform adaptive action to reflect changes in conditions or outcomes. Understanding life cycles and life cycle costing or investment appraisals may offer useful techniques to support investment decisions and priorities for continuous cycles of renewal. The ability to learn from experience and history is important to maximise future actions for improvement. The way in which transitions are managed will impact on the ability to maintain sustainable development paths and continuous improvement. Multiple potential transitions points exist over time and space (e.g. people, territories and administrative boundaries, populations and firms, leadership and decision-making models or actors, organisational, institutional, or governance structures, policies or processes, economies, resources, and technology). Transitions need careful management and focused people development and action to mitigate shocks to the system or misaligned changes in reference systems including conceptual understanding social or cultural norms or habits.

2.4.2.3 Practice Dimension

Pragmatically, the continuous evolution of complex city systems highlights a similar challenge of practicality in the application of systems thinking as with Ulrich's view of holistic thinking for sustainable development (Ulrich 1993, pp. 3–5). Implementing these approaches will necessitate a transformation in systemic, habitual, and professional practice. This change requires a shift in mindset and understanding of the problem and a willingness to engage on the part of many in sharing solutions, neither of which is simple to achieve in complex city systems. This necessitates individual learning skills and abilities to be reflective and reflexive (Schon 1983) as well as individual, institutional, and organisational capacity to engage in different multiple collaborations and decision-making over time, across organisational boundaries and territorial scales. This has implications for education and skills policy and our approach as a society to the development of values and beliefs. Concepts of systems leadership (Van Dyke 2013, pp. 4–6) and systems transformation are highlighted more recently in the literature in contexts where transformational, as distinct from incremental adaption for maintaining development paths, is desired (Lonsdale et al. 2015). For example, this has been considered to enable entrepreneurship (Auerswald 2015), address resource constraints, and enhance services in the NHS and in city devolution contexts (Grant Thornton 2016, pp. 15–19). This could be viewed as innovation at a city systems level to support new or different more sustainable development paths or to overcome major shocks to the system.

2.4.3 Emergent Outcomes Dimension

A shared vision of what constitutes a sustainable city at a point in place-time, informed by diverse worldviews or perspectives is the basis for the analysis of improvement through a learning system approach (Checkland 2011, p. 504). In order to foster action for improvement, there is a need to reflect on judgements, action, process, experience, and outcomes to inform, adapt, and enable continuous improvement. This draws parallels with reflective practice (Schon 1983) with the need to debate and consider reference frames or appreciative systems (Vickers 1965), relevant data and views of the system, and the extent to which sufficient account has been taken of the different and often competing perspectives and interests to inform the necessary priorities and agreed action for transformation and improvement. This emphasis on sustainable city emergent outcomes embraces the integration of the five conceptual dimensions: people, locational, temporal, resource, and interaction and interdependence. Typical sustainable development dimensions such as economic, social, and environmental are reflected in this analysis with an extended construction to include Governance (United Nations General Assembly 2015), Culture (due to the importance of values and mindset), and Education (as a result of the importance of establishing values, mindsets, capacity, and skills for collaborative, inclusive, and reflective practice, and transformational adaption and innovation). This also recognises the need for continuous cycles of investment to at least maintain and with the aim of continuous enhancement of education, skills, infrastructure, or other resources over time. Energy and flows of stocks and resources, ideas and policies, and institutional and societal norms have a parallel in this regard, which need to be understood and taken into account to inform projective action and adaption for resilience and for continuous and inclusive improvement.

2.5 Conclusions

This chapter has offered a contribution to the way in which we might better understand the complexity of sustainable cities and the way in which people and reference frames influence action towards their realisation. The replication globally of urban challenges in the development of sustainable cities over time suggests the need for a different approach to addressing sustainable cities. A pro-sustainability approach, informed by Sustainable City Reference Frame, termed a "Sustainable City Lens" has been explored through the application of systems thinking and methodologies. Such an approach has the potential to change conceptual understanding of sustainable cities and lead to very different solutions to everyday policy and practice. The development of and engagement with reference frames as part of this process provides a richer understanding of the competing priorities and possible options that might support sustainability led action. This analysis has contributed to an explanation of why persistent urban challenges are replicated over time through unsustainable practices and policies being inherited through generations or an inability or lack of capacity to engage in transformational adaptation. Complex systems theories and related systems thinking has enabled a richer understanding of the processes and interdependences at play and provided a reframing, different tools and approaches to policy-making to enhance pro-sustainability decision-making and action.

This chapter has provided a first step in revisiting systems thinking to provide an alternative perspective on the challenges in enabling sustainable cities. This different process of inquiry by decision-makers, policy-makers, and practitioners is proposed as a way to enhance sustainable holistic thinking and increase the potential for transformation of cities globally. Further work will be to extend and refine this "Sustainable City Lens" and the conceptual approach to reference frames through empirical data analysis.

References

Anderson, C., Tornberg, A., & Tornberg, P. (2014). Societal systems—complex or worse? *Futures, 63*, 145–157.

Auerswald, P. E. (2015). *Enabling Entrepreneurial Ecosystems. Insights from ecology to inform effective entrepreneurship policy*. Kauffman Foundation Research Series on City, Metro, and Regional Entrepreneurship. Ewing Marion Kaufman Foundation.

Bachmann, R. (2011). At the crossroads: Future directions in trust research. *Journal of Trust Research, 1*(2), 203–213.

Batty, M., & Marshall, S. (2009). Centenary paper: The evolution of cities: Geddes, Abercrombie and the new physicalism. *TPR, 80*(6), 551–574. doi:10.3828/tpr.2009.12.

Berger, P., & Luckmann, T. (1991). *The social construction of reality*. London: Penguin Books.

Brenner, N. (1998). Between fixity and motion: Accumulation, territorial organization and the historical geography of spatial scales. *Environment and Planning D, 16*(4), 459–481.

Bristow, G., & Healy, A. (2015). Crisis response, choice and resilience: Insights from complexity thinking. *Cambridge Journal of Regions, Economy and Society, 8*(2), 241–456. doi:10.1093/cjres/rsv002.

Bryant, A. (2009). Grounded theory and pragmatism: The curious case of Anselm Strauss. *Qualitative Social Research, 10*(3), Art 2—September 2009, [113 paragraphs]. *Forum Qualitative Sozialforschung/Forum: Qualitative Social Research, 10*(3), Art. 2, http://nbn-resolving.de/urn:nbn:de:0114-fqs090325. [Online]. Retrieved August 15, 2016, from http://www.qualitative-research.net/index.php/fqs/article/view/1358/2850

Charmaz, K. (2006). *Constructing grounded theory: A practical guide through qualitative analysis*. London: Sage.

Checkland, P. (2011). Autobiographical perspectives: Learning your way to 'action to improve'—The development of soft systems thinking and soft systems methodology. *International Journal of General Systems, 40*(5), 487–512.

Checkland, P. (1994). Systems theory and management thinking. *American Behavioral Scientist, 38*(1), 75–91.

Checkland, P., & Scholes, J. (1990). *Soft systems methodology in action*. Chichester, UK: Wiley. http://eu.wiley.com/WileyCDA/WileyTitle/productCd-0471986054.html

Checkland, P., & Winter, M. (2000). Perspectives on people: The relevance of soft systems thinking. *HRDI, 3*(3), 377–383.

Clarke, Adele E. (2003). Situational analyses: Grounded theory mapping after the postmodern turn. *Symbolic Interaction, 26*(4), 553–576. Society for the Study of Symbolic Interaction. [Online]. Retrieved August 15, 2016, from http://sts.ucdavis.edu/summer-workshop/worshop-2008-readings/Clarke%202003%20Situational%20analyses.pdf

Davies, J. S., & Msengana-Ndlela, L. G. (2014). Urban power and political agency: Reflections on a study of local economic development in Johannesburg and Leeds. *Cities, 44*, 131–138.

Eder, K. (1996). *The social construction of nature*. London: Sage.

Ellingsen, W., & Leknes, E. (2012). The city region as concept, object and practice. *Norwegian Journal of Geography, 66*, 227–236.

Emirbayer, M., & Mishe, A. (1998). What is agency? *American Journal of Sociology, 103*(4), 962–1023.

Gibson, R. (1991). Should environmentalists pursue sustainable development? Probe Post, 22–25, cited in Robinson, J. Squaring the circle? Some thoughts on the idea of sustainable development. Ecological Economics 48 (2004) 369–384. Retrieved 29 September 2016 from http://ipidumn.pbworks.com/f/SquaringtheCircleSustainableDevelopment.pdf.

Glaser, B. G. (2001). *The grounded theory perspective: Conceptualization contrasted with description*. Mill Valley, CA: The Sociology Press.

Glaser, B. G. (2002). Constructivist grounded theory? *Forum Qualitative Sozialforschung/Forum: Qualitative Social Research, 3*(3), Art. 12. [Online]. Retrieved August 1, 2016, from http://nbn-resolving.de/urn:nbn:de:0114-fqs0203125

Global Cities Business Alliance (with McKinsey & Company) (2015, June). City-business interaction: Current trends and future outlook. Discussion Paper. Global Cities Business Alliance.

Grant Thornton. (2016, March). *Modelling future care. The NHS under reconstruction*. NHS governance and financial resilience review. Grant Thornton.

Hall, P. (1984, August). Have cities a future? *Futures, 16*(4), 344–350.

Haughton, G., & Counsell, D. (2004). Regions and sustainable development: Regional planning matters. *The Geographical Journal, 170*(2), 135–145.

Healey, P. (2006). Transforming governance: Challenges of institutional adaptation and a new politics of space. *European Planning Studies, 14*(3), 299–320.

Herepath, A. (2014). In the loop: A realist approach to structure and agency in the practice of strategy. *Organization Studies, 35*(6), 857–879.

Holland, J. H. (1987, Summer Fall). Complex adaptive systems: A primer. *Santa Fe Institute Bulletin, 2*(1), 8–10.

Holland, J. H. (1992, Winter). Complex adaptive systems. *Daedalus, 121*(1), 17–30. [Online]. Retrieved July 3, 2016, from http://www.jstor.org/stable/20025416

Holland, J. H. (2006). Studying complex adaptive systems. *Journal of Systems Science and Complexity, 19*(1), 1–8.

Ibrahim, M., El-Zaart, A., & Adams, C. (2015). Paving the way to smart sustainable cities: Transformation models and challenges. *Journal of Information Systems and Technology Management, 12*(3), 559–576. [Online]. Retrieved June 17, 2016, from http://eprints.port.ac.uk/19375/1/Paving_the_way_to_smart_sustainable_cities.pdf

Joas, H. (1996). *The creativity of action*. Chicago, IL: University of Chicago Press.

Joseph, J. (2000). A realist theory of hegemony. *Journal for the Theory of Social Behaviour, 30*, 179–202.

Law, J. (1992). Notes on the theory of the actor-network: Ordering, strategy, and heterogeneity. *Systems Practice, 5*, 379–393. doi:10.1007/BF01059830.

Lee N., Sissons P., Hughes, C., Green, A., Atfield, G., Adam, D., et al. (2014). *Cities, growth and poverty: A review of the evidence*. York: The Work Foundation, Joseph Rowntree Foundation.

Lehmkuhl, D. (2010). Notes of the Translator. In Petkov, V. (Ed.), *Minkowski spacetime: A hundred years later*. Dordrecht: Springer.

Lonsdale, K., Pringle, P., & Turner, B. (2015). *Transformative adaptation: What is it, why it matters and what is needed*. Oxford, UK: UK Climate Impacts Programme, University of Oxford.

Martin, R. (2010). Roepke lecture in economic geography-rethinking regional path dependence: Beyond lock-in to evolution. *Economic Geography, 86*(1), 1–27.

Martin, R., & Sunley, P. (2006). Path dependence and regional economic evolution. *Journal of Economic Geography, 6*, 395–437.

Martin, R., & Sunley, P. (2010). The place of path dependence in an evolutionary perspective on the economic landscape. In R. Boschma & R. Martin (Eds.), *The handbook of evolutionary economic geography*. Cheltenham: Edward Elgar Publishing.

Martin, R., & Sunley, P. (2014). Towards a development turn in evolutionary economic geography? *Regional Studies, 49*(5), 712–732.

Minkowski, H. (1908). Raum und Zeit (Space and Time). In *80th Meeting of German Scientists and Physicians*, Cologne, September 21, 1908. (Translated by Lehmkuhl in Petkov, V. (Ed) (2010) Minkowski Spacetime: A Hundred Years Later. Springer. pp. xiv–xlii)

Mossberger, K., & Stoker, G. (2001). The evolution of urban regime theory, the challenge of conceptualization. *Urban Affairs Review, 36*(6), 810–835.

Norström, A. V., Dannenberg, A., McCarney, G., Milkoreit, M., Diekert, F., Engström, G., et al. (2014). Three necessary conditions for establishing effective Sustainable Development Goals in the Anthropocene. *Ecology and Society, 19*(3), 8. doi:10.5751/ES-06602-190308.

Organisation for Economic Cooperation and Development (OECD) (2012, March). *Promoting Growth in All Regions*. Lessons from across the OECD. OECD.

Paasi, A. (2001). Europe as a social process and discourse: Considerations of place, boundaries and identity. *European Urban and Regional Studies, 8*, 7–28.

Perrings, C. (1998). Resilience in the dynamics of economy-environment systems. *Environmental and Resource Economics, 11*(3–4), 503–520.

Pezzoli, K. (1997). Sustainable development: A transdiciplinary overview of the literature. *Journal of Environmental Planning and Management, 40*(5), 549–574.

Schon, D. A. (1983). *The reflective practitioner: How professionals think in action*. London Temple Smith; New York: Basic Books.

Silverman, D. (1970). The theory of Organisations. London: Heinemann (as cited in Checkland, 1994)

Simmie, J., & Martin, R. (2010). The economic resilience of regions: Towards an evolutionary approach. *Cambridge Journal of Regions, Economy and Society, 3*(1), 1–17, 27–43.

Simonsen, J. (1994, Spring). *Soft systems methodology: An introduction*. Computer Science/ Roskilde University, pp. 1–17. [Online]. Retrieved June 18, 2016, from http://www.jespersimonsen.dk/Downloads/SSM-IntroductionJS.pdf

Stone, C. N. (1989). *Regime politics: Governing Atlanta, 1946–1888*. Lawrence, KA: University Press of Kansas.

Ulrich, W. (1983). *Critical heuristics of social planning: A new approach to practical philosophy*. Bern: Haupt. (Reprint edition, Chichester: Wiley, 1994)

Ulrich, W. (1993). Some difficulties of ecological thinking, considered from a critical systems perspective: A plea for critical holism. *Systems Practice, 6*(6), 583–611. Plenum Publishing Corporation, New York. Post-publication version 2006 published by Ulrich, W. [Online]. Retrieved August 15, 2016, from http://wulrich.com/downloads/ulrich_1993.pdf

Ulrich, W. (1996 and 2014). A Primer to Critical Systems Heuristics for Action Researchers. *The Centre for Systems Studies*, March 1996. Online version of 10 August 2014. The University of Hull. [Online]. Retrieved August 15, 2016, from http://wulrich.com/downloads/ulrich_1996a.pdf

Ulrich, W. (2005). A brief introduction to critical systems heuristics (CSH). *Website of the ECOSENSUS project*, Open University, Milton Keynes, UK, 14 October 2005. [Online]. Retrieved July 3, 2016, from http://ecosensus.info/about/index.html

Ulrich, W., & Reynolds, M. (2010). Critical systems heuristics. In M. Reynolds & S. Holwell (Eds.), *Systems approaches to managing change: A practical guide* (pp. 243–292). London: Springer.

United Nations (General Assembly) (2015). *Transforming our world: The 2030 Agenda for Sustainable Development*. Resolution adopted by the General Assembly, 25 September 2015, A/RES/70/1, United Nations [Online]. Retrieved May 20, 2016, from https://sustainabledevelopment.un.org/post2015/transformingourworld

United Nations Human Settlements Programme (UN-Habitat) (2016). *Urbanisation and Development: Emerging Futures, World Cities Report 2016*. United Nations Human Settlements Programme (UN-Habitat), Nairobi, Kenya. [Online]. Retrieved May 20, 2016, from http://wcr.unhabitat.org/wp-content/uploads/sites/16/2016/05/WCR-%20Full-Report-2016.pdf

Van Dyke, M. (2013). *Systems Leadership: Exceptional leadership for exceptional times*. Source Paper 4a Systems Leadership for Children's Services in the USA. Colebrooke Centre for Evidence and Implementation with the Centre for Health Enterprise, Cass Business School, City University London.

Vickers, G. (1965). *The art of judgement*. London: Chappell & Hall.

Wimsatt, W. C. (1976). Complexity and organization. In M. Grene & E. Mendelsohn (Eds.), *Topics in the philosophy of biology* (pp. 174–193). Dordrecht, Holland: D. Reidel.

World Bank Group (2015). *Mind, Society, and Behavior*. World Development Report 2015, A World Bank Group Flagship Report, International Bank for Reconstruction and Development/ The World Bank Washington.

Chapter 3
A Theoretical Framework for Circular Economy Research in the Built Environment

Francesco Pomponi and Alice Moncaster

Abstract Circular economy is quickly gaining momentum across numerous research fields. The founding principles of circular economies lie in a different perspective on, and management of, resources under the idea that an ever-growing economic development and profitability can happen without an ever-growing pressure on the environment. As such, the built environment has a lot to contribute, being the sector with the greatest environmental impacts. However, the few existing cases of current research in the built environment from a circular economy perspective seem to have just replaced the 3R principle (reduce, reuse, recycle) with the new 'buzz-word'. In this paper, we argue that a significantly different research approach is necessary if the circular economy is to keep up to its promise of being a new paradigm for sustainability. We therefore propose a framework to formulate building research from within a circular economy perspective. The framework is built around six pillars and acknowledges the key role of interdisciplinary research and that of both bottom-up and top-down initiatives to facilitate the transition to 'circular' buildings. Although theoretical in nature, the framework has been tested against current discourse about buildings and circular economies and it has proven a valuable tool to cluster existing initiatives and highlight missing interdisciplinary links. As such it can provide a valuable starting point to contribute to the theoretical foundations of building research from within the new paradigm of circular economies and also shape future research directions.

3.1 Introduction

There are many different schools of thought on the circular economy (CE) (Ellen MacArthur Foundation 2016); however, the shared founding principles lie in the better management of resources. The role of the built, and particularly the urban, environment is crucial. Cities occupy only 3% of the Earth's land but account for

F. Pomponi (✉) • A. Moncaster
Department of Engineering, Centre for Sustainable Development, University of Cambridge, Trumpington Street, Cambridge CB2 1PZ, UK
e-mail: fp327@cam.ac.uk

M. Dastbaz et al. (eds.), *Building Information Modelling, Building Performance, Design and Smart Construction,* DOI 10.1007/978-3-319-50346-2_3

around 70% of energy consumption and carbon dioxide emissions (UN 2016). While buildings provide the essential infrastructure for civilisation and our need for shelter, they also create an ecological threat in terms of resource consumption and depletion, air quality, and pollution of soil and water (Naustdalslid 2014). Buildings are also often unique entities and their long lifespan, and changes of use during their service life, lead to increased uncertainty about future scenarios. Therefore, although buildings are made up of components which are manufactured products, when assembled together those products create an entity which no longer fits into the logic of manufacturing. Since current views on CE tend to focus mainly on manufactured products, the complexities that are inherent within buildings are often neglected. This paper aims to address such gap.

3.2 Building Research and Circular Economies

Two aspects are worth considering when framing building research from within a circular economy perspective. First, solutions devised and engineered for short-lived products are unlikely to be applicable to buildings. A building's 'manufacture' and useful life extend over a significantly longer time span. Evidence of this can be found in figures about the existing building stock. In the UK, for instance, 80% of the buildings that will be standing in 2050 have already been built (Kelly 2008) and the average lifespan is 132 years (Ma et al. 2015). If we are to bring about circularity in buildings, focusing on the new ones may not be enough. Second, buildings are indeed made of more standardly manufactured products but when these are assembled they create a unique, complex, long-lived and ever-transforming entity. The work of Frank Duffy and Stewart Brand (1994) on the shearing layers of buildings qualitatively highlight this aspect particularly well. From a systemic point of view, buildings can be seen as a meso-level, the macro-level being urban agglomerates and the micro-level building components (Fig. 3.1).

For the macro-level, research in terms of 'circular economy' is more advanced if one looks into the concept of eco-cities (inter alia Roseland 1997; Van Berkel et al. 2009), whereas for the micro-level current research on the material dimension (inter alia Braungart et al. 2007; McDonough et al. 2003) and circular Supply Chain Management (SCM) (e.g. Lacy and Rutqvist 2015) would suffice to bring circularity about. As a result, an element of analysis currently lacking is the building as an entity per se. This somewhat contrasts with mere environmental impact assessment research, most often in terms of embodied energy and carbon,[1] which is focused at building level in nearly half of the cases examined in current literature (Pomponi and Moncaster 2016).

[1] Defined as the sum of CO_{2eq} emissions related to all activities and components other than the operational energy consumption related to a building's life. More generally, embodied costs may refer different units such as energy, carbon, water, natural resource depletion, etc.

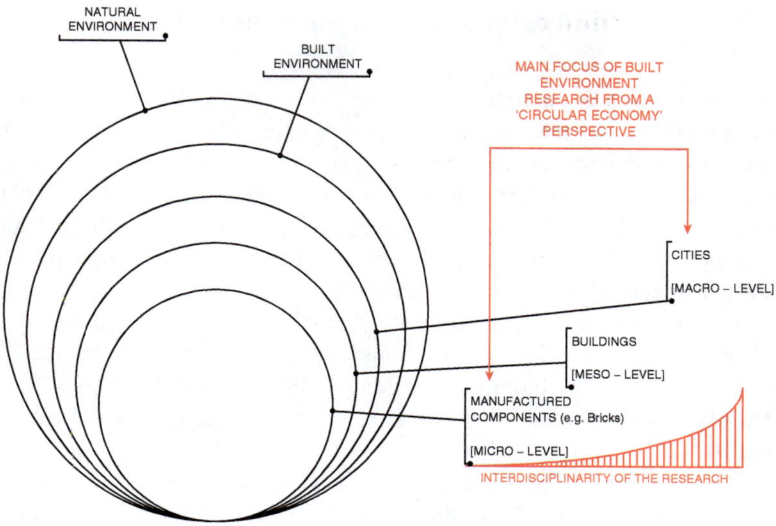

Fig. 3.1. Framing of built environment research

Figure 3.1 also shows a qualitative trend of interdisciplinarity in built environment research. From a methodological point of view, an interdisciplinary, if not a transdisciplinary, approach is necessary since the built environment is not a "discrete discipline with its own standard approaches to philosophy, methodology, and methods" (Knight and Turnbull 2008, p. 72). In fact, however, only research at macro-levels (i.e. cities, neighbourhood, built environment) often acknowledge multiple disciplines. One example in this respect is the huge, though UK-centric, Sustainable Urban Environments programme (EPSRC 2013) which aimed from the outset to have an interdisciplinary focus. At the meso-level (building), the interdisciplinarity is rather moderate, and three main strands are identifiable:

- Post-occupancy evaluation (POE) which considers the effectiveness of occupied environments for humans as users (Zimring and Reizenstein 1980),
- Life cycle assessment (LCA) which could be seen as almost entirely techno-numerical although it aims to understand the impacts of human activities on the environment (Crawford 2011), and
- Recent research trends on operational energy which has started moving from technical to techno-social, by including some thoughts about how people actually live in and use the buildings (Janda 2011).

At micro-level (component), interdisciplinary research is an exception with extremely few cases where the study goes beyond a mere technical point of view (e.g. Forman and Tweed 2014).

In terms of tools for building research in a circular economy, life cycle assessment (LCA) and material flow analysis (MFA) are well-established techniques which have been used also in some CE research (Chen 2009; Genovese et al. 2017; Ghisellini et al. 2016).

3.3 Research Dimensions of Circular Economies

The main innovation of circular economies consists in decoupling resource depletion and growth, under the idea that an ever-growing economic development and profitability can happen without an ever-growing pressure on the environment. Frosch and Gallopoulos (1989) called for a new industrial paradigm that would transform the linear model into a more integrated industrial ecosystem. Their recommendation embedded inherent circularity, for they suggested effluents of industrial processes should serve as raw materials for other processes, so that "[t]he industrial ecosystem would function as an analogue of biological ecosystems" (Frosch and Gallopoulos 1989, p. 144). This principle resurfaced years later in more defined forms known as biomimicry (Benyus 1997) and biomimetics (Bhushan 2009) which, in their simplest meaning, refer to good design inspired by nature (Pawlyn 2011).

Biological analogies are often found as well in the prolific work of William McDonough and Michael Braungart (Braungart et al. 2007; McDonough and Braungart 1998, 2002, 2013; McDonough et al. 2003). Their work is perhaps the form of CE most familiar to the wide public. They identified the source of apparent incompatibility between industrial prosperity, environmental harmony and economic viability in a human-specific activity: design. The role of a different design paradigm can be implicitly found also in Fischer-Kowalski and Hüttler (1998). However when they evaluated the possibility of closing open cycles they concluded that "[u]pon closer scrutiny it is obvious […] that this option applies only to a narrow range of materials and processes" (Fischer-Kowalski and Hüttler 1998, p. 120). A much broader view is instead found in the position of McDonough and Braungart who developed a design framework (best known as cradle to cradle) based on two circular loops, the technical and biological cycles, where resources are kept in for as long as possible, with no or minimal loss of quality and leakage. Their fundamental redesign of industrial flows switches from mainstream eco-efficiency (doing less bad) to eco-effectiveness (doing good).

A similar scepticism over an ever greater efficiency as a pathway to sustainability is argued by Ulanowicz et al. (2009) who used information theory to quantitatively call for caution in maximising efficiency in any field, whether it is physics, economics or ecology. The reasoning underpinning their method of analysis may resemble the biomimicry philosophy at a first look but it bears a fundamental difference. While biomimicry suggests learning from nature to inspire design, Ulanowicz et al. (2009) encourage us to transfer our understanding and modelling of natural elements (such as ecology and ecosystems) to more human concepts such as economies. This is clarified in later work by the same authors (Goerner et al. 2009, p. 76) where a measure called Quantitative Economic Development (QED) is developed to provide a mathematical basis to support "current theory [which] fails to differentiate healthy development from mere growth". Overall, they use System Science as the method of analysis for a sustainable economic development since "similar energy concepts and network analysis methods can be applied to all matter-energy-information flow systems because [...] such systems exhibit strong parallels in behavioural patterns and developmental dynamics" (Goerner et al. 2009, p. 76–77).

Current criticism sees the CE characterised by a sometimes simplistic approach which does not address societal and political challenges or the complexity of human nature. These views resonate with the words of Gregson et al. (2015, p. 219) who argue that academics and practitioners tend to use the CE concept in an "approbatory, uncritical, descriptive and deeply normative" fashion. CE can therefore appear dogmatic (Bakker et al. 2010) in the belief that having devised a solution implicitly means having solved the problem. Such strong faith in the effectiveness of a technical fix (Amelung and Martens 2008) resembles the truth claims of the positivistic philosophy of science which neglects the interdependence between knowledge production and social origins of belief (Dolby 1971). More recent trends, however, see scientific research as a value-laden activity (Gonzalez 2013). One example is the perspective on resource consumption offered by Sauvé et al. (2016, p. 4) who believe that humans "will certainly pursue the means to find and extract increasingly rarer and more expensive resources with ever increasing fervour and cleverness".

A broader view on CE can be found in the work of Naustdalslid (2014) who pairs policy and societal dimensions with the technical one and warns that an excessive focus on materials and their optimisation may underestimate the key role of stakeholder involvement and societal participation to implement CE successfully. The role of society emerges also in a broad discussion on the necessary system perspective for CE by Webster (2013) who emphasises the fundamental part played by education. The need for moral and psychological adjustments was already very clear half a century ago to Boulding (1966) who saw them as indispensable and instrumental in the transition to an embryonic version of CE, which he called closed sphere. As a precursor to the importance of flows and connections underscored by Webster (2013), Boulding (1966) already believed that knowledge (or information) was far more important than matter because, in his view, matter only acquires meaningfulness to humans when it becomes the object of our knowledge. A further dimension seldom considered in framing the CE is the behavioural one. This has been flagged by Smith (2014) who recognises the crucial role of behavioural studies, "even before we arrive at design and repair because it may be the case that people do not want to repair that specific thing". A behavioural dimension for circularity, although in a customer-centric perspective, also emerged in the review of Ghisellini et al. (2016) who reflect on collaborative consumption models.

However, not all scholars see social and behavioural issues as part of circular economies. An example is the framing of Sauvé et al. (2016) who do not see CE as having any social objectives, but rather as a system which focuses on reuse and recycling as substitutes for raw virgin materials. To enable these flows of materials and resources whilst guaranteeing economy growth, some scholars see a key role in the broad spectrum of sustainable supply chain management thus awarding a predominant role to the economic dimension (e.g. Genovese et al. 2017; Lacy and Rutqvist 2015). A further important aspect is the role of policy towards successful circular economies, as discussed by Geng and Doberstein (2008) who identify barriers and challenges in terms of technology and public participation. Barriers are also one of the points considered by Genovese et al. (2017) who see government bodies as facilitators to overcome them in economic and industrial systems. The opportunities for policymakers to implement CE are also discussed in Esposito et al. (2015) who look at the practical levers such as tax, laws and regulatory frameworks within specific industrial sectors or the society at large. Regarding the latter group, however, a very western-centric view emerges in Esposito et al. (2015, p. 2) who see the ultimate goal of a CE as "to preserve *our* current way of life by making it technically viable for the longer term by producing within a closed system" [our italics]. This statement neglects the fact that there is no such thing as a *global* current way of life but rather a very comfortable life in developed countries which we want to hold to as tightly as possible. A concept, which resonates with the views of Gregson et al. (2015, p. 236) who see CE "as a form of geo-political insurance; in a world where rampant economic growth in the developing world threatens the stability of economies long accustomed to having resources their own way". A similar viewpoint comes from Kerschner (2010) who reflected on the popularity that de-growth (*decroissance*) concepts (see e.g. Georgescu-Roegen 1977; Latouche 2007) regained a few years ago and concluded that economic de-growth and growing economy are not mutually exclusive but, in fact, complements, where "de-growth is not a goal in itself, but the rich North's path towards a globally equitable South" (Kerschner 2010, p. 544). Table 3.1 shows a meta-analysis of the dimensions identified in the circular economy literature.

3.4 The Proposed Framework

The previous section has shown how authors from different disciplines view CE differently. In rare cases, the focus on CE was mono-dimensional whereas we often found a link to the three pillars of sustainability: economy, environment and society. Building on the literature reviewed, we however believe that at least three more defining elements are missing from the triple bottom line view and these are the role of governments (i.e. policy), the role of matter (e.g. design, technology, materials) and the role of individuals (i.e. behavioural). All of these are pivotal for the success of a global system such as CE, and in fact all three of them have been in more or less explicit ways mentioned in the current literature on the topic. Figure 3.2 presents our proposal for a frame of reference for building research from within a CE perspective. Our belief behind a 'six pillars' framework is that to meet successfully the goals of today's sustainability research it is necessary to combine the use of different disciplines, such as transdisciplinary research (Kajikawa et al. 2014).

First, the peripheral arrowed arcs represent the need for a holistic approach and a harmonised collaboration of research initiatives in each of the six pillars. Second, the inner dashed lines stress the importance of practical links between each pillar and the others. In some cases, indeed, not all research dimensions may be needed in practice and the framework also allows for sub-groups of two, three, four and five dimensions. We now discuss each dimension in turn with practical examples from current

Table 3.1 Meta-analysis of seminal literature reviewed for this research

	Econ.	Environ.	Tech.	Society	Gov.	Behav.	TOT.
Frosch and Gallopoulos (1989)		x	x				2
Benyus (1997)		x	x				2
Bhushan (2009)			x				1
Braungart et al. (2007)	x	x	x	x			4
McDonough and Braungart (1998)		x	x				2
McDonough and Braungart (2002)		x	x				2
McDonough and Braungart (2013)	x	x	x	x			4
McDonough et al. (2003)	x	x	x				3
Fischer-Kowalski and Hüttler (1998)			x	x			2
Ellen MacArthur Foundation (2013)	x	x	x	x			4
Ulanowicz et al. (2009)	x	x		x			3
Goerner et al. (2009)	x	x		x			3
Webster (2013)	x	x	x	x			4
Huamao and Fengqi (2007)	x	x	x	x			4
Gregson et al. (2015)	x	x	x				3
Sauvé et al. (2016)	x	x	x	x			4
Naustdalslid (2014)	x			x	x		3
Boulding (1966)	x	x	x	x			4
Genovese et al. (2017)	x	x	x		x		4
Lacy and Rutqvist (2015)	x		x		x		3
George et al. (2015)	x	x		x			3
Smith (2014)				x		x	2
Ghisellini et al. (2016)	x	x	x	x			4
Andersen (2007)	x	x	x	x			4
Geng and Doberstein (2008)	x	x	x		x		4
Esposito et al. (2015)	x	x	x		x		4
Totals	19	21	21	15	5	1	

discourse on circular economy and built environment after the authors attended relevant events in London throughout end of 2015/early 2016 themed around the topic.

3.4.1 Governmental Dimension

In discussing the barriers to steel reuse in construction, Roy Fishwick (Corbey et al. 2016) highlighted the role that policy can play as current market prices for steel are so low that steel reuse is hardly economically viable. Additionally, he reported on lack of will at EU regulatory level that could kill steel reuse. At a smaller

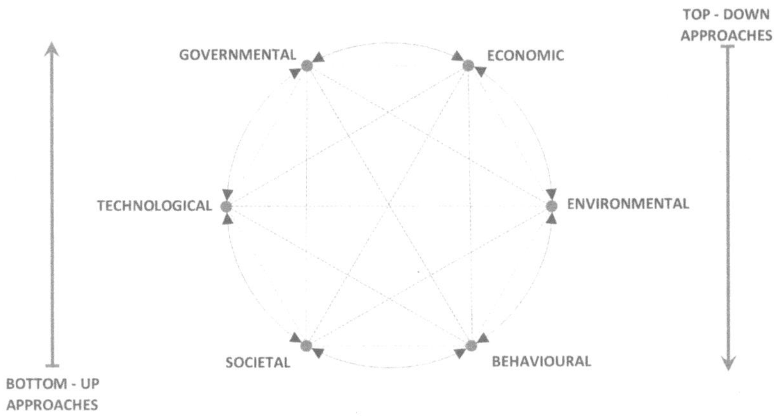

Fig. 3.2 Frame of reference: six dimensions for building research in a circular economy

geographical level, Cécile Faraud (2016) reported on initiatives of planning authorities to achieve circular economy. Her contribution was tailored around the aim of Peterborough in England to become a circular city, and she stressed the difference between Peterborough and the initiatives of worldwide metropolis. Indeed, whereas cities like Amsterdam, Glasgow and Copenhagen are applying circular economy principles to cities, Peterborough believes that a circular city is the pathway to a circular economy (Faraud 2016). She also stressed the need for planning authorities to be locally accurate, i.e. to have strong focus and deep roots in their specific context to make sure they understand the diversity and peculiarity of the challenges ahead (Faraud 2016). Katherine Adams (2016) discussed the importance of tax breaks if a bigger share of reclaimed material use in buildings is to be achieved.

3.4.2 Economic Dimension

The need to change current ownership models and develop a different paradigm for profitability has been a recurring topic in the presentations attended (e.g. Chamberlin 2015; Cheshire 2016; O'Connor 2015). David Cheshire (2016) used the example of lighting systems that are not owned by the building owner/occupier anymore, who just pays for the lighting service through an agreement that also includes performance. Similarly, he discussed other case studies of buildings where collaborative models between all contractors and sub-contractors involved were used from the outset rather than basing the choice of such key actors on the cheapest tenderer at the end of the supply chain (Cheshire 2016). Similarly, Erica Purvis (2015) encouraged more collaborative business models and more openness about relevant data to promote quicker feedback/feed-forward loops.

3.4.3 Environmental Dimension

Research in this respect stresses the lower environmental impacts that reuse has over new products, such as in the cases of steel (Corbey et al. 2016) and wood (Adams 2016). Moreover, most of current research on built environment sustainability focuses on whole life energy and carbon as impact categories (Pomponi and Moncaster 2016) but such an approach might miss out on other, equally crucial, environmental indicators with the risk of shifting environmental burdens from one impact category to another (Pomponi et al. 2016). Therefore, whilst an endless list of environmental indicators is neither desirable nor necessary, a set representative of the majority of environmental impacts should be used nonetheless. (Steinmann et al. 2016).

3.4.4 Behavioural Dimension

The behavioural dimension, seldom mentioned in CE literature, emerged instead as a key element for a breakthrough in built environment sustainability. It was identified as instrumental to succeed in the uptake of recycling (Overbury 2015), energy and carbon reduction (Daly 2015), knowledge on low-carbon buildings and technologies (Fieldhouse 2015) and people's attitude towards reused material (Khoo 2015; Overbury 2015; Adams 2016; Corbey et al. 2016; Owens 2016). Similar issues are also encountered in furniture sharing and reuse (Beavis 2015; O'Connor 2015). Specifically, Roy Fishwick (Corbey et al. 2016) see behavioural issues as one of the two biggest threats to CE uptake and steel reuse in buildings as, in his own words, "people do not want to buy steel for their brand new shiny building from the scrapman". Quite to the contrary, Adams (2016) reported that attractiveness and aesthetic scored as top criteria for people to choose reclaimed wood, which highlights different behavioural patterns depending on the material under consideration. Therefore, there is a strong need to accelerate behavioural research in built environment sustainability as it seems it will eventually be people, not technology, the key to embrace a circular paradigm.

3.4.5 Societal Dimension

Some refer to the circular economy as the sharing economy. Regardless of whether that is fully true or not, it highlights the strong social roots that a circular economy must have. This often involves partnerships and collaboration in building projects (new and existing) and a wider engagement with all involved stakeholders (Daly 2015), networks for resource sharing and reuse (Faraud 2016; Beavis 2015) and a different approach to building's design (Cheshire 2016; Greenfield 2016). In the literature review we have seen that education will have to play a crucial role, this seems specifically important if new building designers have to learn to build with reused and reclaimed materials.

3.4.6 Technological Dimension

Technology emerges as a key aspect to enable circular loops; to connect demand and supplies; and to handle, store and manage huge quantity of data that a circular economy requires. Examples in this latter case are online platforms and web-based apps for resource sharing (Khoo 2015; O'Connor 2015; Owens 2016). Technological innovations in manufacturing and operations can also have enormous impacts, such as mortar-less 3D printed bricks and cardboard ductworks (Cheshire 2016), Design for Manufacture and Assembly (DfMA) (e.g. Laing O'Rourke 2016) or Design for Deconstruction or Disassembly (DfD) (e.g. Densley Tingley and Davison 2011; Adams 2016)—just to name a few.

In addition to the six dimensions discussed, our framework also includes both bottom-up and top-down approaches as boundary conditions. Examples of top-down approaches are CE programmes at EU level (WRAP 2013) or programmes developed by planning authorities at city scale (Faraud 2016). Bottom-up initiatives have equally proven their effectiveness such as the case of grassroots innovations for circular economies (Charter and Keiller 2014; Smith 2014). One further important aspect is that real actions and concrete proposals for a different approach are mostly available from within the technological dimension and, to a lesser extent, in terms of governmental and policy frameworks and environmental assessment metrics. The greatest challenges that lie ahead will deal with the role of people, both as individuals and as society as a whole, and that of new economic models to promote and implement circularity. As such, interdisciplinary research is best placed to succeed in that goal as it will naturally consider multiple dimensions of analysis that switch the narrow focus of technicalities for a wider research basis, without sacrificing depth for breadth. One overall example is the durability of houses and buildings where the problem is not merely technical. In fact, it is not technical at all. The Pantheon was built in 117 AD, and it is still usable and indeed used today. Yet, despite a steady technical development, housing and building construction has severely declined in durability (Boulding 1966). As Boulding (1966, p. 12) worded it, "I suspect that we have underestimated, even in our spendthrift society, the gains from increased durability". Current technology would certainly allow us to build more durable buildings, and the benefits for the environment in terms of resource conservation and waste reduction are undeniable. And yet, there are numerous cases of buildings of 30/40 years that are being demolished (e.g. Cheshire 2016). Building research will have to engage with all relevant stakeholders to understand why it is so, and the reasons behind believing that demolishing such 'baby' buildings is a reasonable choice. Most likely, the answers will be multiple and complex and therefore the contributions that different disciplines can offer are pivotal to achieve a real understanding.

3.5 Conclusions

The built environment is the sector which puts the most pressure on the natural environment and its role in transitioning to a circular economy is pivotal. In framing building research from within a circular economy perspective, we have identified a lack of focus on buildings per se, as unique entities. It also emerged as a decrease in interdisciplinary research related to the scale of analysis. We have therefore framed the problem on a three-tier level: macro (cities and neighbourhoods), meso (buildings) and micro (assemblies and components, e.g. bricks). To understand in which ways building research could be shaped by the circular economy, we have reviewed seminal literature on CEs with an aim of identifying the fundamental dimensions of CE research from within different disciplinary backgrounds. The outcome of the literature review and its meta-analysis has represented the basis for the framework we developed which includes 'six' pillars for building research. These research dimensions have then been tested against recent contributions to the discourse on circular economy and the built environment. All dimensions could be identified in current initiatives, ideas and approaches to achieve more 'circular' buildings therefore showing some robustness of our framework. However, it also emerged a rather poor interdisciplinary underpinning to orchestrate harmoniously and address effectively the six dimensions. We therefore encourage an increase in interdisciplinary research initiatives about buildings and circular economies. Evidence from practical examples has indeed shown that the greatest challenges ahead lie not in further technological innovation but rather in the role of people, both as individuals and as a society.

References

Adams, K. (2016). Important factors to consider for applying circular economy in buildings including a focus on reclamation. #Buildcircular learning hub @ Ecobuild 2016, 8–10 March 2016, London.

Amelung, B., & Martens, P. (2008). Wankele wieg (cradle to cradle). *Milieudefensie Magazine. Amsterdam: Vereniging Milieudefensie, 37*, 28–29.

Andersen, M. S. (2007). An introductory note on the environmental economics of the circular economy. *Sustainability Science, 2*(1), 133–140.

Bakker, C. A., Wever, R., Teoh, C., & De Clercq, S. (2010). Designing cradle-to-cradle products: A reality check. *International Journal of Sustainable Engineering, 3*, 2–8.

Beavis, A. 2015. Remanufacture for the win! The Circular Economy in Organisations—University College London, 11 November.

Benyus, J. M. (1997). *Biomimicry*. New York: William Morrow.

Bhushan, B. (2009). Biomimetics: Lessons from nature–An overview. *Philosophical Transactions of the Royal Society of London A: Mathematical, Physical and Engineering Sciences, 367*, 1445–1486.

Boulding, K. E. (1966). The economics of the coming spaceship earth. Environmental quality issues in a growing economy. In V. D. Lippit (Ed.), *Radical political economy explorations in alternative economic analysis*. London: Routledge.

Brand, S. (1994). *How buildings learn: What happens after they are built*. New York: Viking Penguin.
Braungart, M., McDonough, W., & Bollinger, A. (2007). Cradle-to-cradle design: Creating healthy emissions–a strategy for eco-effective product and system design. *Journal of Cleaner Production, 15*, 1337–1348.
Chamberlin, L. (2015). *Collaboration in the supply chain to deliver circularity—How, what, where and when?* The Circular Economy in Organisations—University College London, London, November 11.
Charter, M., & Keiller, S. (2014). *Grassroots innovation and the circular economy*. The Centre For Sustainable Design ® University for the Creative Arts.
Chen, J. Z. (2009). Material flow and circular economy. *Systems Research And Behavioral Science, 26*, 269–278.
Cheshire, D. (2016). *What principles of the circular economy can be carried across to the built environment?* #Buildcircular learning hub @ Ecobuild 2016, London, 8–10 March.
Corbey, S., Cullen, J. M., Sansom, M., & Fishwick, R. (2016). *Overcoming the barriers to steel re-use in construction*. #Buildcircular learning hub @ Ecobuild 2016, London, 8–10 March.
Crawford, R. H. (2011). *Life cycle assessment in the built environment*. London, New York: Spon Press.
Daly, F. (2015). *Getting things done. Collaborating across departments and sectors*. The Circular Economy in Organisations—University College London, London, 11 November.
Densley Tingley, D., & Davison, B. (2011). Design for deconstruction and material reuse. *Proceedings of the Institution of Civil Engineers-Energy, 164*, 195–204.
Dolby, R. G. A. (1971). Sociology of knowledge in natural science. *Social Studies of Science, 1*, 3–21.
Ellen MacArthur Foundation. (2013). *Towards the circular economy—Economic and business rationale for an accelerated transition*. Author: Cowes.
Ellen Macarthur Foundation (2016). *Circular economy—Schools of thought* [Online]. Retrieved January 29, 2016, from http://www.ellenmacarthurfoundation.org/circular-economy/schools-of-thought/
EPSRC (2013). *Review of the Sustainable Urban Environments (Sue) and Adaptation and Resilience to a Changing Climate (Arcc) Programme* [Online]. Retrieved March 6, 2016, from https://www.epsrc.ac.uk/newsevents/news/sue/
Esposito, M., Tse, T., & Soufani, K. (2015). Is the circular economy a new fast-expanding market? *Thunderbird International Business Review*, n/a–n/a.
Faraud, C. (2016). *Circular economy strategy at a city level*. #Buildcircular Learning Hub @ Ecobuild 2016, London, 8–10 March.
Fieldhouse, E. (2015). Using gamification to change behaviours for ever! The Circular Economy in Organisations—University College London, London, 11 November.
Fischer-Kowalski, M., & Hüttler, W. (1998). Society's metabolism. *Journal of Industrial Ecology, 2*, 107–136.
Forman, T. & Tweed, C. (2014). Solid wall insulation retrofit in UK dwellings: Critical factors affecting management and quality. In: A. B. Raiden & E. Aboagye-Nimo (Eds.) *Association of Researchers in Construction Management (Arcom) 30th Annual Conference*, Portsmouth, UK.
Frosch, R. A., & Gallopoulos, N. E. (1989). Strategies for manufacturing. *Scientific American, 261*, 144–152.
Geng, Y., & Doberstein, B. (2008). Developing the circular economy in China: Challenges and opportunities for achieving 'leapfrog development'. *International Journal of Sustainable Development & World Ecology, 15*, 231–239.
Genovese, A., Acquaye, A. A., Figueroa, A., & Koh, S. C. L. (2017). Sustainable supply chain management and the transition towards a circular economy: Evidence and some applications. *Omega, 66*, 344–357.
Georgescu-Roegen, N. (1977). The steady state and ecological salvation: A thermodynamic analysis. *Bioscience, 27*, 266–270.

George, D. A. R., Lin, B. C.-A., & Chen, Y. (2015). A circular economy model of economic growth. *Environmental Modelling & Software, 73*, 60–63.

Ghisellini, P., Cialani, C., & Ulgiati, S. (2016). A review on circular economy: The expected transition to a balanced interplay of environmental and economic systems. *Journal of Cleaner Production, 114*, 11–32.

Goerner, S. J., Lietaer, B., & Ulanowicz, R. E. (2009). Quantifying economic sustainability: Implications for free-enterprise theory, policy and practice. *Ecological Economics, 69*, 76–81.

Gonzalez, W. (2013). Value ladenness and the value-free ideal in scientific research. In C. Luetge (Ed.), *Handbook of the philosophical foundations of business ethics* (pp. 631–645). Dordrecht: Springer Netherlands.

Greenfield, D. (2016). *Masterplanning for operational resource efficiency—How could we handle recycling and waste more efficiently by considering collection in design?* #Buildcircular Learning Hub @ Ecobuild 2016, London, 8–10 March.

Gregson, N., Crang, M., Fuller, S., & Holmes, H. (2015). Interrogating the circular economy: The moral economy of resource recovery in the EU. *Economy and Society, 44*, 218–243.

Huamao, X., & Fengqi, W. (2007). Circular economy development mode based on system theory. *Chinese Journal of Population Resources and Environment, 5*(4), 92–96.

Janda, K. B. (2011). Buildings don't use energy: People do. *Architectural Science Review, 54*, 15–22.

Kajikawa, Y., Tacoa, F., & Yamaguchi, K. (2014). Sustainability science: The changing landscape of sustainability research. *Sustainability Science, 9*, 431–438.

Kelly, M. J. (2008). Britain's building stock—A carbon challenge. *Communities and local government*. University of Cambridge.

Kerschner, C. (2010). Economic de-growth vs. steady-state economy. *Journal of Cleaner Production, 18*, 544–551.

Khoo, J. (2015). Service design. I don't need a drill i need a hole! The circular economy in organisations—University College London, London, 11 November.

Knight, A., & Turnbull, N. (2008). Epistemology. In A. Knight & L. Ruddock (Eds.), *Advanced research methods in the built environment*. Oxford: Wiley–Blackwell.

Lacy, P., & Rutqvist, J. (2015). *Waste to wealth—The circular economy advantage*. London: Palgrave Macmillan.

Laing O'rourke (2016). *Modular manufacturing* [Online]. Retrieved March 11, 2016, from http://www.laingorourke.com/what-we-do/modular-manufacturing.aspx

Latouche, S. (2007). De-growth: An electoral stake? *The International Journal of Inclusive Democracy, 3*, 14–18.

Ma, J.-J., Liu, L.-Q., Su, B., & Xie, B.-C. (2015). Exploring the critical factors and appropriate polices for reducing energy consumption of China's urban civil building sector. *Journal of Cleaner Production, 103*, 446–454.

McDonough, W., & Braungart, M. (1998). The next industrial revolution. *The Atlantic Monthly, 282*, 82–92.

McDonough, W., & Braungart, M. (2002). *Cradle to cradle—Remaking the way we make things*. New York, NY: North Point Press.

McDonough, W., & Braungart, M. (2013). *The upcycle. Beyond sustainability—Designing for abundance*. New York, NY: North Point Press.

McDonough, W., Braungart, M., Anastas, P. T., & Zimmerman, J. B. (2003). Peer reviewed: Applying the principles of green engineering to cradle-to-cradle design. *Environmental Science & Technology, 37*, 434a–441a.

Naustdalslid, J. (2014). Circular economy in China—The environmental dimension of the harmonious society. *International Journal of Sustainable Development & World Ecology, 21*, 303–313.

O'connor, D. (2015). Every department does not need a laminator! Breaking the rules for circular practices. The Circular Economy in Organisations—University College London, London, 11 November.

Overbury, T. (2015). The circular economy at UCL. The Circular Economy In Organisations—University College London, London, 11 November.

Owens, R. (2016). *How can our industry waste less and ensure re-use? What we've learned at recipro*. #Buildcircular learning hub @ Ecobuild 2016, 8–10 March London.

Pawlyn, M. (2011). *Biomimicry in Architecture*. London: Riba Publishing.

Pomponi, F., & Moncaster, A. M. (2016). Embodied carbon mitigation and reduction in the built environment—What does the evidence say? *Journal of Environmental Management* (Under Review).

Pomponi, F., Piroozfar, P. A. E., & Farr, E. R. P. (2016). An investigation into GHG and non-GHG impacts of double skin façades in office refurbishments. *Journal of Industrial Ecology, 20*, 234–248.

Purvis, E. (2015). Procurement and the standardisation of circular economy products and services. The Circular Economy in Organisations—University College London, London, 11 November.

Roseland, M. (1997). Dimensions of the eco-city. *Cities, 14*, 197–202.

Sauvé, S., Bernard, S., & Sloan, P. (2016). Environmental sciences, sustainable development and circular economy: Alternative concepts for trans-disciplinary research. *Environmental Development, 17,* 48–56.

Smith, A. (2014). *Makers, fixers and circular economies* [Online]. Retrieved January 26, 2016, from http://blogs.sussex.ac.uk/sussexenergygroup/2014/06/25/makers-fixers-and-circular-economies/

Steinmann, Z. J. N., Schipper, A. M., Hauck, M., & Huijbregts, M. A. J. (2016). How many environmental impact indicators are needed in the evaluation of product life cycles? *Environmental Science & Technology, 50*(7), 3913–3919.

Ulanowicz, R. E., Goerner, S. J., Lietaer, B., & Gomez, R. (2009). Quantifying sustainability: Resilience, efficiency and the return of information theory. *Ecological Complexity, 6*, 27–36.

UN (2016). *Sustainable Development Goals—Goal 11: Make Cities Inclusive, Safe, Resilient And Sustainable* [Online]. Retrieved January 25, 2016, from http://www.un.org/sustainabledevelopment/cities/

Van Berkel, R., Fujita, T., Hashimoto, S., & Geng, Y. (2009). Industrial and urban symbiosis in Japan: Analysis of the eco-town program 1997–2006. *Journal of Environmental Management, 90*, 1544–1556.

Webster, K. (2013). What might we say about a circular economy? Some temptations to avoid if possible. *World Futures, 69*, 542–554.

Wrap (2013). *Waste and resource efficiency action programme*. EU Vision 2020. Retrieved from http://www.wrap.org.uk/content/eu-vision-2020

Zimring, C. M., & Reizenstein, J. E. (1980). Post-occupancy evaluation: An overview. *Environment and Behavior, 12*, 429–450.

Part II
Building Information Modelling

Chapter 4
Building Information Modelling (BIM) a Paradigm Shift in Construction

Farzad Khosrowshahi

Abstract The Construction industry in the UK has been recently shaken by a massive BIM (Building Information Modelling) storm, which reached its climax in April 2016 when we reached the British Government's deadline for using BIM for all centrally procured UK Government construction projects. The requirement was pitched at level 2 maturity, which is a managed 3D environment held in separate discipline BIM tools with attached data. BIM has been hailed as a catalyst for a fundamental change in the way the industry conducts its business in a data-intensive and complex environment that significantly relies on effective collaboration of a diverse range of disciplines.

As was the case with Latham Report (1994, Constructing the team. HMSO), and the Egan Report (1998, The Egan Report—Rethinking construction, report of the construction industry taskforce to the Deputy Prime Minister, UK), the industry has the astute ability to welcome the recommendations, but interpret them in the manner suitable for its endurance. Interestingly, in this instance, there is a maturity level which, on the one hand, describes the exact nature of the requirements, but, on the other hand, it allows a degree of interpretation, as to what constitutes BIM capability. However, this time the wave is global, which contain as much collaboration and cooperation, as it imposes competition on moving up the BIM maturity level. Whatever the response of the industry is, the general feeling is that BIM is here to stay. There will be significant "continuous change," but whether it will lead to the well awaited complete "reengineering" of the industry, it remains to be seen. However, what seems certain, is that every step of the process shall leave its worthy mark on the industry.

4.1 Introduction

It is widely acknowledged that the construction industry suffers from inherent inefficient practices, which result in high levels of waste, excessive duplication of processes, poor project co-ordination and delivery, and increased costs. An area where

F. Khosrowshahi (✉)
Victoria University, Melbourne, Australia
e-mail: fk@serenade.org.uk

M. Dastbaz et al. (eds.), *Building Information Modelling, Building Performance, Design and Smart Construction*, DOI 10.1007/978-3-319-50346-2_4

problems have been recognised but not fully understood relates to project information. The 1979 report by the NCC Standing Committee Project Information Group highlighted the impact of poor project information on several areas including time wastages on resolving technical problems (NCCSC Project Information Group 1979). The introduction of National Building Specification (NBS) and SMM7 code of measurement alleviated the issue but only marginally. In fact, over the years, electronic means have not improved the quality of communication and offered little better than paper-based documentation systems (Smith 2014). Indeed, in some ways, new problems have emerged. For instance, the shift from geometric data in CAD to complex parametric data has put excessive demand on the volume of data. This is because the parametric data in BIM define both the objects' size and shape as well as their physical properties and behaviours in relation to other objects (Eastman et al. 2011).

In UK, the adoption of BIM technology has been somewhat slower than those in USA and Scandinavian countries. But, in general, compared to other industries, construction is way behind manufacturing and aerospace. The fragmentation of the industry has been recognised as both the main barrier to the implementation of BIM as well as creating the need to alleviate its adverse nature (Robert and Laepple 2003). Other barriers with significant impact include legacy culture, procurement methods, established work practices, regulatory, legal, and contractual issues, data security, intellectual property rights, and client support (London et al. 2008; Yan and Damian 2009). Furthermore, the extent of the impact of barriers is greatly influenced by the level of the maturity of the organisations making up the supply chain (Aouad et al. 2006). However, as was recognised by the UK government Task force, the adoption of BIM does not have to rely on full and radical changes to the practices. The changes can be incremental and progress in parts on small steps. Already the benefits are visible and measurable: The savings at Heathrow T5 demonstrated the tangible benefit of BIM in UK (International construction intelligence 2009).

4.2 Background

Looking at CAD and BIM from purely technological perspective, it is important not to associate CAD with 2D and BIM with 3D designs. CAD technology can also offer 3D representation. CAD provides a static 2D document that does not relate to the other documents created separately (Ziel et al. 2008). While, in CAD, building elements are represented by lines and geometrical shapes, in BIM the elements hold specifications. For example, a wall definition of the specs include height, width, bearing and non-bearing principle, interior or exterior, fire rating, demolished or new materials such as bricks and boards. BIM offers parametric integrity which relates to the connection and relation between elements which are maintained consistently even when the model is being manipulated Succar 2008.

It has been argued that BIM's inception and evolution is linked all the way back to the ancient Egypt when for the first time the architect and engineer—Imhotep—

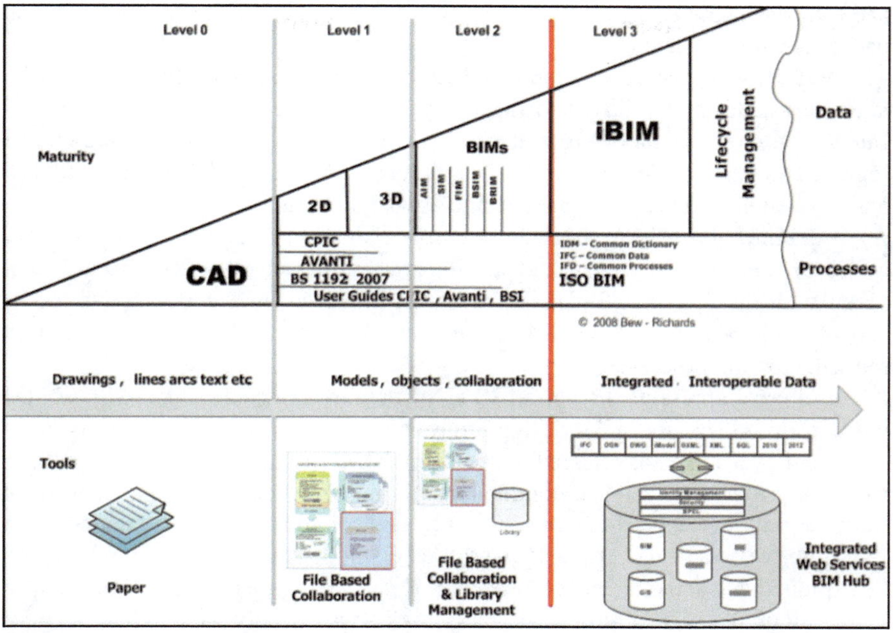

Fig. 4.1 BIM maturity levels (Bew and Richards (2008)

drew lines of ink on papyrus to indicate the outline of a structure. This was then used to communicate the design to the workers who are building it (Hardin 2009). Over time, the inefficiencies of hand-drafting (such as time and cost wastage of alterations to the design) highlighted the need for better coordination and representation. By this time, technological advancements were more appealing than the need to work in a truly collaborative environment. Modern BIM has its roots in the 1960s when Computer-aided Design (CAD) laid the foundation for a major technological breakthrough in this area. With the introduction of 2D geometry and its conversion into 3D in the 1970s, the automated task of drafting entered a new era. These advances were primarily due to designers' ability to shift the focus from pen and paper to graphical interaction with the computer, albeit, in a somewhat limited fashion. Another leap was facilitated through the introduction of object-oriented CAD systems (OOCAD), which facilitated the incorporation of 'intelligence' to the relationship between building elements (Howell and Batcheler 2004).

Bew and Richards (2008) recognised that the definition and implementation of BIM is linked to a defined level of maturity that ranges from Level 0 to Level 3. These are depicted in Fig. 4.1 and described accordingly.

The maturity levels depicted by Bew and Richards are defined as follows (BIM Industry Working Group – DBIS 2011, pp. 16–17):

Level 0: Unmanaged CAD probably 2D, with paper (or electronic) as the most likely data exchange mechanism.

Level 1: Managed CAD in 2 or 3D format using BS1192:2007 with a collaboration tool providing a common data environment, possibly some standard data structures, and formats. Commercial data managed by standalone finance and cost management packages with no integration.

Level 2*: Managed 3D environment held in separate discipline "BIM" tools with attached data. Commercial data managed by an Enterprise Resource Planner (ERP). Integration on the basis of proprietary interfaces or bespoke middleware could be regarded as "pBIM" (proprietary). The approach may utilise 4D programme data and 5D cost elements as well as feed operational systems.

**Mandatory for all UK public sector projects from April 2016*

Level 3: Fully open process and data integration enabled by "web services" compliant with the emerging IFC/IFD standards, managed by a collaborative model server. Could be regarded as iBIM or integrated BIM potentially employing concurrent engineering processes.

Level zero represents the use of 2D CAD drawings in conjunction with written specifications. The use of 3D design information, by individual members, is introduced at Level 1. This is referred to as 'lonely BIM', because members use BIM in isolation rather than using a common platform for collaboration and sharing of information. Level 2 is where the UK Government aimed at, as the starting point for its public projects from April 2016. At this level, model checking software tools can be exploited and some degree of coordination can be achieved. Integration can take place, but on the basis of proprietary interfaces or use of bespoke middleware. This level tends to replicate the traditional practice of bringing independent drawings together, but using discipline-specific models. This will allow problem solving by walkthrough, clash detection, and design scrutiny (Nisbet and Dinesen 2010). It is only at level 3 where a single project model is used as a platform for collaboration. There are some concerns that the integrated working at BIM level 3 generates confusion as who is responsible and who owns the model. This will have an impact on contracts and insurances (BIM Industry Working Group 2011). Also, at level 3, there are potential issues relating to the provision of conflicting information from different models and liability for design. These are likely to affect professional indemnity insurance and intellectual property rights (Barnes and Davies 2014). Other areas of concern include problems associated with data loss due to interoperability inefficiencies (Kramer et al. 2012).

4.3 What is BIM?

BIM is defined in many different ways and tends to mean different things to different people. In one extreme, BIM is purely a technical enabler in form of a sophisticated software, and at the other extreme, it offers a philosophical framework that offers a paradigm shift within the construction sector. In effect, BIM is both of these extremes and everything that comes in between them.

The early definitions of BIM always placed the digital nature of BIM at its core. According to AGC (2006), BIM is "The development and use of a technology to simulate the construction and operation of a facility from which views and data appropriate to various user needs can be extracted and analysed. These data are then used to generate information for making decisions that improve the process of delivering the facility". National Institute of Building Sciences (2007) defines BIM as "A Building Information Model, or BIM, utilizes cutting edge digital technology to establish a computable representation of all the physical and functional characteristics of a facility and its related project/life-cycle information, and is intended to be a repository of information for the facility owner/operator to use and maintain throughout the life-cycle of a facility." More directly, London et al. (2008) stated that "Building Information Modelling (BIM) is an IT enabled approach to managing design data in the AEC/FM industry." Autodesk (2002) views collaborative nature of BIM from technology window. They become further technical by noting BIMs have three main features: Create and operate on digital databases for collaboration; Manage change via the databases to ensure a change to any part of the database is coordinated in all other parts; Capture and preserve information for reuse by adding industry-specific applications. A more all-rounded definition is offered by Eastman et al. (2008) as "A modelling technology and associated set of processes to produce, communicate, and analyse building models. Building models are characterised by:

1. Building components that are represented with intelligent digital representations (objects) that 'know' what they are, and can be associated with computable graphic and data attributes and parametric rules.
2. Components that include data that describe how they behave, as needed for analyses and work processes, e.g., take-off, specification, and energy analysis
3. Consistent and non-redundant data such that changes to component data are represented in all views of the component.
4. Coordinated data such that all views of a model are represented in a coordinated way."

The emphasis by the Computer Integrated Construction research group, C.I.C. (2010), p. 1) goes beyond just digital representation: according to them "Building information model is a shared knowledge resource for information about a facility forming a reliable basis for decisions during its lifecycle; defined as existing from earliest conception to demolition."

The UK's Construction Project Information Committee (CPIC) echo RIBA that BIM model is a focal point for the sharing of information throughout the assets life cycle and places the emphasis on the feature of BIM that enables sharing of knowledge, as follows; "*Building Information Modelling is a digital representation of physical and functional characteristics of a facility creating a shared knowledge resource for information about it forming a reliable basis for decisions during its life cycle, from earliest conception to demolition.*"(RIBA BIM Overlay Report 2012).

The recognition of the increasing need to place "I" at the heart of the definition of BIM gave rise to a new set of definitions. Succar (2008) suggested that BIM is "a set of interacting policies, processes and technologies producing a methodology to manage the essential building design and project data in digital format throughout the building life-cycle." The interesting point in this definition is the indication of a paradigm shift in the ways of business and operational process of construction industry. Moreover, BIM is not restricted to a certain level of achievement in expertise. Process is considered more valuable than use of a certain level of technology or a certain level of product. He further discussed in detail the technology, process, and policy as three main fields of BIM.

The product view of BIM promoted a new wave of definitions. Kymmell (2008) suggests that BIM is "A tool, process and/or product that develop virtual intelligent models linked to other construction management tools (i.e. schedule, estimates) that promote collaboration, visualization and constructability reviews to benefit all stakeholders throughout the lifecycle of the facility."

An industry's attempt by SKANSKA recognises the 'intelligence' inherent in building elements and the relationship between them. They define BIM as *"A method to describe a project and its spaces, structures, components and materials with their essential information and properties. The model is a container for the information."* (Stagg 2011, p. 3).

After emphasizing what BIM is not *"BIM is not CAD. BIM does not have to be 3D. BIM is not application oriented. BIM is not a single building model or a single database. BIM is not Revit (or ArchiCad, or Bently), BIM is not a replacement for people and will not automate you out of existence. BIM is not perfect."*, Jernigan (2007) then suggests that BIM is simply about *managing information to improve understanding.*

4.4 Potential Benefits

In 2011, DBIS predicted that the use of BIM in the UK can generate a net saving of about £1–2.4 bn/annum (DBIS 2011). Kennet (2010) used the 'Heathrow Airport Terminal 5' as a benchmark case study where a saving of £210 m was achieved (5%). The potential for benefits is particularly noted for reducing waste: these include reduction of the number of field coordination errors through early integration of the design models of the main disciplines (Cohen 2010); time and cost savings due to the visualisation of digital model during design and Construction (Lichtig 2005); timely provision of accurate information exchange open to all participants leading to reduced level of unknowns in contract documents (Carbaso 2008); and minimise waste due to reduced uncertainties, guesswork, and inefficiency in preconstruction estimating (Cohen 2010). Figure 4.2 illustrates the measured benefits of 'managed' BIM mapped on the RIBA Plan of Work stages:

BIM offers advances through better visualisation, better coordination, and better management. The benefits of BIM have been categorised into tangible, semi-tangible, and intangible (Becerik-Gerber and Rice 2010). Based on a number of case studies, Manning and Messner (2008) summarised the benefits of BIM at early design stages into the following six areas: Rapid visualization, better decision support upstream in the project development process, rapid and accurate updating of changes, reduction of man-hours, increased communication, and increased confidence in completeness of scope.

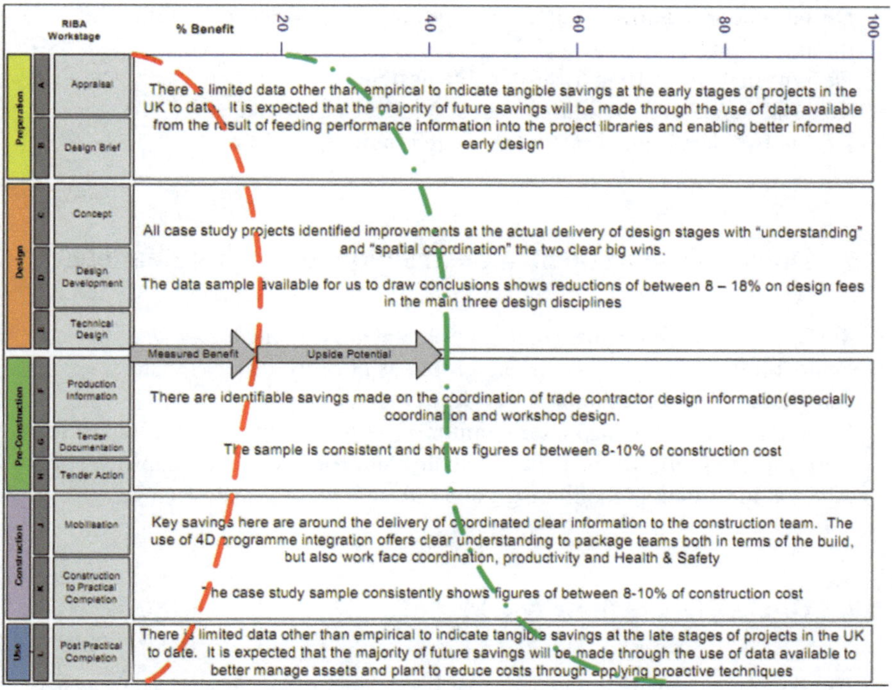

Fig. 4.2 Benefits of BIM(M) against RIBA stages DBIS (2011, p. 92)

Due to platform for early collaboration, BIM facilitates accurate and consistent drawing sets, early collaboration, synchronized design and construction planning, model-driven fabrication, as well as the enabling the implementation of lean techniques, and improved supply chain integration. The technical capabilities of BIM that starts with visualisation and coordination of architectural design and construction work is complemented with numerous tools that apply to all levels of construction lifecycle. These range from clash detection, requirement management, energy calculation (Ashcraft 2008), comfort simulation, environmental analysis, lighting simulation, facilities management and space management, to lifecycle cost evaluation of alternative building design or alternative choice of building components and elements.

The use of BIM would yield different benefits at different phases of the overall process, starting with the conception all the way to operation and Facilities management phases. Some of these benefits are outlined below:

4.5 During Concept & Design

- Visualisation and VR leading to informed and improved decisions and H&S management.

- Design Coordination leading to optimisation; clash detection and waste minimisation
- Design Analysis leading to sustainable design
- Material Schedules, facilitating fast and error-free schedules
- Design Efficiency, allowing to test design options

4.6 During Procurement, Construction, & Commissioning

- 4D BIM, linking schedule to design of maintenance schedule
- Buildability and Logistics, solving problems prior to site
- Clash Management, enabling clash-free construction
- Stage Payment, allowing better monitor of progress & payment mechanism
- Impact of building works, hence, reducing site congestion and improve H&S
- As-built information, facilitating good information for handover

4.7 During Operations & FM

- Facilitating incremental population of FM Database for fast and accurate use
- Providing full knowledge of how it was built for operations and building management
- Safe and efficient management of facility over time for use in new works and changes
- Better product Information by linking to supplier product information
- Facilitating cross-organisation and cross-project knowledge Management
- Safer decommissioning using structure & material information

4.8 BIM and Sustainability

As noted by Autodesk (2006), the use of CAD (or object CAD) for evaluating building performance requires significant human intervention which makes it too laborious, time-consuming, and costly. Further, CAD solutions offer very little in the way of achieving sustainability objectives. BIM, on the other hand, contains all the data needed for supporting sustainable design of projects throughout its whole life cycle, particularly when coupled with relevant performance analyses tools. What makes the process more effective is BIM's ability to provide simultaneous and real-time solution to a diversity of what-if scenarios, all empowered by powerful visual representation of choices.

Sustainability is broadly defined in social, environmental, and economic terms. There are numerous ways by which BIM can contribute to the sustainability agenda. Energy modelling, building orientation (saving energy), lifecycle evaluation, building massing (optimize the building envelope), daylighting analysis, water harvesting, and sustainable materials (to reduce material needs and to use recycled materials) are only a few examples where all three sustainability parameters come together.

While CAD drawings can be used for the certification of Energy and Environmental Design (LEED1), BIM offers a massive leap in this direction. All of these are due to the added advantage yielded by automatic parametric control. However, the most powerful way by which BIM can contribute to sustainability is through minimisation of waste that arises from collaborative working within a unified and integrated supply chain. This is particularly true about the decisions at the design phase where the lifecycle implication of critical decisions can be evaluated collectively, including the client.

4.9 Information Manager

A stand-alone role of Information Manager has been identified in order to look after the principle functions of Common Data Environment Management, Project Information Management and Collaborative Working, Information Exchange and Project Team Management (CIC 2013). The Information Manager also creates and implements the Project Information Plan and Asset Information Plan. This will provide the information at each stage, as well as details of the format of information and level of detail required (CIC 2013). The Information Manager's responsibility does not cover design, though they are expected to facilitate the exchange of information.

Since the Information Manager is central to information communication among architecture, contractors, structural engineer, and building services engineering, they need to have wider knowledge of the construction process. In conjunction with the need for technical knowledge of BIM, the Information Manager also requires soft skills such as communication and power of persuasion. Some argue that the attributes and skills required of the information manager fit those of Quantity Surveyors, which makes them appropriate candidate for the role. Others argue the case for a key player within design consultancy fields such as architecture or engineering.

4.10 Construction Disciplines

There is a need for re-examination of roles within the overall network of supply chain (Henrik and Linderoth 2010). Architecture practices' familiarity with CAD might give the impression that the BIM is likely to affect them less than it does to contractors. This can only represent the technical view of BIM, whereas it cannot be further from the truth, when it comes to fostering collaborative work through BIM. The traditional role of the architects in leading the team that represents the client needs to undergo a significant cultural and procedural development. Naturally, the technological marvels of BIM will assist the designer in this endeavor, but there also must be a will on the part of the designer to embrace the new way of collaborative working with the range of disciplines directly and indirectly involved in the realization of the project. The real-time engagement of all contributory disciplines might be the gateway to the BIM promise-land, but it also offers serious challenges to the current prevailing culture. Similarly, the participative nature of supply chain integration has a serious diminishing influence on the authority of architects, but it empowers them to generate better products and more efficient processes.

In contrast to architectural companies, contractors tend to lack capabilities on CAD management. This may give rise to the need for BIM roles within the company. Indeed, the skill changes required for BIM implementation may well exceed that of the introduction of one or two roles. BIM is likely to restructure contractors' role in changing contractual procedures and emerging procurement routes.

Contractors' specific interest in BIM is driven by the higher demand on interoperability among teams and among software packages. Contractors are also inspired by host of benefits such as better tools for evaluation and simulation, and benefits such as collisions detection; visualisation of the design; fewer errors and corrections in the field; higher reliability; increased use of offsite prefabrication; "what if" scenarios; product visualisation by the client; fewer call backs; scoping capabilities during bidding and purchasing; further analyses such as value engineering; construction sequencing; project demonstration and constructability (Associate General Contractors [AGC] 2006).

RICS's survey of its members revealed that Quantity Surveyor's focus is on Measurement and Cost Estimation and using BIM to link construction schedule data to the BIM model in order to extract quantities from the BIM model (RICS 2011). Monteiro and Martins (2013) reiterate that despite significant automation, QS's skills will still be needed to interpret and examine cost information. In fact, Norcliffe (2013) suggests that the role of PQS will be enhanced by removing them from trivial calculation, thus allowing them to focus on offering better value through enhanced services such as life cycle costing and value management.

BIM is also likely to have a profound effect on project management practices. Equally, the full potential of BIM will also be contingent upon adopting changes in the work tasks and skill sets of the project participants (Froese 2009).

Guidelines for improving the efficiency of government procurement projects included removal of wastes associated with the practice of generating multiple designs for one project. Two models were presented that supported the integrated approach. In the first model, if framework contractors' submission does not meet the cost benchmark, the tender will open up to outside the framework. The second model uses guaranteed maximum price, underwritten by insurance. (Cabinet Office 2011).

Three procurement routes were suggested by The Procurement/Lean Client Task Group for trial on a public sector construction projects. All three models address waste control through early contractor involvement, integration, and transparency. These were The Cost Led Procurement and Integrated Project Insurance model, and Two Stage Open Book model (Procurement/Lean Client Task Group 2012).

The first model relies on a supply chain framework agreement using a set of criteria which promotes collaboration and delivery against cost targets. In the Integrated Project Insurance model, design competition is the basis for the creation of an integrated project team. The criterion for success is based on experience, capability, and fees. Cost overruns, up to a maximum liability cap, are covered by insurance. As for the third model—the Two Stage Open Book model is also based on framework arrangement of contractor consultant teams. Framework bidding is based on a client brief and benchmarked cost target, though the main selection criteria relate to the supply chain strength. The second stage is based on open book.

An area of great uncertainty is the contractual, legal, and insurance issues. However, it has been stipulated that contractual complexity is linked to the level of BIM maturity. There is guidance by both the JCT and the NEC on how to incorporate BIM into the contracts. (JCT 2011).

It is envisaged that at least for up to level 2, there will be little issues relating to incorporation of BIM into standard forms. The forms of contract suitable for public sector projects were NEC3 option C, ACA PPC2000, and JCT Constructing Excellence (Procurement/Lean Client Task Group 2012).

4.11 Supply Chain

Currently, the members of most supply chains in the industry aim to maximise their reward with minimum risk. Production process is geared to lowest cost rather than best value; the bidding processes encourage opportunism and allow bullying; profit is typically improved by reducing quality and through variations; there is little signs of collaborative problem solving; and profits margins tend to suffer due to material, labour, and management wastages as well as design inefficiencies. Moreover, poor communication among the members tends to create misunderstanding, mistrust, and tension, inevitably leading to low quality and low productivity and ultimately blame culture. It is therefore clear that benefits of BIM are intrinsically linked to the level of the supply chain maturity. Five level of maturity apply to supply chain.

Level 1 is referred to as ‘Ad-hoc’: it is unstructured and ill-defined. Operations are carried out more or less sequentially, with very little concern about other members of supply chain. Negotiations are often based on win-lose and there is low awareness and concerns about the interest of the client, leading to high cost and low customer satisfaction.

Level 2 is called ‘Defined’ because supply chain processes are defined and documented. The cost is still high, but customer satisfaction is slightly improved.

Level 3, referred to as 'Linked', shows initial signs of breakthrough in cooperation, resulting in some cost savings and marked improvement in customer satisfaction. Here, processes begin to link to the chain. Organised attempts are made to create coordination and optimisation. Indeed, there are trade-offs of opportunities.

Level 4 is 'Integrated' where cooperation takes place at the process level and advanced supply chain management practices such as collaborative forecasting and risk management are applied. At this level, cost savings and customer satisfaction are significant.

At Level 5, namely, 'Extended' system of processes and functions are established. An elevated level of collaboration with common processes and goals is achieved, thus leading to highest level of cost savings and customer satisfaction. The competition is with external supply chains rather than company against company and organisations within the own supply chain.

4.12 Building Down Barriers

In 1997, a study was sponsored by the MOD, DETR, AMEC, and Laing under the title of Bringing Down Barriers. It was an idealist process-driven model of collaboration that paid a great deal of attention to processes but very little to the technology side of collaboration. Its aim was to take the waste out of:

- Design, through lifecycle and sustainability considerations.
- Construction, through clarity coordination, buildability, and reduced rework.
- Maintenance, through informed lifecycle choices and just-in-time maintenance

This was to be achieved through single point responsibility, simultaneous engineering, and early involvement of all involved and participants drawing from common pools of trades, material, and skills.

BDB was an excellent work with a deep insight into what could today be referred to as the 'spirit of BIM'. However, the aspirations of BDB were never pursued nor realised because its success relied on intent, goodwill, trust, and volunteered process re-engineering by all organisations. Its expectation of the supply chain includes:

- Efficient; resourceful; reliable; financially secure;
- Existence of close and continuous relationship between all parties
- Existence of seamless processes (no disruptions)
- To think and operate outside local needs
- High dependency on soft trust
- Collective involvement in risk management
- Contractor supporting supply chain: Making experts & expertise available; training; improvement plan; investment in competence development; standardisation.

While these seemed like a tall order in 1997, today the advents of BIM have prepared the groundwork for the achievement of these objectives on a wider scale.

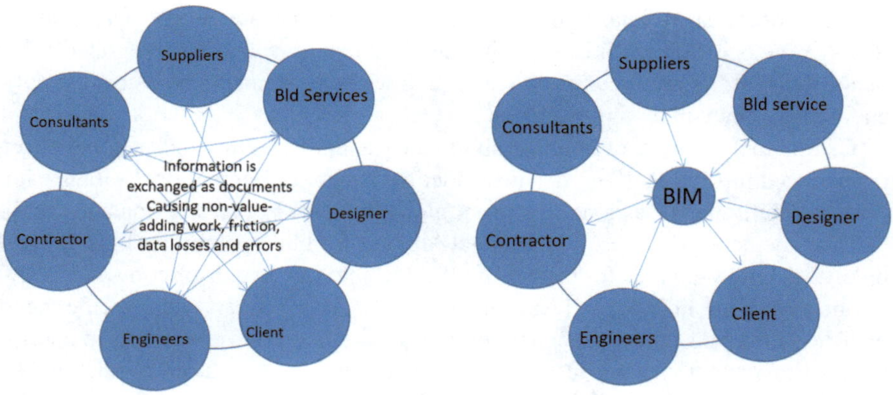

Fig. 4.3 Supply chain communication—Conventional vs. BIM-based interoperability

4.13 Collaboration

Much of the inefficiencies in the construction performance is reflected and attributable to its communication structure. This is depicted in the communication structure in Fig. 4.3. The figure also shows that, in contrast, BIM offers a digital model providing seamless communication using a common database holding all project information. It also facilitates seamless processes, dealing with inadequate or misinterpreted information.

At the heart of BIM lies the way building information is managed and shared by all involved in the project. Based on this premise, 'Interoperability' becomes the core issue for transferability and reusing of information. According to IEEE, Interoperability is the "ability of two or more systems or elements to exchange information and to use the information that has been exchanged". By application, therefore, seamless collaboration among the construction team relies on the ability of their systems to interact and exchange data and information. Therefore, in essence, Interoperability is *"The fundamental characteristic of tools that are designed to work together as part of an integrated system to complete tasks"* (Smith and Tardif 2009, p. 146).

Construction projects involve collaborative contribution of several disciplines throughout the whole lifecycle of buildings. These disciplines tend to use a diversity of software for simulation, calculation, operation, and management of projects. Their interactions need not be sequential, but are often in parallel and need to be seamless. However, the process is significantly hindered due to lack of compatibility between these systems. The cost of inefficient interoperability was studies in 2002 by the likes of National Institute of Standards and Technology (NIST), though some argue that the exact impact has yet to be evaluated (Gallaher et al. 2004; Nisbet and Dinesen 2010). The adverse effects are not simply limited to data re-entry, processing delays, or incompatibility in interpretations. The gravest damage relates to the inability of the supply chain to act in an integrated way, hence their efforts are focused on competing with the members of its own supply chain rather than other supply chains.

Central to the subject of interoperability are standards. To this end, a global effort commenced in 1995 by 12 companies under the banner of International Alliance for Interoperability (IAI), as a not-for-profit global alliance of construction and facilities management industries. International Alliance for Interoperability (IAI) focuses on establishing standards for the use of object technology in construction and facilities management. In 2005, IAI was rebranded to Building Smart with specific focus on BIM, covering wider concepts pertaining to collaborative working including contracts, payment systems, insurance, education, and training. Their mission is "To provide a universal basis for information sharing and process improvement in the construction and facilities management industries". Some of the common file exchange formats and standards used for AEC applications are:

4.14 File Formats

- IFC (Industry Foundation Classes).
- DXF-DWG (Autocad Drawing).
- PDF (Portable Document Format).
- XML (Extensible Markup Language).
- Other native CAD file formats.

4.15 Standards

- ISO PAS 16739 (contains core part of IFC)
- IFC2×4 (2008 release, includes an interface to GIS data)
- IFD Library (International Framework Dictionary) [ISO 12006-3]; for creating uniform object libraries
- IDM (Information delivery manual) [ISO 29481-1]; for guidance on how & when to provide information during project

4.16 Electronic Trading

The manufacturing industry has implemented interoperability in the widest possible extent consisting of numerous automated interrelated tools performing tasks ranging from design visualisation and processing such as measuring, cutting, painting, and packing in a lean integrated system (Smith and Tardif 2009). Following significant inroads by the manufacturing industries, construction industry realised the need to explore the potential use of electronic trading, focusing particularly on interactions between contractors, sub-contractors, and suppliers. Due to construction industry's limitations, caused by fragmentations, electronic trading has not been exploited to its full potential. Its benefits have been better realised more in tendering; requisitions; orders; invoices; acknowledgement; delivery; statements; and remittance.

'Electronic Data Interchange' (EDI) aims to provide sharing of information across multiple software programmes, from one native format to another, while avoiding inefficient and error-ridden task of data re-entry (Faulkner 2006). As yet, there is no single software application that could facilitate seamless information sharing across building design and construction (Eastman et al. 2008).

In 1995, Construction Industry Trading Electronically (CITE) was formed with the mission "To develop and promote the adoption of e-business standards in the construction and facilities management industries".

4.17 Standards

Central to the implementation of BIM are the standards, particularly those associated with data exchange. These are primarily addressed by BuildingSMART (formally known as the IAI), which is an international alliance of industry bodies, software producers, government, and academic institutions with a common goal of achieving open data exchange. In particular, the focus of BuildingSMART has been on the production of a series of object-based exchange file formats called Industry Foundation Classes (IFC). Taking an object view of all construction physical and abstract entities (e.g. process and space), IFC data models have been developed to facilitate sharing and communication between applications. While not every BIM is readily IFC-based, currently IFC offers the greatest potential for exchange (Nisbet and Dinesen 2010). As a neutral file format, IFCs provide a common denominator for software vendors to provide seamless exchanges between differing applications. The IFC format has been recognised by the ISO and is in the process of becoming an official International Standard under ISO 16739 (BuildingSMART 2014). While there have been significant inroads in the IFC development, the work is frequently criticised for not being complete. Alternative solutions include trade-specific neutral file formats such as CIS/2 for the structural steelwork industry (Crotty 2012).

Within the UK, a powerful complementary parallel movement has been the work of British Standards BSI555 group who are primarily the same champions driving BuildingSMART UK. BSI - British Standards Institution (2010). Their output includes British Standard BS1192, which is a code of practice for the collaborative production of architectural, engineering, and construction information. A PAS version (PAS 1192-2:2013) was developed for the design and delivery phases of projects in order to support the UK Government's mandate of achieving Level 2 BIM on all public sector procurement (BSI 2013, p. 3). This covers six phases, namely, assessment and need, procurement, post-contract award, mobilisation, production, and asset information model maintenance (BIM Task group). The Specifications from BSI include:

- PAS 1192-2 Specification for information management for the capital/delivery phase of construction projects using Building Information Modelling. http://shop.bsigroup.com/en/forms/PASs/PAS-1192-2
- PAS 1192-3:2014 – Specification for information management for the operational phase of assets using building information modelling. http://shop.bsigroup.com/upload/Construction_downloads/PAS1192-3%20final%20bookmarked.pdf
- BS 1192-4: Collaborative production of information. Part 4: Fulfilling employer's information exchange requirements using COBie – Code of practice.
- http://shop.bsigroup.com/upload/Construction_downloads/BS1192-4_Collaborati
- PAS 1192-5: Specification for security-minded building information management, digital built environments, and smart asset management. http://shop.bsigroup.com/forms/PASs/PAS-1192-5

These are supported by a BIM Execution Plan (BEP) and, later, a Master Information Deliver Plan (MIDP), reflecting the information delivery requirements and the delivery plan by the supply chain. The roles and responsibilities and timings and criteria of information outputs are also defined. The Common Data Environment (CDE) is defined as a "single source of information for any given project, used to collect, manage and disseminate all relevant approved project documents for multi-disciplinary teams in a managed process" (BSI 2013, p. 3).

Reflecting the pending needs of the industry, COBie was introduced for the delivery of project. COBie provides non-geographical asset data such as product data sheets, warranties, and maintenance requirements in a simple Spreadsheet format. It can also be used to transfer asset data to facilities management to manage the operational phase of the buildings life (Barnes and Davies 2014).

References

Aouad, G., Wu, S., & Lee, A. (2006). nDimensional modeling technology: Past, present and future. *Journal of Computing in Civil Engineering, 20*(3), 151–153.

Ashcraft, H. (2008). Building information modelling: A framework for collaboration. *Construction Lawyer, 28*(3), 5 .bu American Bar Association.

Associate General Contractors – AGC. (2006). *The contractors' guide to BIM, Edition 1. The Associated General Contractors of America*. Las Vegas: AGC.

Autodesk. (2002). *Building information modelling, white paper*. Retrieved February 20, 2010, from http://www.laiserin.com/features/bim/autodesk_bim.pdf

Autodesk. (2006). *A model decision HOK pledges support for building information modelling and Autodesk Revit Building*. Retrieved March 3, 2010, from http://images.autodesk.com/apac_grtrchina_main/files/aec_customer_story_en_v30.pdf

Barnes, P. T. & Davies, N. (2014). *BIM in principle and practice.* Institute of Civil Engineering – ICE Publishing, – 4 Feb 2014.
Becerik-Gerber, B., & Rice, S. (2010). The perceived value of building information modelling in the U.S. building industry. *Journal of Information Technology in Construction (ITcon), 15,* 185–201. Retrieved March 13, 2010, from http://www.itcon.org/2010/15
Bew, M., & Richards, M. (2008). *Bew-Richards BIM Maturity Model, CPIC, 2007.* Avanti—Report from The Construction Research Programme—Project Showcase [online]. Retrieved May 5, 2013, from http://www.cpic.org.uk/en/publications/avanti/index.cfm
BIM Task Group. Retrieved from www.bimtaskgroup.org.
BIM-Industry-Working-Group. (2011). Strategy Paper for the Government Construction Client Group.
British Standards Institution [BSI]. (2010). *BS 6079-1:2010 Project management. Principles and guidelines for the management of projects.* London: British Standards Institution.
British Standards Institution [BSI]. (2013). *PAS 1192-2:2013.* London: BSI.
Cabinet Office. (2011). *Government construction strategy – May 2011* [Online]. Retrieved June 28, 2011, from http://www.cabinetoffice.gov.uk/sites/default/files/resources/Government-Construction-Strategy.pdf
Carbaso, T. (2008). *Integrated project delivery improves efficiency, streamlines construction.* Retrieved from wwww.tradelineinc.com/reports/0A03D1C0%2DB525%2D85702BCEDF900F61/print.
CIC (2013). *Outline scope of services for the role of information management* (1st ed. 2013). CIC/INF MAN/S, Construction Industry Council.
Cohen, J. (2010). *Integrated project delivery: Case studies.* In: Group, A.C.C.I.S.C.A.A.N.I. (ed.).
Computer Integrated Construction research group. (2010). *Building information modelling execution planning guide.* The Computer Integrated Construction research group, The Pennsylvania State University. University, Retrieved from http://www.engr.psu.edu/ae/cic/bimex/
Crotty, R. (2012). *The impact of building information modelling, transforming construction.* Spons Press. isbn:978-0-415-60167-2.
Eastman, C., Teicholz, P., Sacks, R., & Liston, K. (2008). *BIM handbook: A guide to building information modeling for owners, managers, designers, engineers and contractors.* New York: Wiley.
Eastman, C., Teicholz, P., Sacks, R., & Liston, K. (2011). *BIM handbook: a guide to building information modeling for owners, managers, designers, engineers and contractors* (2nd ed.). Wiley. isbn:978-0-470-54137-1.
Faulkner, L. (2006). Interoperability for the Steel Industry [Online]. *Structure Magazine,* Feb(2006). Retrieved July 29, 2011, from http://www.structuremag.org/archives/2006/February-2006/C-Technology-Feb-06.pdf
Froese, T. M. (2009). *The impact of emerging information technology on project management for construction, Automation in Construction, AUTCON-01064*; p. 8
Gallaher, M. P., O'connor, A. C., Dettbarn, J. L., & Gilday, L. T. (2004). Cost analysis of inadequate interoperability in the U.S. Capital Facilities Industry. *National Institute of Standards and Technology, 2004,* 1–210.
Hardin, B. (2009). *BIM and construction management: Proven tools, methods and workflows.* Indiana: Wiley.
Henrik, C. J., & Linderoth. (2010). Understanding adoption and use of BIM as the creation of actor networks. *Automation in Construction, 19,* 66–72.
Howell, I., & Batcheler B. (2004). *Building information modelling two years later – Huge potential, some success and several limitations* [Online]. Retrieved July 24, 2011, from http://www.laiserin.com/features/bim/newforma_bim.pdf
International Construction Intelligence. (2009). *Building information modelling – Coming of age?. 21*(2) – 2nd Quarter 2009 – U.S. Edition, Retrieved February 1, 2010, from http://www.fgould.com/media/resources/files/ICI-US-Qtr2-2009.pdf
JCT. (2011). *Public sector supplement, fair payment, transparency and building information modelling.* Joint Contracts Tribunal Limited, Publisher Sweet & Maxwell, 100 Ave Road London NW3 3PF, part of Thomson Reuters UKL Ltd.

Jernigan, F. (2007). *Big BIM little bim*. Salisbury, MD: 4Site Press.
Kennett, S. (2010). *The use of force: Building information modelling*. Building.co.uk. Retrieved October 22, 2010, from http://www.building.co.uk/the-use-of-force-building-information-modelling/5007579.article
Kramer, M. E., Klein, J., & Steel, J. R. H. (2012). *Building specifications as a domain-specific aspect language*. ACM
Kymmell, W. (2008). *Building information modelling: planning and managing construction projects with 4D CAD and simulation*. New York: McGraw Hill.
Lichtig, W. A. (2005). Sutter health: Developing a contracting model to support lean project delivery. *Lean Construction Journal, 2*(1), 105–112.
London, K., Singh, V., Taylor, C., Gu, N., & Brankovic, L. (2008). Building information modelling project decision support framework. *Proceedings of 24th Annual ARCOM (Association of Researchers in Construction Management) Conference, Cardiff* (pp. 665–673).
Manning, R., & Messner, J. I. (2008). Case studies in BIM implementation for programming of healthcare facilities. *ITcon, 13*, 446.
Monteiro, A., & Martins, J. P. (2013). A survey on modeling guidelines for quantity takeoff-oriented BIMbased design. *Automation in Construction, 35*, 238–253.
National Institute of Building Sciences. (2007). *National Building Information Modelling Standard—Version 1.0—Part 1: Overview, principles and methodologies*. National Institute of Building Sciences.
NCC Standing Committee. (1979). *On Computing & data co-ordination, project information, its content and arrangement*. A report and proposals on the way forward. Project Information Group, Department of the Environment
Nisbet, N., & Dinesen, B. (2010). *Constructing the business case: Building information modelling*. London: British Standards Institution.
Norcliffe, S. (2013). *BIM: Is this the end of the PQS?* Leeds Metropolitan University, 2013, p. iv, 53 leaves.
Procurement/Lean Client Task Group Report. 2012. Retrieved from https://www.gov.uk/government/publications/government-construction-task-groups.
RIBA. (2012). *BIM overlay to the RIBA plan of work*.
RICS. (2011). *Building information modelling survey report, BCIS – Building cost information service*. London: Royal Institute of Chartered Surveyors www.bcis.co.uk.
Robert, E., & Laepple, E. S. J. (2003, December). *Digital innovation and organizational change in design practice* (CRS Center Working Paper no. 2). CRS Center, Texas A&M University.
buildingSMART. (2014). Developing the open standards, tools and training that will drive the successful uptake of BIM, *buildingSMART Bulletin*, Bulletin 15, January 2014.
Smith, K. D., & Tardif, M. (2009). *Building information modelling: A strategic implementation guide for architects, engineers*. New Jersey: Wiley.
Smith, P. (2014). BIM & the 5D Project Cost Manager. Proceedia – Social And Behavioural Science. *119(Selected papers from the 27th IPMA (International Project Management Association)), World Congress, Dubrovnik, Croatia, 2013*, 4754–484. Doi:10.1016/j.sbswpro.2014.03.053
Stagg, M. (2011). CPA Presentation; Major Contractors and BIM, Skanska UK. Retrieved February 17, 2011, from http://www.howittconsulting.co.uk/download/skanska%20presentation.pdf
Succar, B. (2008). Building information modelling framework: A research and delivery foundation for industry stakeholders. *Automation in Construction, 18*, 357–375.
Yan, H., & Damian, P. (2009). *Benefits and barriers of building information modelling*. Loughborough: Department of Civil and Building Engineering, Loughborough University.
Ziel, R., Haus, A., & Tulke, A. (2008). Quantification of the pore size distribution (porosity profiles) in microfiltration membranes by SEM, TEM and computer image analysis. *Journal of membrane science, 323*(2), 241–246.

Chapter 5
Using Agile Project Management and BIM for Improved Building Performance

Mohammad H. Sakikhales and Spyros Stravoravdis

Abstract The early design stage is the most crucial stage to achieve sustainability targets because this is when major design decisions that affect sustainability performance are taken. A way to facilitate more effective decisions in this stage is to explore as many design alternatives as possible and analyse them to find the optimum one. However, the traditional architecture workflow does not support an iterative design process that allows for the exploration and evaluation of many design alternatives. On the contrary, design decisions are made without supporting performance analysis, which is typically done at the end of the process. On the other hand, other industries such as IT and more specifically software development realised the importance of early analysis and feedback and understood the potential of using an iterative design process for improving product performance. To facilitate this, agile project management was developed which focuses on the incremental development of a product by implementing an iterative process of designing and testing and delivering a better quality product in each iteration. Although the advantages of the agile project management have been proven in the IT industry, this method has hardly been used in the construction industry and the applicability of it has not been studied.

This chapter explains the advantages of agile project management through an extended literature review and analyses the potential benefits from the adoption of this methodology in the construction industry and sustainable design process. It introduces an iterative design framework for the design phase of construction projects, using agile principles. Then, it explores how Building Information Modelling (BIM) can facilitate the implementation of this framework to achieve improved building performance.

M.H. Sakikhales (✉) • S. Stravoravdis
Built Environment Department, University of Greenwich, London, UK
e-mail: M.H.Sakikhales@greenwich.ac.uk; S.stravoravdis@greenwich.ac.uk

M. Dastbaz et al. (eds.), *Building Information Modelling, Building Performance, Design and Smart Construction*, DOI 10.1007/978-3-319-50346-2_5

5.1 Introduction

The early design stage has the most impact on achieving sustainability targets, as effective decisions are usually made at this stage (Azhar et al. 2009). However, many factors such as building mass, building orientation, surface area-to-volume ratio, thermal insulation, natural lighting and natural ventilation, and so on can affect the building performance (Smeds and Wall 2007). These aspects have the potential to significantly decrease the energy requirements and improve occupant thermal comfort. For instance, optimising a building orientation, shape, insulation, and ventilation can decrease the heating load of a building up to 80% (Jansson et al. 2013). In order to optimise a building design at the early design stage to take advantage of all these potential savings, it is necessary to evaluate the contribution of each factor before selecting the optimum one as the design evolves. Nevertheless, traditional architecture workflow does not support enough opportunities to analyse design alternatives at the early design stage. Traditionally, as a building design evolves at the early design stage, many changes are implemented in the design due to practical, regulatory, aesthetic, cost, or other considerations and rarely any consideration on other criteria such as energy performance. Therefore, any performance analysis usually is done at the later stage and only for the finalised alternative.

However, the engineering industry has realised the benefits of using the iterative design process to analyse and explore many design alternatives at the earliest stages in order to improve the specification of a product (Ulrich and Eppinger 2012). For instance, the Information Technology (IT) industry faced the same problems a few years ago when most of their products did not perform as designed (Highsmith 2009) mainly because they were tested only after the design and coding processes were completed. Therefore, agile management was developed to support an iterative design process and improve the performance of the products. Experts in the IT industry realised that using such an iterative process can maximise cooperation, which results in delivering a higher quality product more quickly than with traditional methods (Frank Cervone 2011). This led them to develop a new product development method called agile project management.

On the other hand, Building Information Modelling (BIM) as a set of policies, processes, and tools has the potential to facilitate iterative design to achieve improved building performance (Sakikhales and Stravoravdis 2015). However, current research on integrating building performance and BIM has put a lot of emphasis on improving software capabilities, and there is little research on how BIM processes can influence project sustainability outcomes (Wu and Issa 2015).

This chapter explains the advantages of agile project management through an extended literature review and analyses the potential benefits from the adaption of this methodology in the construction industry and sustainable design process. It introduces an iterative design framework for the design phase of construction projects, using agile principles. Then, it explores how Building Information Modelling (BIM) can facilitate the implementation of this framework to achieve improved building performance.

5.2 Method

In order to assess the appropriateness of agile project management for improved building performance, firstly, an extended literature review was conducted to assess the agile approach and its benefits in the IT industry. Secondly, to assess the potential, the advantages, the limits and barriers of implementing this approach in construction projects, a literature review through case studies was conducted, while a mapping exercise followed, where agile principles were analysed so that the agile phases can be mapped to the design work stages. The findings from the literature review and analysis of agile principles and phases were then used to propose an iterative design framework for the design phase in construction projects.

5.3 What Is an Agile Project Management?

5.3.1 Classic Project Management

In the software development industry, traditional project management or waterfall methodology is a well-known approach which is a sequential process of life-cycle model. As can be seen in Fig. 5.1, in this methodology every step must be completed before the next step can start. The waterfall methodology has been embraced by the software development industry, but also is well known in other industries including construction. However, there is an argument that this method is robust in simpler

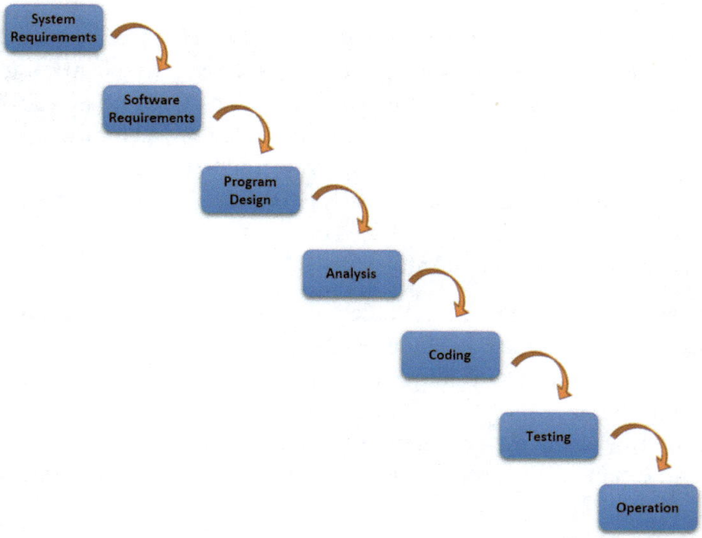

Fig. 5.1 Waterfall software development workflow reproduced from Royce (1970)

domains when cause and effect has a known and direct relationship; but when variability is high and change is inevitable the waterfall system does not work efficiently (Measey and Radtac 2015) because feedback is usually received only at the end stage when it is expensive, too late, and difficult to return to previous stages. For instance, a development team may spend 18 months building a service that doesn't meet users' needs at the end (Agile delivery Community 2016).

Some voices were raised against this method in the 1980s and 1990s (Owen and Koskela 2006) which subsequently led to developing the agile management method.

5.3.2 Agile Project Management

The agile project management idea can be traced back to ideas from a paper by Takeuchi and Nonaka in the January 1986 issue of the Harvard Business Review. However, the agile concept did not gain attraction until Jeff Sutherland and Ken Schwaber discussed the first agile method for software development in 1995 (Frank Cervone 2011). Later, in 2001, seventeen software developers gathered together in Utah and wrote the Agile Manifesto.

"*We are uncovering better ways of developing software by doing it and helping others do it. Through this work we have come to value:*

Individuals and interactions	*over*	*processes and tools*
Working software	*over*	*comprehensive documentation*
Customer collaboration	*over*	*contract negotiation*
Responding to change	*over*	*following a plan*

That is, while there is value in the items on the right, we value the items on the left more" (Beck et al. 2001a).

These concepts inspired twelve agile principles which cover customer satisfaction, welcoming changes, delivering working software frequently, face-to-face communication, and regular intervals team reflection (Beck et al. 2001b). These characteristics can help project team members to quickly adapt the unpredictable and rapidly changing requirements during project timeline (Frank Cervone 2011). Agile management uses these changes as opportunities to enhance the value of the final product (Owen and Koskela 2006). The five phases of agile project management which are shown in Fig. 5.2 are:

- Envision: Determining project objectives, constraints, and community
- Speculate: Developing a feature-based release plan
- Explore: Running and testing features in a short iteration to reduce the risk of the project
- Adapt: Reviewing the delivered results and adapt as necessary
- Close: Conclude the project and review key learning (Highsmith 2009)

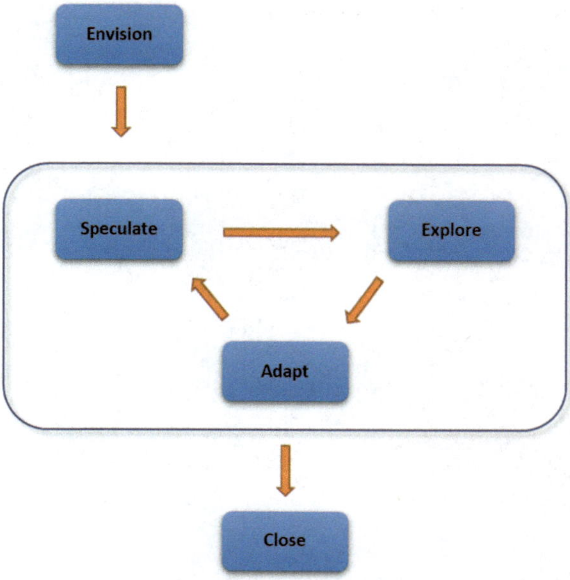

Fig. 5.2 Agile workflow reproduced from Highsmith (2009)

5.3.3 Benefits of Agile Management

Agile is focused on delivering business value through the frequent iteration in a shorter time framework (Measey and Radtac 2015). Many surveys and studies have been conducted to show benefits of using agile management (Highsmith 2009; Vijayasarathy and Turk 2008; Owen and Koskela 2006). As can be seen in Table 5.1, the majority of respondents agreed on substantial improvement in productivity, quality, and business satisfaction in their project. Moreover, cost reduction is also reported as one the important agile benefits. In addition to these qualitative studies, a series of quantitative studies by Michael Mah pictures more tangible benefits of the agile management. Figures 5.3 and 5.4 summarise two of these studies (Highsmith 2009).

Figure 5.3 illustrates improvements in performance at Scientific Instrument Company between agile and non-agile projects which shows 83% improvement in cumulative defects. Figure 5.4 also shows general improvements in BMC software projects versus the industry norms. The most significant number is 58% improvement in the project schedule. Moreover, the chaos manifesto report from the Standish Group in 2011, as illustrated in Fig. 5.5, compares the results of using waterfall and agile management in the software development projects from 2002 to 2010 (Measey and Radtac 2015) which shows agile projects are three times more successful than waterfall projects.

However, implementing agile management has some difficulties. This method is relatively difficult to people who used to work with the traditional method. (Bowes 2014; Agile delivery Community 2016). The traditional waterfall method is a linear

Table 5.1 Improvement in agile project management

	Version One (Highsmith, 2009)	**Shine technology** (Owen and Koskela, 2006)	**Vijayasarathy & Turk** (Vijayasarathy and Turk, 2008)
Increase Productivity	89% of respondents	93% of respondents	6 out of 7
Reduce Cost	66% of respondents	49% of respondents	5.34 out of 7
Improve quality	84% of respondents	88% of respondents	6.18 out of 7
Better customer satisfaction	-	83% of respondents	6.28 out of 7

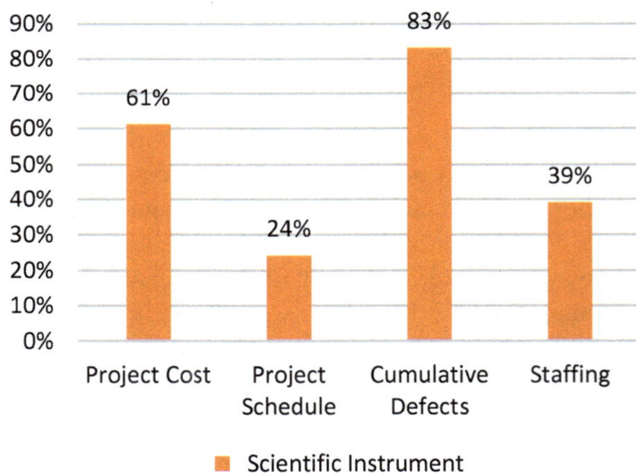

Fig. 5.3 Improvement in agile project management against other method adapted from Highsmith (2009)

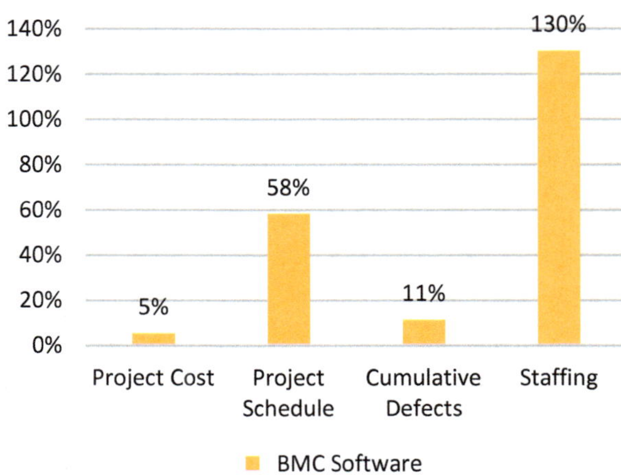

Fig. 5.4 Improvement in agile project management against industry norm adapted from Highsmith (2009)

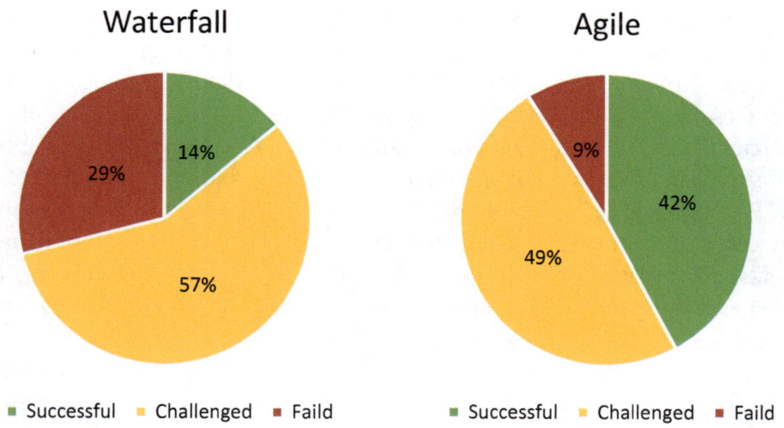

Fig. 5.5 Comparing success rate in agile and waterfall projects adapted from Measey and Radtac (2015)

and sequential method, easy to understand and implement. On the contrary, if an agile method is implemented inappropriately, it can introduce extra inefficiencies (Bowes 2014). Research shows that organisational resistance, management apathy, and inadequate training are the most important barriers towards implementing agile project management (Vijayasarathy and Turk 2008).

Benefits of using agile have been proven in the IT industry and now in the UK, it is mandatory to use the agile approach to build and run government digital services (Agile delivery Community 2016). Having established that agile management offers significant improvements in project delivery, it is worth exploring the application of agile in the construction industry.

5.4 Application of Agile in the Construction Industry

Many other types of projects resemble the same characteristics of IT projects such as creativity, iterative design and testing, dynamic development environment, and constant change (Conforto et al. 2014). Therefore, agile practices, techniques, and tools can be adapted to these types of projects. While there is an extensive research on the effectiveness of agile management in the software industry, there is a lack of empirical studies on the use of agile management in the other industries (Conforto et al. 2014). A recent survey by the Software Advice Company (Leslie 2015) indicates that agile is not common in architecture and construction industries with only 10% of their project managers use this method. The construction industry is famous for adapting tools and techniques from other industries, such as lean methodology from the manufacturing industry. However, there are not many examples of implementing agile management in the construction industry. Recently, however, some firms have understood the importance of this methodology and have tried to implement it in their projects, such as Proving ground, a consultancy which specialises in data-driven design and construction that has published on the benefits and use of agile on their project (Nate 2015). Moreover, from 2012 to 2014, Centrus Energy Corp and the US Department of Energy executed a cooperative research, development, and demonstration programme including a commercial nuclear plant. The project adopted many agile principles to complete, on time, under budget and safely. For instance, instead of a normal sequential construction, the project team implemented component, system, subsystem, and integrated test before delivered any functional system to the customer. Although they did not officially use agile management, informal adoption and use of the agile concepts was the key player to the project's success. (Straçusser 2015)

Researchers also have begun to explore the application of agile management in construction. Demir et al. (2012) combine agile and lean concepts to present a potentially new framework, called "AgiLean PM", which illustrates how a project can react to change through agile management. Suresh Kumar and McArthur (2015) propose the application of agile project management to BIM in order to optimise the process of collaboration, documentation, and life-cycle data management. Tomek and Kalinichuk (2015) introduce a hybrid scheduling approach which is a synthesis of the agile project management and Building Information Modelling and explore its practical application in the construction industry.

5.4.1 Potential of Agile Management for the Construction Projects

The Royal Institute of British Architects (RIBA) Plan of Works is the most well-known workflow of the design and construction projects in the UK. This workflow is a descriptive and point-to-point process and each stage starts when the previous one has finished and as a result, like a waterfall methodology it cannot represent the iterative process. (Pektasx and Pultar 2006). Figure 5.6 compares the RIBA plan of works with the waterfall method from the IT industry. Although they may have different stages, overall they follow the same logic.

Implementing this stage to stage workflow increases the chance of discarding design alternatives at the early stage before their capacities are truly explored (Sakikhales and Stravoravdis 2015). Therefore, it is important to introduce a new workflow that supports iterative design. On the other hand, there is an argument that

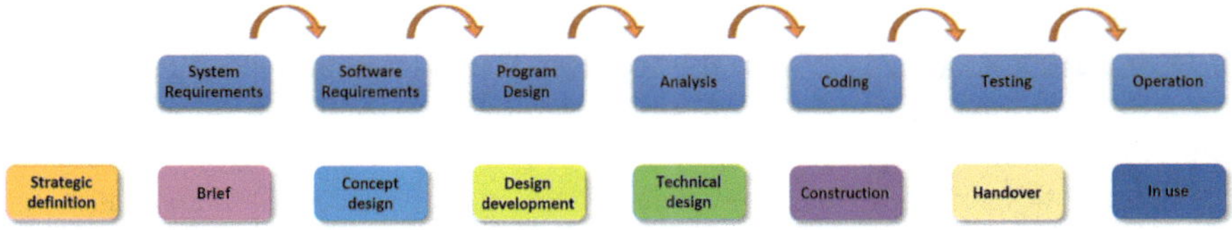

Fig. 5.6 Comparing RIBA plan of work and waterfall method

construction projects are not a good candidate for an iterative process as it has sequential nature and changes are expensive as projects move forward (Straçusser 2015). However, the construction process consists of a broad variety of activities with different nature which may need a different methodology to work best. Based on their nature, these activities can be categorised in four phases: pre-design, design, construction, and operation. The characteristics of activities in these four phases are different; therefore, suitable management systems and tools can be implemented for each one, as needed. For instance, while design phase has to be finished in a short time (e.g. a few months) and involves lots of inspiration and creative activities, the operation phase is a long-term process (e.g. 50 years) with managing day-to-day activities. Owen et al. (2006) analyse the potential of agile management for the pre-design, design, and construction phases based on criteria such as organisational attitudes and practices, planning and execution, and conclude that:

- The agile management has the potential to improve the pre-design phase because it can address the new constantly emerging opportunities and risks and ability to devise the creative solutions.
- The adoption of agile principles in the design phase is very appropriate. It allows the development of high quality and complex products at lowest possible cost. It is particularly appropriate when solutions evolve or requirements change during the project.
- Implementation of agile management in the construction phase is expected to be harder because the consequences of changes may cause big impacts on the project objectives.

The in-use phase involves facility management in general and operation and maintenance in particular. The nature of facility management activities is more similar to the business management method (Cotts et al. 2009). Planning and programming are longer (2–10 years) in this phase, and it is harder to find the benefits of using agile management.

There are other factors that can be considered for adapting an agile management approach including the number of people in the team, the level of education in team members, the culture of team members towards change and client involvement. Considering these factors in all four different phases, design is the most compatible one to agile, as the number of team members is relatively small, design team members are usually well educated, they are more flexible to change and finally client involvement in the design phase is common. Therefore, the design phase can get more benefits from an agile approach.

5.5 An Agile Framework for Iterative Design

Based on the definition of agile management phases, it is possible to map these phases with the activities during the design process. Some of the phases can be entirely mapped to the RIBA plan of works stages. For instance, the explore phase which includes running and testing can be mapped into design stages (e.g.

Table 5.2 Mapping agile management phases and design work stages

Agile management phases	Design work stages
Envision	Strategic definition
Speculate	Brief & preparation
Explore	Design alternatives, Sustainability assessment
Adapt	Review and feedback
Close	Final documentation for the construction

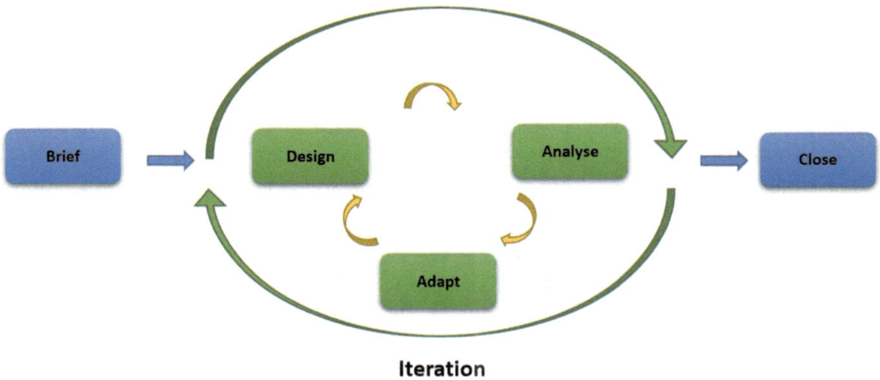

Fig. 5.7 Agile iterative design framework

conceptual design and design development) from the RIBA plan of works. However, for some phases such as adapt, it is harder to find a direct relationship to any work stage. Table 5.2 shows the proposed mapping process for these processes.

Based on the agile project management workflow (Fig. 5.2) and the above mapping process, the following iterative design framework for integrating design and building performance has been proposed which is illustrated in Fig. 5.7. This framework allows to explore many design options and evaluate them to find the optimum alternative.

Brief: in the brief phase, project objectives including quality, budget, and sustainability targets should be developed. It is important to address all environmental requirements for the project. If the project needs to follow a specific time frame, it is vital to have clear sustainability targets for each milestone to achieve. Iteration time should be developed based on the timeline, complexity of the project, and sustainability targets at this stage. The first iteration is usually longer than the others as the project team has to develop the initial outline, for instance, 1 month for the first iteration and then 1 or 2 weeks for the later iterations. The collaboration method

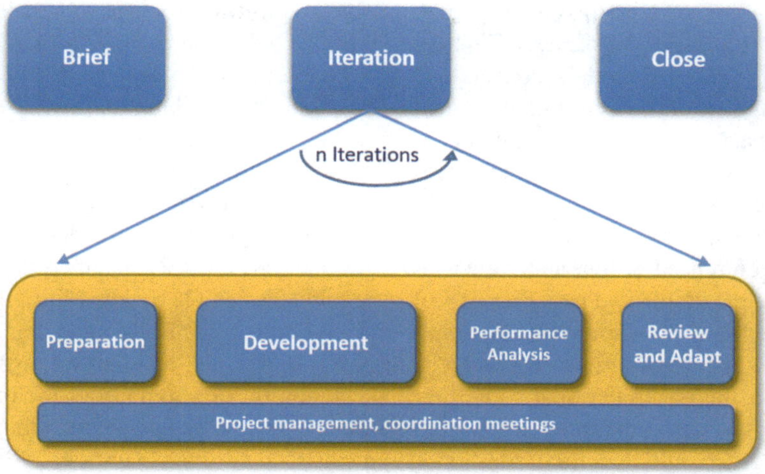

Fig. 5.8 Agile Iterative design framework life cycle

including any data exchange format, quality assurance, and communication system has to be developed in this phase.

Iteration: this phase consists of three steps: design, analyse, and adapt. This includes determining project characteristics, searching for the solutions, exploring and developing design alternatives in accordance with the brief. Performance analysis mainly includes 3D simulation analysis but it may also include physical modelling. This performance analysis can be done for sun and shadow, solar, lighting, wind, energy, and so on. In this phase, it is important to create a collaborative project community for all team members to communicate and work collaboratively which could be facilitated by the project leader (Highsmith 2009). The adapt is the most important phase of this cycle. The designers may be familiar with the process of constantly changing and reviewing their concepts but here they need to review their work not only from aesthetics but also from sustainability and performance perspectives. The feedback from this stage can be used to modify the design in the next iteration.

Close: the key objective of this phase is to finalise the project and review key learning points and passing the project to the next team which is the construction team. Figure 5.8 shows the agile iterative design framework life cycle.

5.6 Discussion

Construction projects traditionally perform in a linear continuous workflow. Many new approaches such as Integrated Project Management (IPM), Integrated Project Delivery (IPD), and Building Information Modelling (BIM) have been introduced to overcome the barriers of this workflow. Agile and BIM have many core principles in common. Similar to agile management objectives, BIM promises to reduce project cost and delivery time and increase its productivity and quality (Azhar 2011). Using traditional models, conventional CAD and drawing to analyse building performance is costly and time-consuming (Autodesk 2005) which makes it the most important barrier to implementing sustainable design and construction (Azhar, Brown, and Farooqui 2009). However, BIM as a data-rich, object-oriented, intelligent and parametric digital representation of a building has the potential to integrate building performance analysis during the iterative design process in order to explore more optimised design alternatives. Kryegiel and Nies (2008) specified that BIM can help with the two major aspects of sustainable design (1) building form including building orientation, building mass, and daylighting, (2) Building systems such as energy modelling, waste management, water harvesting, renewable energy, and sustainable materials. Beyond a 3D representation of the geometry of the project, BIM can carry important information to support sustainable design including location, material thermal properties, operational hours, and so on. This greatly simplifies the often cumbersome and difficult analysis and provides immediate feedback on design options early on in the design process (Azhar, Brown and Farooqui 2009) which makes it appropriate for an iterative process. Schemes such as IFC and gbXML have been developed to facilitate data exchange process between design analysis software and to reduce the need to remodel the same project in analysis software (Bahar et al. 2013) that could result in a significant time saving.

BIM documents such as Employer's Information Requirements (EIR) and BIM Execution Plan (BEP) can cover important information in the brief phase including project scope, project objective, communication and collaboration method, exchange information format, and so on. However, too much emphasis on documentation without having the agile mindset might create conflict.

The adapt phase is probably the most challenging part for BIM. One the one hand, BIM promises better collaboration and communication among all parties involved in a project which is aligned to agile principles; on the other hand, most of BIM communication and collaboration platforms are cloud based which contradicts with the face-to-face communication in the agile management process. However, with the advent of Augmented Reality (AR), there is a chance that in the near future web-based communication may become more effective and more similar to the face-to-face communication.

5.7 Conclusion

The early design stage is the most critical stage to achieve building performance targets as this is when the important decisions are made. However, the traditional architectural workflow does not allow analysis of many design alternatives at the earlier stages to find the optimum one, and it is typically done only for the finalised alternative. On the other hand, other industries such as IT with the similar problem understood the significant impact of early analysis and feedback on the product performance. Therefore, agile project management was developed to, unlike the traditional method, support iterative design and testing process. Implementing agile management has brought significant improvement in the IT projects and resulted in reducing cost, increasing productivity and improving delivery. As the design phase in the construction industry has the same characteristics of a software development project, application of agile can be applicable. This chapter introduces the agile iterative design framework which supports incremental development in each iteration to achieve higher building performance. In this framework, each iteration has three phases: design, analyse, and adapt which allows exploring, evaluating, and developing design alternatives. Building information modelling can facilitate the implementation of this approach by improving the data exchange process to enable faster and more accurate data transferring and also by providing important documents on collaboration and facilitating communication among team members.

5.7.1 Limitations and Future Work

The application of agile project management is relatively new to the construction industry, and there is not enough proof yet to support the potential benefits of implementing this method. This chapter presented a theoretical framework for using agile project management, BIM, and iterative design to achieve improved building performance. However, the application of this framework has not been evaluated in practice or in specific situations. The outlined framework will form the basis for a series of case studies to analyse the effectiveness of this framework in different types of projects. The legal and contractual aspects of using this approach are beyond the scope of this chapter and will be investigated in the future.

References

Agile delivery Community (2016). *Agile government services introduction.* Retrieved April 20, 2016, from https://www.gov.uk/service-manual/agile-delivery/agile-government-services-introduction

Autodesk, I. (2005). *Building Information Modeling for Sustainable Design*. Autodesk, Inc. White Paper.

Azhar, S. (2011). Building information modeling (BIM): Trends, benefits, risks, and challenges for the AEC industry. *Leadership and Management in Engineering, 11*(3), 241–252.

Azhar, S., Brown, J., & Farooqui, R. (2009). BIM-based Sustainability Analysis: An Evaluation of Building Performance Analysis Software. *45th Associated Schools of Construction International Conference*, Gainesville.

Bahar, Y. N., Pere, C., Landrieu, J., & Nicolle, C. (2013). A thermal simulation tool for building and its interoperability through the building information modeling (BIM) platform. *Buildings, 3*, 380–398.

Beck, K., Beedle, M., van Bennekum, A., Cockburn, A., Cunningham, W., Fowler, M., et al. (2001a). *Agilemanifesto.org*. Retrieved March 15, 2016, from http://agilemanifesto.org/

Beck, K., Beedle, M., van Bennekum, A., Cockburn, A., Cunningham, W., Fowler, M., et al. (2001b). *Principles behind the Agile Manifesto*. Retrieved April 15, 2016, from http://www.agilemanifesto.org/principles.html

Bowes, J. (2014). *Agile vs waterfall: Comparing project management methods*. Retrieved April 15, 2016, from https://manifesto.co.uk/agile-vs-waterfall-comparing-project-management-methodologies/

Conforto, E. C., Salum, F., Amaral, D. C., Silva, S. L. d., & Almeida, L. F. M. d. (2014). Can agile project management be adopted by industries other than software development? *Project Management Journal, 45*(3), 21–34.

Cotts, D. G., Roper, K. O., & Payant, R. P. (2009). *The facility management handbook*. New York: Amacom.

Demir, S. T., Bryde, D. J., Fearon, D. J., & Ochieng, E. G. (2012). Re- conceptualising agile for lean construction: The case for "agilean" project management. In *Proceedings of the 28th Annual ARCOM Conference*, Edinburgh, pp. 1013–1023.

Frank Cervone, H. (2011). Understanding agile project management methods using Scrum. *OCLC Systems & Services: International Digital Library Perspectives, 27*(1), 18–22.

Highsmith, J. (2009). *Agile project management: Creating innovative*. Boston: Pearson Education.

Jansson, G., Schade, J., & Olofsson, T. (2013). Requirements management for the design of energy efficient building. *Journal of Information Technology in Construction, 18*, 321–337.

Krygiel, E., & Nies, B. (2008). *Green BIM: Successful sustainable design with building information modeling*. Indianapolis: Willy.

Leslie, J. (2015). *Agile Project Management Software User Report–2015*. Retrieved March 25, 2016, from http://www.softwareadvice.com/resources/agile-project-management-user-trends-2015/

Measey, P., & Radtac. (2015). *Agile foundation*. London: BCS, The Chartered Institute for IT.

Nate (2015). *What building teams can learn from scrum*. Retrieved April 20, 2016, from https://provingground.io/2015/08/03/what-building-teams-can-learn-from-scrum/

Owen, R. L., & Koskela, L. (2006). Agile Construction Project Management. In *6th International Postgraduate Research Conference in the Built and Human Environment*, Delft, pp. 22–33.

Owen, R., Koskela, L., Henrich, G., & Codinhoto, R. (2006). Is agile project management applicable to construction? In *14th Annual Conference of the International Group for Lean Construction*, Santiago, pp. 51–66.

Pektasx, S. T., & Pultar, M. (2006). Modelling detailed information flows in building design with the parameter based design structure matrix. *Design Studies, 27*, 99–122.

Royce, W. W. (1970). *Managing the development of large software systems*. In *Proceedings of IEEE WESCON*, pp. 1–9.

Sakikhales, M. H., & Stravoravdis, S. (2015). Using BIM to facilitate iterative design. *Building Information Modelling (BIM) in Design. Construction and Operations, 149*, 9–19.

Smeds, J., & Wall, M. (2007). Enhanced energy conservation in houses through high performance design. *Energy and Buildings, 39*, 273–278.

Straçusser, G. (2015). *Agile project management concepts applied to construction and other non-IT fields*. Retrieved from http://www.pmi.org/learning/agile-software-applied-to-construction-9931

Suresh Kumar, S., & McArthur, J. J. (2015). Model, Streamlining Building information creation using Agile project management. In *Building Information Modelling (BIM) in Design, Construction and Operations*, Bristol, pp. 229–240.

Tomek, R., & Kalinichuk, S. (2015). Agile PM and BIM: A hybrid scheduling approach for a technological construction project. *Procedia Engineering, 12*, 557–564.

Ulrich, K. T., & Eppinger, S. D. (2012). *Product design and development* (5th ed.). New York: Mc Graw Hill.

Vijayasarathy, L. R., & Turk, D. (2008). Agile software development. *Journal of Information Technology Management, 19*(2), 1–8.

Wu, W., & Issa, R. R. A. (2015). An integrated green BIM process model (IGBPM) for BIM execution planning in green building projects. In R. R. A. Issa & S. Olbina (Eds.), *Building Information Modeling*, American Society of Civil Engineers, Reston, VA, pp. 135–165.

Chapter 6
Use of Simulation Through BIM-Enabled Virtual Projects to Enhance Learning and Soft Employability Skills in Architectural Technology Education

Barry J. Gledson and Susan Dawson

Abstract Traditional teacher-led methods in higher education such as the customary use of lectures can result in passive learning behaviour being adopted by students. Academic theorists consider alternative approaches that encourage active learning more desirable. This chapter reports on the use of a simulated virtual project designed to introduce project-based learning into the classroom environment. Over a 2-day duration, several pairs of second year undergraduate Architectural Technology students developed competing designs across multiple predetermined work stages. Starting from an initial strategy briefing, students were able to rapidly progress through to a more developed design stage because of the use of BIM technology and processes. Data were collected from participants to measure perceptions of the various areas of learning and skills development that had occurred because of their participation in the virtual project. Findings indicate that students believe they had gained greater depth of subject understanding and developed a range of personal, interpersonal, self-management, and initiative and delivery skills. The conclusion is that further introduction of project-based learning via simulation using virtual projects could be of high value across a range of built environment programmes taught across the higher education sector.

B.J. Gledson (✉) • S. Dawson
Faculty of Engineering and Environment, Northumbria University,
Newcastle upon Tyne NE1 8ST, UK
e-mail: barry.gledson@northumbria.ac.uk

M. Dastbaz et al. (eds.), *Building Information Modelling, Building Performance, Design and Smart Construction*, DOI 10.1007/978-3-319-50346-2_6

6.1 Introduction

> *"College is a place where a professor's lecture notes go straight to the students' lecture notes, without passing through the brains of either"* (attributed to Edwin Emery Slossan, circa 1910)

Traditional teacher-led classroom activity, including the prevailing lecture type delivery common to Universities continues to attract criticisms along the lines of the above. In a single academic year spent studying in a Higher Education Institution (HEI) that packages related classes into modules, a student taking six typical 20-credit modules, each employing a 1-h weekly lecture and a 1-h accompanying seminar delivery pattern would spend nearly 4 weeks in lectures alone. This time can largely be classified as *"passive learning"* in which the only activities engaged in are listening and note taking (Exley and Dennick 2009). Built environment education has conventionally been delivered through similar "chalk and talk" methods of teaching, which has not optimised student learning (Forsythe et al. 2013). Project-based learning (PBL) where active learning occurs through student endeavour focusing on problem-solving and the application of taught skills has been recognised as a valuable alternative in built environment education (Demian 2007). "Simulated projects", that replicate real-life problems within a safe environment, are suitable vehicles for PBL in built environment teaching and for enhancing learning and soft employability skills. Researchers in education have also recognised that the use of computer technologies can support project-based learning (Peterson et al. 2011). Recently, the introduction of Building Information Modelling (BIM) into built environment education has also enabled "virtual projects" to be used for facilitating learning through such project simulations. The aim of this research was to introduce a simulated virtual project within the second year of the Architectural Technology undergraduate degree programme in order to enhance learning about the AT discipline, BIM, and augment key softer employability skills that are desired by industry employers.

6.1.1 Project-Based Learning and the Use of Simulated Projects in Built Environment Education

Project-based learning anchors learning to the solving of predetermined problems that are generated within time-limited resource-constrained projects. This form of teaching activity is strongly linked with the more "open-ended" problem-based learning, which allows creativity and deeper understanding of concepts to be tested. Allison and Pan (2011) identified the difficulty in distinguishing between Project-based and Problem-based learning, and note the work of Graaff and Kolmos (2003) who argue that projects are always problem based in nature. To avoid confusion, PBL is hereafter used to refer to Project-based learning in this chapter. Demian (2007) notes the validity of PBL as an alternative to traditional built environment teaching methods and argues that PBL facilitates the development of interdisciplinary teamworking skills whilst also encouraging students to take ownership of the learning process. Other benefits include the opportunities to introduce additional impromptu situational learning activities into the project in order to further broaden the knowledge base and improve and enhance student communication and leadership skills. Such project simulations are recognised as a means of developing professional practice within professionally oriented courses, which help students develop useful real-world competencies and, importantly improve their employability skills (Sambell et al. 2012).

6.1.2 Categorising and Defining Soft Employability Skills

"*Basic employability skills are transferable core proficiencies that represent essential functional and enabling knowledge skills and abilities required to succeed at all levels of employment in the 21st century workplace*" (Overtoom 2000, as cited in Rosenberg et al. 2012). In addition to harder "technical" transferable skills such as literacy and numeracy, softer employability skills are concerned with personal and social skills. From an appraisal of previous practice reports, Blades et al. (2012) were able to categorise 15 separate soft employability skills and attributes within four distinct areas: *personal skills; interpersonal skills; self-management-skills*; and *initiative and delivery skills.* Their categorisations of these employability skills are listed in Table 6.1.

In the present work, the research team was interested to observe how the use of PBL could help build career confidence and develop softer employability skills within Architectural Technology education. 14 of the 15 original skills identified by Blades et al. were used as constructs for the design of a research instrument, and the definitions of these 14 constructs used are reproduced in Table 6.2.

More detail on the design of the research instrument is provided further below, but first, as the use of a BIM-enabled virtual project was considered to be a suitable vehicle for purposes of PBL and the development of soft employability skills, it is worth briefly introducing BIM and commenting upon its current application in built environment education.

Table 6.1 Categorisation of soft employability skills and attributes (adapted from Blades et al. 2012)

Personal	Interpersonal	Self-management	Initiative and delivery
Confidence	Social/Interpersonal skills	Self-control	Planning
Self-esteem	Communication skills	Reliability	Problem-solving
Motivation	Teamwork	Positive attitude	Prioritising
Self-efficacy	Assertiveness	~~Presentation~~ (not used)	

Table 6.2 Definitions of soft employability skills and attributes (adapted from Blades et al. 2012)

Skill/attribute	Definition
Self-confidence	Belief in oneself or one's own abilities
Self-esteem	A positive or negative orientation towards oneself; an overall evaluation of one's worth or value
Motivation	Interest/engagement, effort and persistence/work ethic
Self-efficacy	Belief in one's ability to succeed in a particular situation
Social/interpersonal skills	Ability to interact appropriately with other people, without undue conflict or discomfort
Communication skills	Ability to convey information effectively so that it is received and understood; appropriate verbal/nonverbal communication with colleagues, managers, and customers/others
Teamwork	Ability to work cooperatively with others
Assertiveness	Ability to confidently express views or needs without either aggression/dominance/undue submissiveness towards others
Self-control	Ability to control own emotions and behaviour, particularly in difficult situations or under stress
Reliability	Attendance, time-keeping, consistent standards
Positive attitude	Keen to work, learn, accept feedback, and take responsibility
Planning	Ability to plan tasks and monitor progress
Problem-solving	Ability to identify problems and devise solutions
Prioritising	Ability to identify and focus on priority tasks

Table 6.3 Ranking of impact of VP on learning about industry and disciplinary-related aspects

Undertaking the virtual project helped me gain greater depth of understanding about…	Min.	Max.	Sum.	Mean
BIM	4	5	96	4.57
4D construction simulation	4	5	89	4.24
COBie and data drops	3	5	87	4.14
Design coordination through clash detection	2	5	87	4.14
Common Data Environments (CDEs)	2	5	86	4.10
5D cost modelling	3	5	85	4.05
General Architectural Technology concepts	3	5	85	4.05
5D quantity take-off processes	3	5	84	4.00
Design review in BIM	3	5	84	4.00
The use of massing models for the conceptual design stage	3	5	84	4.00
The project design briefing stage	3	5	83	3.95
The potential of BIM in facilities management	2	5	82	3.90
The NBS BIM Toolkit	3	5	82	3.90
Rendering and optimisation of visual production methods	3	5	82	3.90
Design authoring in BIM	3	5	81	3.86
Employers Information Requirements (EIR's)	2	5	80	3.81
Design coordination through model federation	2	5	80	3.81
BIM Execution Plans (BEP's)	3	5	79	3.76

6.1.3 Building Information Modelling in Higher Education

Building Information Modelling (BIM) is a model-focused methodology to planning, creating, and preserving assets in the architectural engineering and construction (AEC) industry. Gledson (2016) describes BIM as the most "*prominent radical, transformative and disruptive innovation to hit construction industry*". In AEC education, Forsythe et al. (2013) note that BIM can be used to "*facilitate a more integrated and visual mode of teaching* [that] *provides a new basis for developing problem based learning – one that has the potential to allow students to aggregate their learning around a central project whilst enabling problems to be scaled at different levels of complexity*". Despite the importance of BIM to the AEC sector, HEIs have responded variously, and in some cases, relatively slowly to it. Woo (2006) argued there was no widespread strategy for teaching BIM in AEC curricula. Tasked by the Government BIM Task Group, a recent survey undertaking by the BIM Academic Forum (BAF) found there were clear distinctions between the higher and lower performing Higher Education Institutions (HEIs) in relation to UK HEI BIM readiness (Underwood et al. 2015). At the researchers home institution, Northumbria University, use of BIM and earlier 3D computer modelling precursors such as Virtual Reality (VR) have been embedded into the academic curriculum since 2003 (Horne and Thompson 2008). The current research team however, did not begin to use BIM for the purposes of simulating virtual projects for collaborative student group work to embed PBL until more recently, which presented the current research opportunity.

6.2 Research Method

A virtual project was used to facilitate project-based learning and ran over a 2-day duration in March 2016. The project opened with an initial discussion sessions that helped to contextualise and reinforce learning around several important BIM concepts detailed in PAS 1192-2 (BSI 2013). These included the importance of BIM Execution Plans (BEPs), and Common Data Environments (CDEs), the capture of Employers' Information Requirements (EIRs), and the use of Construction Operations Building Information Exchange (COBie) data drops. Pairs of students worked together with hardware and software provided by the tutors. The students were provided with a virtual site model and several partially completed architectural, structural, and services models. The VP was loosely structured around an existing UK industry process delivery model (RIBA 2013) and over the course of day 1, these small student groups worked to develop and enhance the existing designs in alignment with three COBie data drop stages (for details, see Love et al. 2014). During day 2, the students progressed to the aggregation of model files and producing BIM output in relation to 4D (time), 5D (cost) and facility and asset management data.

6.2.1 Research Instrument

A questionnaire survey was constructed to be able to gather sufficient data for analysis. The questionnaire consisted of 53 separate questions. An initial section included demographic questions to establish information about cohort composition and to be able to determine variances between individual groups. These questions focused on sex/gender, mode of study, and level of previous practical experience working within the Architectural Engineering Construction (AEC) industry.

Subsequent sections were included to measure the effectiveness of the virtual project on industry and discipline-specific learning; and measure the development of *Personal; Interpersonal; Self-management;* and *Initiative and delivery* skills. The constructs measured were identified from a review of literature. Use was made of an online web-based questionnaire to administer the survey and collect data from the cohort. From the 29 students who participated in the virtual projects, 21 students completed the questionnaire to the end meaning a response rate of 72.0 % was achieved. An additional four "abandoned" responses were also received although these were excluded from the analysis because of their incompleteness.

6.3 Results and Analysis

The majority of students were Male (71.4 %), and studying Full-Time (81.0 %). Two questions were included to determine relevant prior work experience although these were the only two questions not to receive full responses. Of the 19 students to complete Q6, a majority (88.2 %) confirmed that they did not undertake a placement year in the AEC industry. Similarly, Q7 was worded "*Have you had any work experience at all in the Architectural Engineering Construction (AEC) industry?*" The most frequent response from the 15 students that completed this question was "*Yes*" with 60.0 % of respondents selecting that option.

Questions 8–25 were designed to measure how undertaking the virtual project had helped the student gain a greater depth of understanding across a range of industry and disciplinary-related aspects. 5-point Likert scale questions were used and the available response options were listed as: *Strongly disagree* (1) *Disagree* (2) *Neutral* (3) *Agree* (4) *Strongly agree* (5). Table 6.1 shows the mean score and sum score of these aspects in descending order. This table highlights that the virtual project was most useful for gaining greater depth of understanding about the following aspects: *BIM; 4D construction simulation; COBie and data drops; Common Data Environments; 5D cost modelling concepts*; and *general Architectural Technology concepts*, all of which received a mean score of more than 4.00 out of 5.00.

Questions 26–53 were designed to measure the impact of the virtual project across a range of key soft employability skills and comprised a series of 14-paired, "before and after" questions that focused on the development of *Personal,*

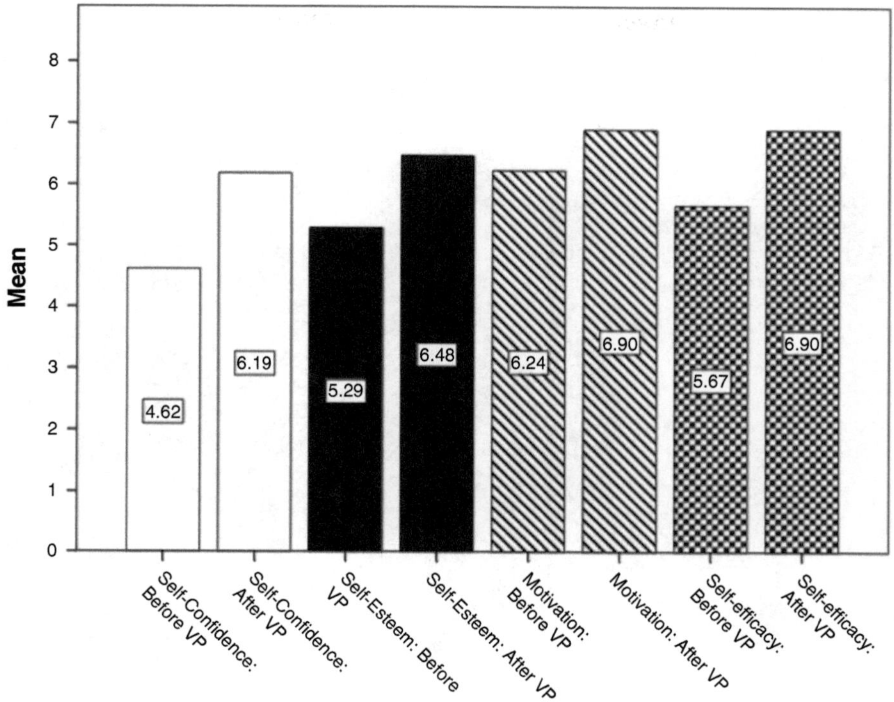

Fig. 6.1 Personal skills—mean scores before and after virtual project—all 21 responses

Interpersonal, Self-management; and *Initiative and delivery skills*. The students were first asked to reflect upon their ability before having undertaking the virtual project in relation to an employability skill and then score this skill on a scale from 0–10. One such example of this type of question is reproduced from Q38: "***Teamwork skills:*** *Please indicate on the sliding scale from 0-10, your ability to work cooperatively with others,* ***before*** *undertaking the virtual project*." After each "before" question, the students were then asked to self-reflect on the same skill after having undertaken the virtual project, and then score their subsequent ability in relation to this skill on the same 0–10 scale. One such example of this type of question is reproduced from Q39: *"**Teamwork skills:** Please indicate on the sliding scale from 0-10, your ability to work cooperatively with others,* ***after*** *having undertaking the virtual project"*.

The first group of questions focused on four separate *Personal skills*. These were: *Self-confidence; Self-esteem; Motivation;* and *Self-efficacy*. Figure 6.1 shows a rise across each measure with the joint largest mean scores of **6.90** being recorded in *Motivation* and *Self-efficacy* after having undertaken the virtual project. This group also records the greatest mean increase of all employability skills of **1.57** occurring in *Self-confidence* because of having undertaken the virtual project.

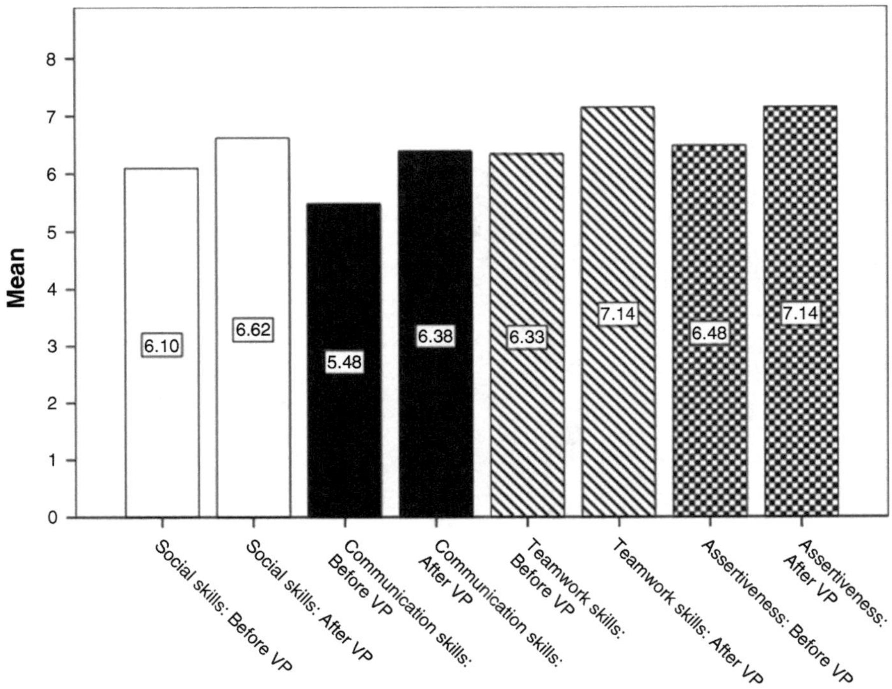

Fig. 6.2 Interpersonal skills—mean scores before and after virtual project—all 21 responses

The next group of questions focused on four separate *Interpersonal skills*. These were: *Social skills; Communication skills; Teamwork skills;* and *Assertiveness*. Figure 6.2 shows a rise across each measure with the joint largest mean scores of **7.14** being recorded in *Teamwork skills* and *Assertiveness* after having undertaken the virtual project. The greatest mean increase of **0.90** occurred in *Communication skills* because of having undertaken the virtual project.

The next group of questions focused on three separate *Self-management skills*. These were: *Self-control, Reliability,* and *Positive attitude*. Figure 6.3 shows a rise across each measure with the largest mean score of **7.24** of *Positive attitude* being recorded after having undertaken the virtual project. The greatest mean increase of **0.72** also occurred in *Positive attitude*.

The final group of questions focused on three separate *Initiative and delivery skills*. These were: *Planning, Problem-solving,* and *Prioritising*. Figure 6.4 shows a rise across each measure with the largest mean score of **7.57** being recorded in *Problem-solving skills* after having undertaken the virtual project. This was also the largest mean score across all employability skills recorded. The greatest mean increase of **0.90** also occurred in *Planning* because of having undertaken the virtual project.

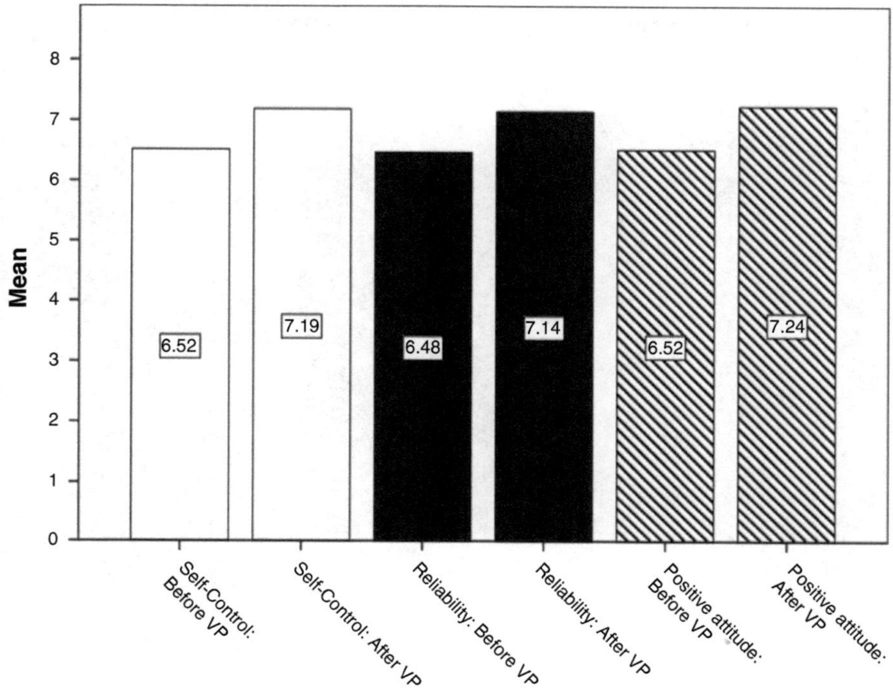

Fig. 6.3 Self-management skills—mean scores before and after virtual project—all 21 responses

To provide a recap: across the entire dataset of all students and in all four categories, the largest recorded mean scores all occurred after having undertaken the virtual project. These were *Problem-solving skills* (7.57); *Positive attitude* (7.24); *Self-control* (7.19); *Teamwork skills, Assertiveness,* and *Reliability* (all 7.14). The greatest increases (when comparing "before" and "after" mean scores) occurred in *Self-Confidence* (1.57) followed by *Self-Efficacy* (1.24), *Self-Esteem* (1.19) then *Communication skills* and *Planning* (both 0.90).

Examination of differences between groups of students with and without prior work experience in the AEC industry then occurred in relation to the key soft employability skills. For this analysis to occur, data could only be used from the subset of 15 students who responded to Q7 "*Have you had any work experience at all in the Architectural Engineering Construction (AEC) industry?*" Table 6.4 presents these filtered data in response to questions 26–53.

The five largest areas of variance between groups of students with and without prior relevant work experience ***before*** undertaking the virtual project occurred in the following categories: *Self-Esteem* (2.33); *Problem-solving* (2.11); *Planning* (2.06); *Self-Confidence* (2.00); and *Positive attitude* (1.94). Surprisingly in each of these categories, the mean self-perception of students without prior relevant

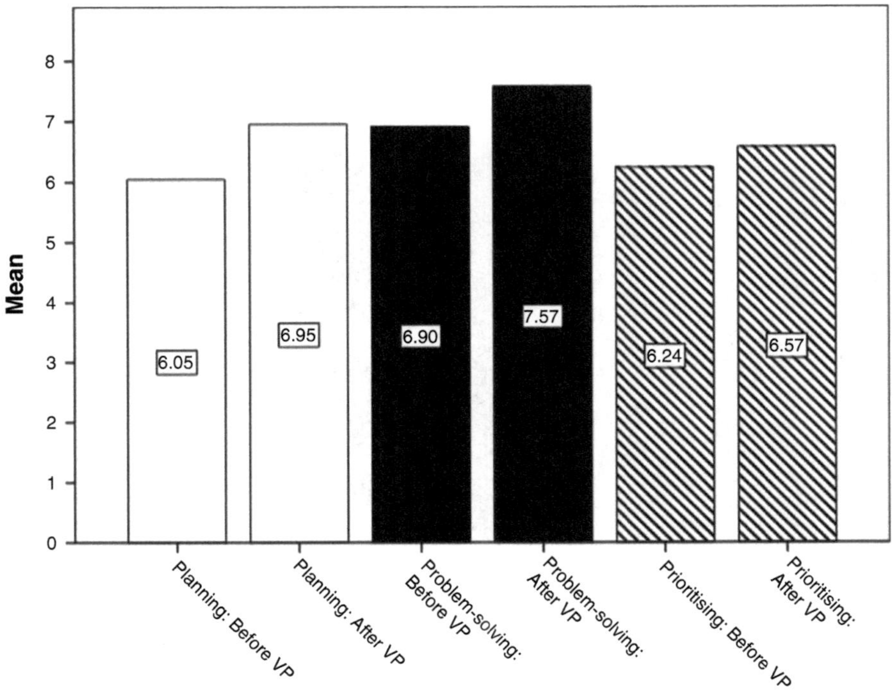

Fig. 6.4 Initiative and delivery skills—mean scores before and after virtual project—all 21 responses

work experience was higher in comparison with students who had relevant work experience.

The five largest areas of variance between groups of students with and without prior relevant work experience ***after*** undertaking the virtual project occurred in the following categories: *Motivation; Self-Efficacy; Self-Esteem; Planning;* and *Communication Skills* (all 1.44). Again, in each of these categories, the mean self-perception of students without prior relevant work experience was higher in comparison with students who had relevant work experience.

The five largest growth areas in skills development for students with prior work experience occurred in the following categories: *Self-Esteem* (1.89); *Self-Confidence* (1.67); *Self-Efficacy* (1.56); *Planning* (1.44); *and Problem-solving* (1.22).

The five largest growth areas in skills development for students without prior work experience occurred in the following categories: *Self-Efficacy* (1.50), *Self-Confidence; Self-Esteem; Motivation* and *Teamwork skills* (all 1.00).

In this subset of students, remaining consistent with the analysis of data across the entire data set, the largest mean score overall was again recorded in the category of an increase in *Problem-solving skills* after having undertaken the virtual project (7.87). Other categories where notable increases were recorded include

Table 6.4 Employability skills—differences between students with and without prior work experience in the AEC industry (subset of 15 responses)

	Yes		No		Variance	Total	
Any prior work experience in the AEC industry?	Mean	N	Mean	N	in groups	Mean	N
Self-confidence: Before virtual project	4.00	9	6.00	6	*2.00*	4.80	15
Self-confidence: After virtual project	5.67	9	7.00	6	*1.33*	6.20	15
Impact of VP	*1.67*		*1.00*		*-0.67*	*1.40*	
Self-esteem: Before virtual project	4.00	9	6.33	6	*2.33*	4.93	15
Self-esteem: After virtual project	5.89	9	7.33	6	*1.44*	6.47	15
Impact of VP	*1.89*		*1.00*		*-0.89*	*1.53*	
Motivation: Before virtual project	5.33	9	6.67	6	*1.33*	5.87	15
Motivation: After virtual project	6.22	9	7.67	6	*1.44*	6.80	15
Impact of VP	*0.89*		*1.00*		*0.11*	*0.93*	
Self-efficacy: Before virtual project	4.67	9	6.17	6	*1.50*	5.27	15
Self-efficacy: After virtual project	6.22	9	7.67	6	*1.44*	6.80	15
Impact of VP	*1.56*		*1.50*		*-0.06*	*1.53*	
Social/interpersonal skills: Before virtual project	5.67	9	6.50	6	*0.83*	6.00	15
Social/interpersonal skills: After virtual project	6.33	9	7.00	6	*0.67*	6.60	15
Impact of VP	*0.67*		*0.50*		*-0.17*	*0.60*	
Communication skills: Before virtual project	4.33	9	6.17	6	*1.83*	5.07	15
Communication skills: After virtual project	5.44	9	6.83	6	*1.39*	6.00	15
Impact of VP	*1.11*		*0.67*		*-0.44*	*0.93*	
Teamwork skills: Before virtual project	6.00	9	6.67	6	*0.67*	6.27	15
Teamwork skills: After virtual project	6.78	9	7.67	6	*0.89*	7.13	15
Impact of VP	*0.78*		*1.00*		*0.22*	*0.87*	
Assertiveness: Before virtual project	5.22	9	6.83	6	*1.61*	5.87	15
Assertiveness: After virtual project	6.33	9	7.50	6	*1.17*	6.80	15
Impact of VP	*1.11*		*0.67*		*-0.44*	*0.93*	

(continued)

Table 6.4 (continued)

	Yes		No		Variance	Total	
Any prior work experience in the AEC industry?	Mean	N	Mean	N	in groups	Mean	N
Self-control: Before virtual project	6.00	9	7.33	6	*1.33*	6.53	15
Self-control: After virtual project	7.00	9	8.00	6	*1.00*	7.40	15
Impact of VP	*1.00*		*0.67*		*-0.33*	*0.87*	
Reliability: Before virtual project	5.67	9	7.33	6	*1.67*	6.33	15
Reliability: After virtual project	6.67	9	7.83	6	*1.17*	7.13	15
Impact of VP	*1.00*		*0.50*		*-0.50*	*0.80*	
Positive attitude: Before virtual project	5.56	9	7.50	6	*1.94*	6.33	15
Positive attitude: After virtual project	6.67	9	8.00	6	*1.33*	7.20	15
Impact of VP	*1.11*		*0.50*		*-0.61*	*0.87*	
Planning: Before virtual project	4.78	9	6.83	6	*2.06*	5.60	15
Planning: After virtual project	6.22	9	7.67	6	*1.44*	6.80	15
Impact of VP	*1.44*		*0.83*		*-0.61*	*1.20*	
Problem-solving: Before virtual project	6.22	9	8.33	6	*2.11*	7.07	15
Problem-solving: After virtual project	7.44	9	8.50	6	*1.06*	7.87	15
Impact of VP	*1.22*		*0.17*		*-1.06*	*0.80*	
Prioritising: Before virtual project	5.56	9	7.00	6	*1.44*	6.13	15
Prioritising: After virtual project	6.33	9	7.33	6	*1.00*	6.73	15
Impact of VP	*0.78*		*0.33*		*-0.44*	*0.60*	

Self-control (7.40); *Positive attitude* (7.20); *Teamwork skills*; and *Reliability* (both 7.13) all of which were measures of skills having developed after having undertaken the virtual project.

Across this subset of students the joint greatest mean increase of all employability skills was recorded in the *Personal skills* and *Interpersonal skills* categories of *Self-Esteem* and *Self-Efficacy* (both 1.53), followed by *Self-Confidence* (1.40); and *Motivation, Communication skills,* and *Assertiveness* (all 0.93).

6.4 Conclusion

Having undertaken the virtual project, students perceived that they had gained greater depth of conceptual understanding, subject-specific knowledge and improved their employability skills. Analysis of the data reveals that actual gains had been recorded across all of the aspects. The virtual project was most useful for gaining greater depth of understanding about the following aspects: *BIM; 4D construction simulation; COBie and data drops; Common Data Environments; 5D cost modelling concepts*; and *general Architectural Technology concepts*, all of which received a mean score of more than **4.00** out of **5.00**. There was positive correlation between self-perception of their existing skill level before the virtual project and the gains they had made because of having undertaken the virtual project. The largest mean score recorded was in *Problem-solving skills* after having undertaken the virtual project. This was consistent even when filtering the subset of students who responded to a question enquiring about prior relevant work experience. Other categories that remained consistently high across all students even when filtering for differences in prior relevant work experience included the categories of *Positive attitude*; *Self-control*; *Teamwork skills* and *Reliability.* Similarly, the largest mean increases of all employability skills because of undertaking the virtual project were recorded in *Personal skills* categories of *Self-Confidence; Self-Esteem;* and *Self-Efficacy.*

6.4.1 Limitations, Consequences, and Opportunities for Future Research

One possible limitation in the research method used relates to the timing of when the students were presented with the survey questionnaire. In the present research, students answered both the "before" and "after" questions at the same time, on completion of the virtual project. It is clear that administering the "before" and "after" questions at those respective project stages may have produced different results. Having observed such student learning and skills development as a result of undertaking the virtual project however, the implications of this research are clear.

There is high educational value in providers of built environment education further absorbing and embedding Building Information Modelling into their curriculum for the purposes of introducing simulated virtual projects as vehicles to facilitate project-based learning. A similar virtual project is planned for the next academic year for purposes of research replication. In the next VP, quantitative data will be augmented with qualitative interview data for purposes of triangulation.

References

Allison, J., & Pan, W. (2011). Implementing and evaluating the integration of critical thinking into problem based learning in environmental building. *Journal for Education in the Built Environment, 6*(2), 93–115.

Blades, R., Fauth, B., & Gibb, J. (2012). *Measuring employability skills: A rapid review to inform development of tools for project evaluation*. London: National Children's Bureau.

BSI (2013). *Specification for information management for the capital/delivery phase of construction projects using building information modelling* (PAS 1192–2:2013).

Demian, P. (2007). Theory and practice of project-based learning in built environment education: A CEBE case study on innovative design project work. A case study by the Centre for Education in the Built Environment, pp. 1–9.

Exley, K., & Dennick, R. (2009). *Giving a lecture: From presenting to teaching*. New York, NY: Routledge.

Forsythe, P., Jupp, J., & Sawhney, A. (2013). Building information modelling in tertiary construction project management education: A programme-wide implementation strategy. *Journal for Education in the Built Environment, 8*(1), 16–34.

Gledson, B. J. (2016). Hybrid project delivery processes observed in constructor BIM innovation adoption. *Construction Innovation Emerald, 16*(2), 229–246.

Graaff, E. D. E., & Kolmos, A. (2003). Characteristics of problem-based learning. *Int J Engng Ed, 19*(5), 657–662.

Horne, M., & Thompson, E. M. (2008). The role of virtual reality in built environment education. *Journal for Education in the Built Environment, 3*(1), 5–24.

Love, P. E. D., Matthews, J., Simpson, I., Hill, A., & Olatunji, O. A. (2014). A benefits realization management building information modeling framework for asset owners. *Automation in Construction, 37*, 1–10.

Peterson, F., Hartmann, T., Fruchter, R., & Fischer, M. (2011). Teaching construction project management with BIM support: Experience and lessons learned. *Automation in Construction, 20*(2), 115–125.

RIBA (2013). *RIBA Plan of Work 2013*. Retrieved from http://www.architecture.com/TheRIBA/AboutUs/Professionalsupport/RIBAOutlinePlanofWork2013.aspx.

Rosenberg, S., Heimler, R., & Morote, E.-S. (2012). Basic employability skills: A triangular design approach. *Education + Training , 54*(1), 7–20. Emerald Group Publishing Limited

Sambell, K., McDowell, L., & Montgomery, C. (2012). *Assessment for learning in higher education*. London: Routledge.

Underwood, J., Ayoade, O., Khosrowshahi, F., Greenwood, D., Pittard, S., & Garvey, R. (2015). Current position and associated challenges of BIM education in UK higher education. BIM Academic Forum.

Woo, J. H. (2006). BIM (building information modeling) and pedagogical challenges. In *Proceedings of the 43rd ASC National Annual Conference*, pp. 12–14.

Chapter 7
Development of a Building Information Modelling (BIM)-Based Real-Time Data Integration System Using a Building Management System (BMS)

Muhammad Umar Khalid, Muhammad Khalid Bashir, and Darryl Newport

Abstract Building Information Modelling (BIM) has become a very important part of the construction industry, and it has not yet reached its full potential, but the use of BIM is not just limited to the construction industry. The aim of BIM is to provide a complete solution for the life cycle of the built environment from the design stage to construction and then to operation. One of the biggest challenges faced by the facility managers is to manage the operation of the infrastructure sustainably; however, this can be achieved through the installation of a Building Management System (BMS). Currently, the use of BIM in facilities management is limited because it does not offer real-time building data integration, which is vital for infrastructure operation. This chapter investigates the integration of real-time data from the BMS system into a BIM model, which would aid facility managers to interact with the real-world environment inside the BIM model. We present the use of web socket functionality to transmit and receive data in real-time over the Internet in a 3D game environment to provide a user-friendly system for the facility managers to help them operate their infrastructure more effectively. This novel and interactive approach would provide rich information about the built environment, which would not have been possible without the integration of BMS with BIM.

M.U. Khalid (✉) • D. Newport
Sustainability Research Institute, University of East London, London E16 2RD, UK
e-mail: m.u.khalid@uel.ac.uk

M.K. Bashir
Le Meriden Makkah, King Abdulaziz Road, Ajyad Area, Makkah 13700, Saudi Arabia

M. Dastbaz et al. (eds.), *Building Information Modelling, Building Performance, Design and Smart Construction*, DOI 10.1007/978-3-319-50346-2_7

7.1 Introduction

The Building Information Modelling (BIM) framework specifies using the BIM model after the infrastructure is constructed and handed over to the operators, in order to take advantage of the rich information provided by the BIM model (Motawa and Almarshad 2015). The BIM guidelines do not specify a methodology for adapting or incorporating the BIM model in an operational built environment; therefore, this research will describe one of the methodologies that could be adapted in order to implement this into the workflow of a facilities manager and Building Management System (BMS) designer. The objective of the BIM is to promote collaboration between engineers and designers from the design stage to completion of the project. However, to implement BIM up to level 3 there is a requirement to also manage the lifecycle of the project (Paterson et al. 2015). This chapter evaluates the potential of implementing a BIM model into a BMS. The current BMS uses 2D vector graphics to aid facility managers with the management of the building; but the BMS is not fully interactive and can only be manipulated by a trained operator (Reeser et al. 2015).

The BMS is used daily by facility managers; therefore, it is logical and advantageous to combine a BIM model with the existing BMS to improve their standard operations. One method to achieve this would be through the introduction of an information-rich BIM model, providing real-time information through BMS rather than static information that is normally generated by the BIM model. Hence, this will allow the facility managers to interact with the built environment in real-time and provide a better user interface as compared to that of a traditional BMS.

7.2 Literature Review

7.2.1 Building Information Modelling

Building Information Modelling (BIM) is constantly being developed and improved through the addition of new features, in order to maximise the potential use of an intelligent BIM model. The concept of BIM was first illustrated in research by Douglas C. Englebart in 1962 (Engelbart 1962). There were many papers then published in the 1960s and 1970s relating to the BIM concept, but at that time it was very difficult to implement because the concept required more computing power than was readily available (Eastman et al. 2011). The first software based on this concept was released in 1980 called ArchiCAD (Walliss and Rahmann 2016)—before the term BIM was used to describe it. More recent improvements in computing power and substantial benefits to the designers, contractors, and the clients have enforced the use of BIM in the construction industry (Eynon 2013). BIM currently has three levels: BIM up to level 2 is very well defined but BIM level 3 is still under development. The BIM level 2 promotes collaboration between different teams using common file formats (Mordue et al. 2015) and provides 5D information which includes project scheduling and project costing (Morledge and Smith 2013). The

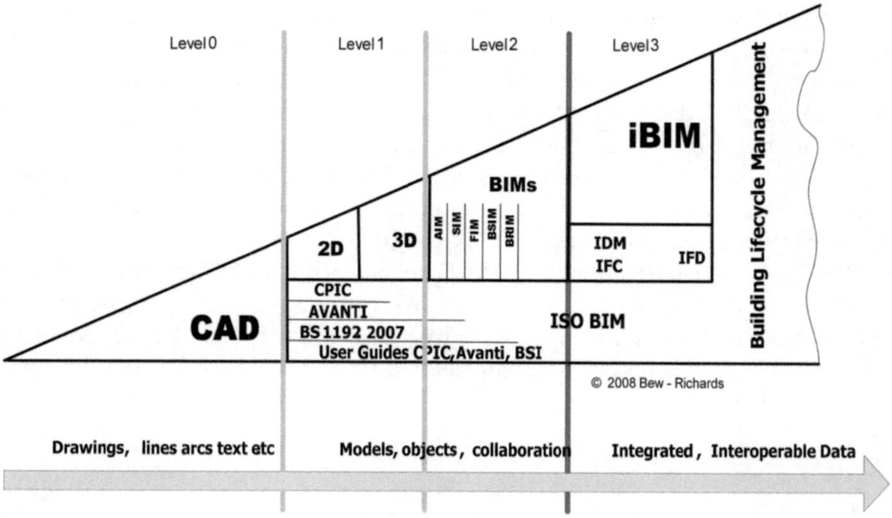

Fig. 7.1 Different levels of BIM (BIMtaskgroup 2011)

information requirement at each BIM level can be seen in Fig. 7.1. The main requirement of BIM level 3 is integrated BIM (iBIM) and Lifecycle Management (Goodhew 2016); thus, integration of a BMS into BIM would help achieve this next level of BIM.

7.2.2 *Building Management System*

The Building Management System (BMS) is used to monitor and control all the electrical and mechanical components of the built environment, and it is a computer-based system (Reddy 2011). The BMS connects the security system; lighting system; fire system; access control system; heating, ventilation, and air-conditioning (HVAC); elevators; and many other electrical and mechanical components of infrastructure (Capehart and Capehart 2007) and aids the facilities manager to manage the built environment more efficiently (Levermore 2013). The BMS is connected to electrical and mechanical components using a Programmable Logic Controller (PLC) (Levermore 2013). The PLC is programmed to monitor and extract information from these electrical and mechanical components. The extracted information is saved and sent to the computer aiding the facility managers for real-time monitoring (Bolton 2015).

The PLC uses many different types of transmission protocols, the most widely used are SOAP, Profibus, BACnet, LonWorks, Modbus, XML, and JSON (Capehart and Capehart 2007). In this chapter, only the use of the latter two—XML and JSON—have been evaluated. The transmitted data is controlled and monitored using a Supervisory Control and Data Acquisition (SCADA) system. The SCADA system controls and monitors all the components connected to the PLC (Macaulay and Singer 2012) and reads the data from the PLC, returning any control commands if needed. The SCADA system manages the communication (at hardware level) between the PLC and the computer (Krutz 2015). And it is capable of managing many different types of transmission technologies, such as Ethernet, RS485, GSM, 3G, or 4G (Radvanovsky and Brodsky 2016). The SCADA is hosted onto a server, and this server can then be hosted on a Local Area Network (LAN) or Wide Area Network (WAN) (Radvanovsky and Brodsky 2016).

The Human-Machine Interface (HMI) collects the data from the SCADA and displays it to the facility manager. The HMI uses 2D vector graphics to aid the user, providing the location and the position of the data in the built environment (Krutz 2015). In this research, an information-rich BIM model replaces this conventional 2D HMI.

7.2.3 Cloud Computing and Internet of Things (IoT)

Cloud computing is a paradigm that involves hosting different computing services over the Internet (Mahmood and Saeed 2013). The services hosted within the cloud can be divided into three different categories, namely, Software as a Service (SaaS), Platform as a Service (PaaS), and Infrastructure as a Service (IaaS) (Yang et al. 2014). The major advantage of the cloud is that it allows for multi-tenancy, enabling a user to share resources across a large number of users (Wang 2012). In our case, the data from the BMS can be shared with multiple users.

7.2.3.1 Infrastructure as a Service (IaaS)

This cloud service model provides virtual machines with pools of hypervisors to provide cloud users with physical computing power (Jamsa 2013). These pools of hypervisors can be scaled up or down according to the user requirement. This service provides storage, Virtual Local Area Network (VLAN) with IP addresses, firewall, load balancers, software applications, and user data (Sosinsky 2011).

7.2.3.2 Platform as a Service (PaaS)

This cloud service model provides a computing platform that can include a webserver, database, operating system, or a programming language execution environment such as PHP, Python, or any other programming language (Jamsa 2013). This type of cloud platform provides developers with the solution to test, run, and deploy their applications on the cloud without the cost and complexity required to setup different programming environments and hardware (Chandramouli 2011). This cloud service is also scalable according to the needs of the users (Jamsa 2013).

7.2.3.3 Software as a Service (SaaS)

This cloud service provides users with access to web-based software applications such as email, browser-based games, or any other type of web-based software service (Sabharwal and Wali 2013). This cloud service is installed and operated by the cloud service provider, and the users of this type of cloud can only use this service (Buyya et al. 2013).

The data from the BMS can be hosted onto the cloud by connecting the BMS to the Internet. This would make the BMS a part of an Internet of Things (IoT). The IoT refers to connecting different embedded devices to the existing Internet infrastructure (Castellani et al. 2010) and is a technology that opens new possibilities to create machine-to-machine communication. This would help monitor, maintain, and manage different embedded devices, i.e. indoor climate control system and/or lighting system (Zanella et al. 2014) that are linked using a wireless or wired connection. The data from these devices can be hosted on the cloud, and the cloud can provide advance analysis of the data (Jin et al. 2014). This would enable the facilities manager to analyse the data of the built environment and run it more efficiently.

7.3 Methodology

A simple data sensor network was established to mimic the data produced by the BMS. This was achieved by using five Arduino microcontrollers, connected to the server using Wi-Fi. The purpose of this chapter is to illustrate the management side of the BMS with a strong focus on the software. Elements of the hardware have not been included as it was beyond the scope of this work.

7.3.1 Design Methodology

The design methodology used in this research is shown in Fig. 7.2. The data generated by the sensors is communicated to the server, and then the server manages all information, communicating this data to the webpage and the database. The webpage sends the data to the game engine using a web socket connection. The server provides the real-time data directly to the web page, but if the user needs historical data then the request is sent via the game engine to the webpage using web socket connection. The webpage requests the data from the server and then the server requests the data from the database, which in response transmits it back to the server and then to the webpage. The webpage then feeds the requested historical data to the game engine. In this chapter, a Unity 3D game engine was used to test the methodology. The Unity 3D game engine is programmed to have a series of trigger points in the 3D model to trigger the extraction of data and display it in the game to improve

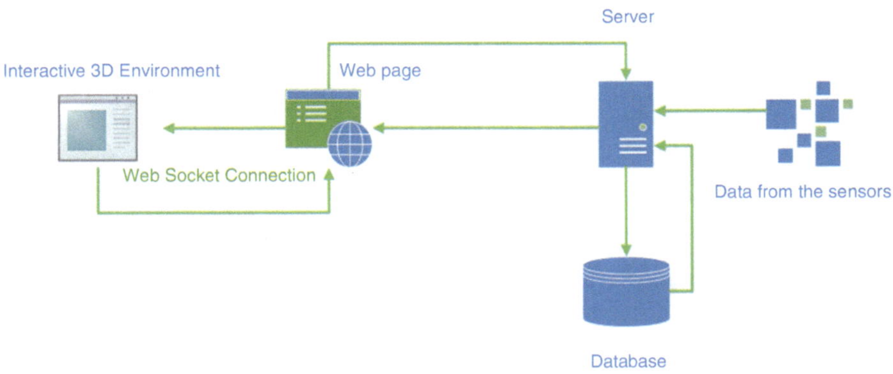

Fig. 7.2 The flow of data from BMS to BIM

the performance of the 3D game environment as this reduces the load on the computer.

The connection between the webpage and the game engine is established using the web sockets instead of polling; this is because web sockets provide full duplex communication (Pimentel and Nickerson 2012). The performance of the web socket is also greater than polling as it provides 500:1 reduction in unnecessary HTTP header traffic and 3:1 reduction in latency (Lubbers et al. 2011). Furthermore, the use of web sockets reduces the work load on the server and hence allows for better integration of real-time data (Gutwin et al. 2011).

7.3.2 *Database Evaluation*

In this project, two different types of databases have been tested for their scalability, namely, MongoDB and MySQL database. The MongoDB is an NoSQL database and MySQL is an SQL database. Their scalability has been investigated by adding new columns into an existing table that has already been populated with 50 rows of sensor data—this was to mimic the expansion of the BMS.

7.3.3 *XML and JSON Evaluation*

The XML and JSON formatted data was sent to the server to evaluate the efficiency of both transmission formats, and then these were assessed for their performance. The data set was increased from 100 to 2500 data points in increments of 100 and was replicated five times to account for any fluctuations in the readings. A similar testing approach was also adapted by (Lin et al. 2012).

7.3.4 3D Environment Performance Evaluation

The performance of the Unity 3D game engine was evaluated by implementing two simple game setups to record frame rate and number of vertices. In both the setups, the number of cubes were increased to increase the number of vertices inside the Unity 3D game engine. In the first setup, each cube had 245,000 vertices and in the second setup, each cube had 100 vertices. The frame rate after each increment was recorded. The experiment was repeated five times to account for fluctuations in the readings.

7.4 Results and Discussion

7.4.1 Database Evaluation

For the MySQL-tested database, it was found that when new columns were added into the SQL database, in order to mimic the expansion of the BMS, the SQL database corrupted. The SQL database had 50 rows of data in the existing table corrupted on adding the new column; the first 50 rows of the new column had no data and could not be resolved. In this case, the solution would be to assign "NULL" value to these rows. An alternative solution would be to make a completely new empty table. Meanwhile, the NoSQL database, which has no schema, can add additional new columns as new data becomes available. Thus making NoSQL database better suited for the BMS as additional sensors can be added to the built environment after restructure or renovation. The NoSQL database easily adapts to the test due to its agility and its dynamic schema, compared to SQL database (Cattell 2010), and can deal with a large volume of structured data, unstructured data, or semi-structured data (Tiwari 2011). This flexibility also means that they are much easier to programme in an object-oriented paradigm (McCreary and Kelly 2013) and are more efficient as they have a scale-out architecture, compared to the SQL database's monolithic architecture (Stonebraker 2010). These findings also support the existing literature available on the SQL and NoSQL databases.

7.4.2 XML and JSON Evaluation

The JSON data exchange format is more beneficial than XML because JSON can be directly parsed by the NoSQL database. It is supported by the NoSQL database by default (Dede et al. 2013), whereas the XML data first has to be parsed by the server to convert it to JSON and only then can it be sent to the NoSQL database; making the use of XML computationally expensive (Lin et al. 2012). The use of JSON is recommended because it is self-descriptive, and there is no overhead for parsing the XML tags. The XML also requires more space to store and parse as

XML	JSON	Number of Character Comparison between XML to JSON
<sensordata> <name>MasterBedRoom</name> <temperature>25</Temperature> <light25>ON</light25> <light26>OFF</light26> </sensordata>	{"sensordata":{ "name": "MasterBedRoom", "temperature":25, "light25":"ON", "light25":"OFF"}}	123:87

Fig. 7.3 Number of character comparison between XML to JSON

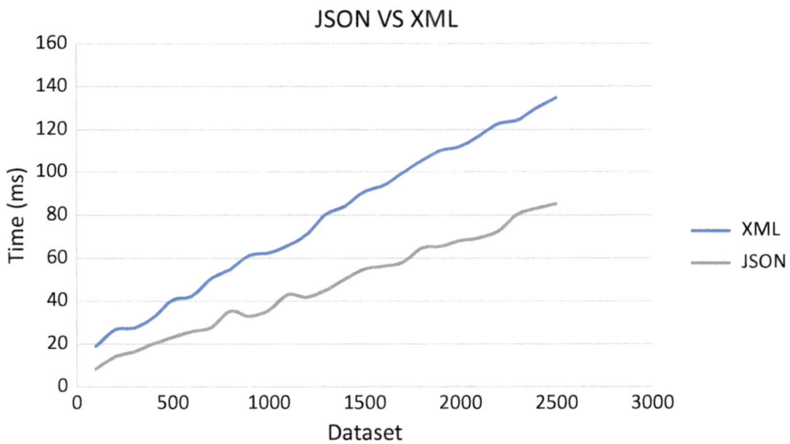

Fig. 7.4 Data transmission speed

shown in Fig. 7.3. Because it uses more characters to represent similar information to that in JSON. This extra memory requirement is important, as memory usage is a limiting factor in BMS, which deals with enormous amounts of data. JSON would make the system faster and more efficient. The transmission efficiency of JSON is also much better than that of XML as shown in Fig. 7.4.

7.4.3 3D Environment Performance Evaluation

The Unity 3D game engine is very efficient in handling large amounts of vertices in a scene (Watkins 2012). The performance test conducted suggested that Unity 3D game engine can efficiently handle large number of vertices in a single scene as shown in Figs. 7.5 and 7.6. All the tests were performed on a computer with an Intel

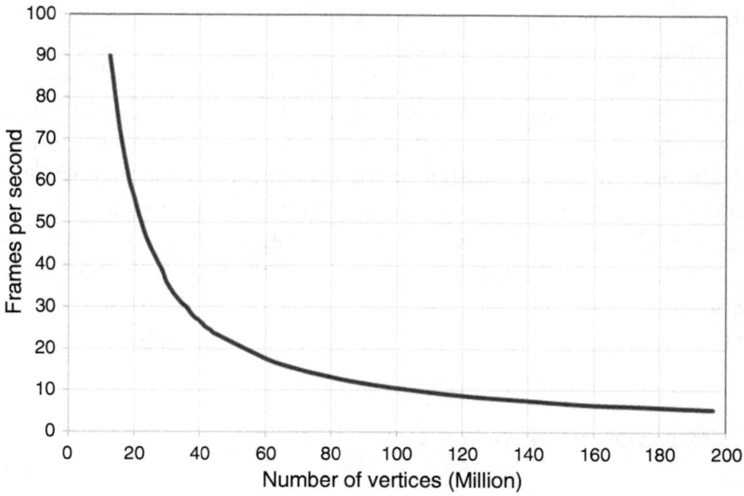

Fig. 7.5 The performance of the Unity 3D game engine using a cube, each with 245,000 vertices

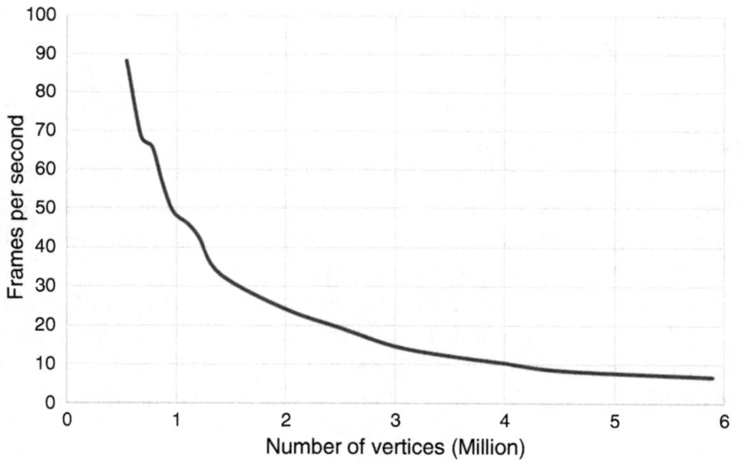

Fig. 7.6 The performance of the Unity 3D game engine using a cube, each with 100 vertices

i7 (2.6 GHz), 16 GB Ram and NVIDIA GeForce GTX 970M 3GB graphic card, the experiments were repeated five times and the average readings were taken to produce Figs. 7.5 and 7.6. It can be seen in Fig. 7.5 that the Unity 3D game engine can handle 43 million vertices at 25 frames per second (fps) using the first setup. In the second setup, the Unity 3D game engine was only able to handle 1.8 million vertices at 25 frames per second (fps) as can be seen in Fig. 7.6. There is a reduction in the performance of the game engine in the second setup because the game engine had to make more draw calls to increase the number of vertices in the scene as each cube would increase 100 vertices in the scene hence contributing to the reduction of performance of the game engine. Therefore, it is recommended to join different 3D elements of the BIM model into one object as this would reduce the number of draw calls and hence would improve the performance of the Unity 3D game engine. The number of vertices supported by different computers would be different therefore while designing the software system it is important to take into consideration the hardware specification of the computer. The Unity 3D game engine can handle very large number of vertices, but it is best to use a trigger-based model generation in a 3D scene to control the visibility of the non-essential 3D elements because the BIM model can become very complex especially when using 3D information relating to the electrical wiring, HVAC, and other types of 3D information.

7.4.4 The Benefit of the Implementation

The integration of BMS into BIM would aid the facility managers to plan the maintenance of the built environment by helping them inspect the built environment inside the game engine. This is most apparent for areas such as the plant room and the high voltage generator room. It is impossible to complete this type of inspection within the conventional HMI interface. The 3D interactive environment also provides rich BIM model information, which would assist the facilities manager to identify hidden built environment elements such as electrical wiring, ducts, and the network of pipes. There are many additional benefits for the facility managers which includes planning maintenance and assessing the safety of the environment by using an information-rich BIM model to carefully examine the safety of the site. The maintenance process can also be animated and saved into the system by the built environment designers or the facilities managers to help them train new technicians and produce a knowledge base of the built environment.

This methodology promotes additional collaboration between the BMS design engineer and the built environment designers. The collaboration provides positive knowledge transfer and provides an opportunity for the BMS design engineer to use an intelligent BIM model designed by the built environment designers.

7.5 Conclusion and Recommendation

It is very difficult to pinpoint the best approach to establish a 3D game-based Building Management System (BMS). The best approach identified in this project is to use JSON to send the data to the server from the BMS. The JSON transmission format is recommended over XML because it uses less memory and has a faster transmission speed. In addition, the JSON is the standard format used to send data to the NoSQL databases such as MongoDB; therefore, the data transmitted in the JSON format can be saved to the database without pre-processing. The use of the

NoSQL database is recommended to store the data as compared to the traditional SQL database because it offers improved scalability. The web sockets are used to transmit data from the webpage to the game engine and vice versa. The use of a web socket is recommended instead of polling as it provides full duplex communication; therefore, the performance of the web socket is very high allowing the data to flow in real time. The adaption of this methodology would aid facility managers to manage their infrastructure more efficiently as it provides real-time information from the BMS in an information-rich BIM model. This provides both static and real-time information to the facility managers hence greatly improving operability. The implementation of this system can be further investigated by reviewing its integration with the facility managers and other stakeholders to test and validate the system for its robustness and ease of use.

References

BIMtaskgroup. (2011). *A report for the Government Construction Client Group Building Information Modelling (BIM) Working Party Strategy Paper*. Retrieved May 2, 2014, from http://www.bimtaskgroup.org/wp-content/uploads/2012/03/BIS-BIM-strategy-Report.pdf.

Bolton, W. (2015). *Programmable logic controllers*. Amsterdam: Elsevier Science.

Buyya, R., Vecchiola, C., & Selvi, S. T. (2013). *Mastering cloud computing*. New Delhi: McGraw Hill.

Capehart, B. L., & Capehart, L. C. (2007). *Web based enterprise energy and building automation systems*. Lilburn: Fairmont Press.

Castellani, A. P., Bui, N., Casari, P., Rossi, M., Shelby, Z., & Zorzi, M. (2010). Architecture and protocols for the internet of things: A case study. In*8th IEEE international conference on pervasive computing and communications workshops, PERCOM workshops 2010* (pp. 678–683).

Cattell, R. (2010). Scalable SQL and NoSQL data stores. *SIGMOD Record, 39*(4), 12–27.

Chandramouli, R. (2011). Cloud service feature driven security policies for virtualized infrastructures. In*WMSCI 2011 – The 15th world multi-conference on systemics, cybernetics and informatics, proceedings* (pp. 166–171).

Dede, E., Govindaraju, M., Gunter, D., Canon, R. S., & Ramakrishnan, L. (2013). Performance evaluation of a MongoDB and Hadoop platform for scientific data analysis. In*ScienceCloud 2013 – Proceedings of the 4th ACM workshop on scientific cloud computing* (pp. 13–20).

Eastman, C., Teicholz, P., Sacks, R., & Liston, K. (2011). *BIM handbook: A guide to building information modeling for owners, managers, designers, engineers and contractors*. Indianapolis, IN: Wiley.

Engelbart, D. C. (1962). *Augmenting human intellect: A conceptual framework*. Air Force Office of Scientific Research, AFOSR-3233.

Eynon, J. (2013). *The design manager's handbook*. Indianapolis, IN: Wiley.

Goodhew, S. (2016). *Sustainable construction processes: A resource text*. West Sussex: Wiley.

Gutwin, C. A., Lippold, M., & Graham, T. C. N. (2011). Real-time groupware in the browser: Testing the performance of web-based networking. In*Proceedings of the ACM conference on computer supported cooperative work, CSCW* (pp. 167–176).

Jamsa, K. (2013). *Cloud computing SaaS, PaaS, IaaS, virtualization, business models, mobile, security and more* (1st ed.). Burlington, MA: Jones & Bartlett Learning.

Jin, J., Gubbi, J., Marusic, S., & Palaniswami, M. (2014). An information framework for creating a smart city through internet of things. *IEEE Internet of Things Journal, 1*(2), 112–121.

Krutz, R. L. (2015). *Securing SCADA systems*. Indianapolis, Indiana: Wiley.

Levermore, G. (2013). *Building energy management systems: An application to heating, natural ventilation, lighting and occupant satisfaction*. Boca Raton, FL: Taylor & Francis.

Lin, B., Chen, Y., Chen, X., & Yu, Y. (2012). Comparison between JSON and XML in applications based on AJAX. In*International conference on computer science and service system, CSSS 2012, Nanjing* (pp. 1174–1177).

Lubbers, P., Albers, B., Salim, F., & Pye, T. (2011). *Pro HTML5 programming. The expert's voice in web development* (2nd ed.). New York: Apress.

Macaulay, T., & Singer, B. L. (2012). *Cybersecurity for industrial control systems: SCADA, DCS, PLC, HMI, and SIS*. Boca Raton, FL: Taylor & Francis.

Mahmood, Z., & Saeed, S. (2013). *Software engineering frameworks for the cloud computing paradigm. Computer communications and networks series*. New York: Springer.

McCreary, D., & Kelly, A. (2013). *Making sense of NoSQL: A guide for managers and the rest of us*. Connecticut: Manning Publications Company.

Mordue, S., Swaddle, P., & Philp, D. (2015). *Building information modeling for dummies*. Hoboken, NJ: Wiley.

Morledge, R., & Smith, A. (2013). *Building procurement*. Chichester, West Sussex: Wiley.

Motawa, I., & Almarshad, A. (2015). Case-based reasoning and bim systems for asset management. *Built Environment Project and Asset Management, 5*(3), 233–247.

Paterson, G., Harty, J., & Kouider, T. (2015). *Getting to grips with BIM: A guide for small and medium-sized architecture, engineering and construction firms*. London: Routledge.

Pimentel, V., & Nickerson, B. G. (2012). Communicating and displaying real-time data with WebSocket. *IEEE Internet Computing, 16*(4), 45–53.

Radvanovsky, R., & Brodsky, J. (2016). *Handbook of SCADA/control systems security*. Boca Raton, FL: CRC.

Reddy, K. P. (2011). *BIM for building owners and developers: Making a business case for Using BIM on projects*. Indianapolis: Wiley.

Reeser, J., Jankowski, T., & Kemper, G. M. (2015). Maintaining HMI and SCADA systems through computer virtualization. *IEEE Transactions on Industry Applications, 51*(3), 2558–2564.

Sabharwal, N., & Wali, P. (2013). *Cloud capacity management. The expert's voice in information technology*. New York: Apress.

Sosinsky, B. A. (2011). *Cloud computing bible*. Indianapolis: Wiley.

Stonebraker, M. (2010). SQL databases v. NoSQL databases. *Communications of the ACM, 53*(4), 10–11.

Tiwari, S. (2011). *Professional NoSQL*. Indianapolis: Wiley.

Walliss, J., & Rahmann, H. (2016). *Landscape architecture and digital technologies: Re-conceptualising design and making*. Taylor & Francis.

Wang, L. (2012). *Cloud computing: methodology, systems, and applications*. Boca Raton, FL: CRC.

Watkins, A. (2012). *Creating games with unity and maya: How to develop fun and marketable 3D games*. Taylor & Francis.

Yang, W. J., Luo, Y. J., Zhao, H. Y., & Li, X. T. (2014). Research on information resources sharing patterns based on cloud computing. *Applied Mechanics and Materials, 644–650*, 3012–3015.

Zanella, A., Bui, N., Castellani, A., Vangelista, L., & Zorzi, M. (2014). Internet of things for smart cities. *IEEE Internet of Things Journal, 1*(1), 22–32.

Chapter 8
Procurement Route and Building Information Modelling (BIM) Implementation Effect on Sustainable Higher Education Refurbishment Projects

Mohammad Alhusban, Salam Al bizri, Mark Danso-Amoako, and Mark Gaterell

Abstract Procurement routes currently applied in higher education (HE) predate the use of life cycle Building Information Modelling (BIM) for the delivery of construction projects. To date, little has been done to align the various procurement routes used in refurbishing HE buildings with the novel opportunities offered via BIM. The importance of HE estate that it is nearly as large as the total National Health Service (NHS) estate and that both of these are significantly larger than the Government's mandated estate. This chapter is a part of an on-going research project to determine whether procurement approaches influence the ability to use BIM techniques to deliver sustainable refurbished buildings in the higher education (HE) sector.

8.1 Introduction

Current concerns about energy consumption, natural resources, increasing oil prices, and other factors affecting the global environment have put pressure on making any development sustainable. At the same time, issues pertaining to global warming and carbon emissions make sustainability a priority area for concern (Sheth 2011).

Sustainable development is an ambiguous concept, with a meaning that is complex and contested (Carter 2007). It has been defined, interpreted, or used in a variety of ways by different groups (planning, academia, business, or environmental policy) to suit their own goals (Redclift 2005). The most frequently quoted definition is from the Brundtland Report (WCED 1987): "Development that meets the needs of the present without compromising the ability of future generations to meet their own needs".

M. Alhusban (✉) • S. Al bizri • M. Danso-Amoako • M. Gaterell
School of Civil Engineering and Surveying, University of Portsmouth, Portsmouth PO1 3AH, UK
e-mail: mohammad.alhusban@port.ac.uk

M. Dastbaz et al. (eds.), *Building Information Modelling, Building Performance, Design and Smart Construction*, DOI 10.1007/978-3-319-50346-2_8

The construction sector is critical in delivering sustainable development, as this is important to everyone's quality of life (Burgan and Sansom 2006; Morrell 2010). "Sustainable built environment" or "sustainable construction" is a subset of sustainable development that integrates effectively low-energy design materials whilst maintaining ecological diversity (Edwards 1999). The sustainable construction goal has been defined as "the creation and management of a healthy built environment based on resource efficient and ecological principles"(Publication et al. n.d.).

Buildings are responsible for almost 50% of the country's CO_2 emissions (Carbon Trust 2013). University buildings are important for the UK's sustainable development strategy as the HE sector comprises 26.6 million square metres (non-residential space) [Association of University Directors of Estates (AUDE) 2014], with around £400 million in annual energy costs and three million tonnes of CO_2 emissions per year (Carbon Trust n.d.).

Although universities have committed to building sustainably through their estate department strategies in both the short and long term, as each HE institution has produced carbon management and implementation plans to achieve absolute CO_2 emission reduction, including capital projects and actions to embed carbon management within the institution [Higher Education Funding Council for England (HEFCE) 2015]. JM Consulting Ltd., which was consulting on behalf of HEFCE, reported a lack of sustainable performance in HE stock. This report noted that over 50% of the UK's HE estates had exceeded their original design life cycle (Crowe 2013).

Modernizing the ageing HE stock can be carried out by refurbishment or total demolition and rebuilding. In 2008, AUDE developed a toolkit that can be used to assist higher education institutions (HEIs) to decide whether to demolish and rebuild or to refurbish. This tool was applied to structures built between 1960 and 1980, as they represent 40% of the current total HE stock. The project found that refurbishment would have a much smaller environmental impact (AUDE 2008). Moreover, sustainable refurbishment is often considered to be a more practical, viable, and potentially affordable solution compared to complete demolition and rebuilding. However, refurbishment projects are more complicated than new buildings and are more prone to go over schedule or budget (or both) (Painting et al. 2014).

Designing sustainable non-domestic buildings through refurbishment must be considered from inception to minimize environmental impact throughout the building life cycle. This is helped by several fairly recent innovations in design and builds process, such as BIM and Integrated Project Delivery (IPD) (Schneider Electric and O'Mara 2012). BIM is defined by Succar (2009) as "*a methodology to manage the essential building design and project data in digital format throughout the building life-cycle*". IPD is defined by the American Institute of Architects (AIA) as:

"*a project delivery approach that integrates people, systems, business structure and practices into a process that collaboratively harnesses the talents and insights of all participants to optimise project results, increase value to the owner, reduce water consumption and maximize efficiency through all phases of design, fabrication and construction*" (AIA CA Council 2007).

BIM implementation by HE estate departments can assess, design, and deliver sustainable HE buildings. BIM can be used in the design phase to enhance the building's sustainability by, for example, evaluating various building skin options and performing daylight (Azhar and Brown 2009). On the other hand, BIM can assess to deliver the required sustainable design via its ability to continuously evaluate buildings' performance over their whole life cycle, which in turn assesses the gap reduction between the design intent and the in-use performance in order to deliver the required whole life performance and to achieve financial savings for the universities' estates when operating their facilities (Kapogiannis et al. 2015). However, BIM implementation in the HE sector is limited (Mair 2013).

BIM implementation requires profound process changes (Volk et al. 2014), particularly to deliver the necessary collaborative platform that brings together multiple stakeholders over the various project phases of the project life cycle (Laishram n.d.). Therefore, clients are likely to change the way that they procure buildings when implementing BIM to ensure a more integrated and collaborative working process (Foulkes 2012). This will unlock the usefulness of BIM to HE estate departments by treating BIM as a shared resource for the facilities over their whole life cycle, from earliest design conception through construction; operation phases; and through additions and adaptive reuse and any alterations until the end of their useful operating life cycle (Laishram n.d.).

The issue with the above is that, to date, procurement types were not chosen on their ability to deliver collaborative environments; therefore, BIM has been used in a relatively isolated way, with limited collaboration between designers and other professionals within the projects (Stirton and Tree 2015).

8.2 Literature Review

8.2.1 Sustainable Development and Higher Education Sector

In the HE sector, the Learning and Skills Council (LSC) and HEFCE published their own sustainable development strategies for HE sectors to promote and support sustainable development, as shown in Table 8.1 (UK Parliament 2005).

HEFCE published "Sustainable development in higher education: Consultation outcomes" at the end of 2014, with 69 responses from HE–HEFCE-funded institutions (50 responses), other HE providers, representative bodies, consultants, statutory or regulatory bodies, and students' advisors. It is being argued that the sector does not operate in a sustainable way; however, the sector is in the process of sustainable development (Smith 2014).

Cost, performance, and quality issues are the focus of traditional design and construction (Latham 1994), where shifting to sustainable design and construction adds the issues of creating a healthy and comfortable built environment and minimizing resource consumption and environmental degradation (Sev 2009). Therefore, sustainable buildings are those that are designed to meet the above issues and are environmentally benign, socially acceptable, and economically viable. However, the presence of a gap between design intent and in-use performance can hinder achieving these sustainable principles (Brett 2012).

Table 8.1 . Sustainable development guiding principles

Sustainable development guiding principles	Learning and Skills Council (CSC) (Council & Council, 2008)	HEFCE (England, n.d.)
1	Putting people at the centre	Living within environmental limits
2	Taking a long-term perspective	Ensuring a strong and healthy society
3	Combating poverty and social exclusion	Achieving a sustainable economy
4	Respecting environmental limits	Using sound science responsibly
5	Using scientific knowledge and transparency information	Promoting good governance
6	Participation and access to justice	

8.2.2 Higher Education Buildings and Refurbishment

It has been estimated that more than half of already built buildings will still be standing in 2050. Despite the focus on reducing the emissions from new buildings, existing buildings remain largely untouched, missing opportunities to reduce emissions and deliver zero or near-zero carbon buildings (Carbon Trust 2008).

According to AUDE (2008), the substantial growth of universities in the late 1950s and 1960s, with a typical life cycle of 30 years, lead to the refurbishment and replacement of these buildings, becoming a significant part of many universities' capital programmes (Smith et al. 2011).

However, recent research on refurbishing HE buildings concluded that the process was not easy or straightforward; this was due to the complexity of client briefs, the difficulties of timetabling, the desire for democratic processes in decision-making, and the inevitable compromises resulting from these often conflicting interests and demands. A single case study including a questionnaire and interviews was used to obtain the data on engaging stakeholders in the process and the buildings whilst they are being refurbished (Painting et al. 2014).

8.2.3 BIM Implementation and Procurement Routes

Practitioners believe that BIM can achieve sustainable construction outcomes more effectively than what is currently being achieved in the industry (McGraw Hill 2010). Political pressure in the UK influences BIM implementation through the construction industry (Cabinet Office 2011), as the use of BIM for public projects (> £50 M) will be mandatory by 2016.

BIM can be used as a sustainable tool to manage and design construction projects via its seven dimensions: 1D: program data, 2D: design-based upon the data, 3D: modelling, 4D: time, 5D: costs, 6D: procurement, and 7D: sustainability. It has been argued that the sustainability dimension (7D) could impact the other six dimensions of the BIM concept (Kapogiannis et al. 2015), as, for example, BIM can make the required information for sustainable design, certification, and analysis consistently available, leading to cost (5D) reduction associated with sustainability analyses (Paper 2003).

Bynum et al. (2013) aimed to investigate the perception of using BIM to design and construct sustainably among designers and constructors. An online survey was distributed to different types of professionals in the USA, and 123 questionnaires were completed. The majority of the respondents believed that IPD and design/build are the optimal procurement routes to integrate BIM as a sustainability tool. However, only a quantitative method was used to obtain the data, and the study concentrated on new buildings (Bynum et al. 2013).

Further research in the Netherlands about the "legal and regulatory aspects of BIM" (Sebastian 2011) discussed the appropriateness of four procurement types that were applied to BIM-based projects: design–bid–build, Dutch building team, design and build, and IPD. The researcher categorized the legal aspects that affect the success of BIM implementation into two types, firstly, the general aspects concerning regulation and law on information and communication, and secondly, the project-specific aspects concerning the procurement type for a project. This research focused on the effect of BIM usage on the division of tasks, the formal agreements on the parties' roles, and changing the working process between the parties involved. Based on the literature and real practices, the research concluded that BIM can be used in any building type (public and private, small and large, residential and non-residential, and domestic and non-domestic). Moreover, BIM can be used in combination with any procurement type. However, BIM's greatest impact can occur when used in complex projects where stakeholders collaborate in an open and integrated way through the building life cycle. The research recommends further investigation into the most effective procurement for applying BIM for a certain typology of building project (Sebastian 2011).

Research conducted by Crowe (2013) on English HE estate departments resulted in a framework to evaluate the collaborative practices through procurement routes in HE property and estate. This research depends on a real-world problem and focuses on universities' deliverables; therefore, the purpose of this framework is to help directors of estates in the HE sector to improve the performance of their construction supply chains. The framework consists of an implementation theme that relates to working efficiency, risk to achieving performance requirements, and motivation to exceed performance requirements. However, buildings' performance and BIM were not a central focus in the research. Data was obtained through interviews and documentation (qualitative) (Crowe 2013).

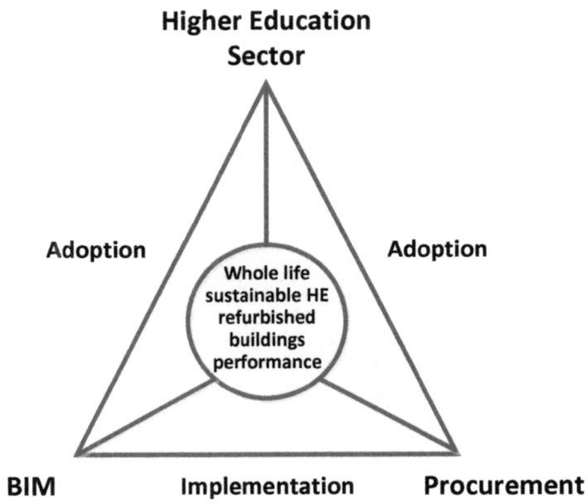

Fig. 8.1 Research framework

8.3 Research Review and Methodology

As discussed above, the main issues to be addressed regarding HE stock of non-domestic buildings is how to modernize their high percentage of ageing, low performance, and non-sustainable facilities. Limited information about the construction practices in the HE sector was available from the published literature, particularly the practices of refurbishing existing facilities. On the other hand, the available literature suggested that BIM has the potential to address sustainability and building performance issues when implemented effectively in refurbishment projects; however, for effective implementation of BIM, procurement approaches that facilitate the necessary collaborative environment are required. Therefore, the research framework highlights the main themes discussed above: BIM, procurement routes, and sustainability within the HE estates context (see Fig. 8.1).

8.3.1 *Gaps of Knowledge*

Reviewing the literature according to the research themes leads to find the following gaps:

- Most of the previous studies on BIM were on new buildings.
- Most of the previous studies were on the impact of BIM implementation for refurbishment in the environmental phase (CO_2 emission), with limited consideration for the ecological and social phases of buildings' sustainability.
- Many studies have been done on incorporating BIM into universities' educational and training courses and programmes to fill in the construction industry skill gaps, with very little research done regarding BIM's impact on sustainable construction practices in the UK's HE sector.
- Limited research has been conducted on the effect of procurement routes on project performance in the HE sector. Previous studies concentrated on the HEI's supply chain performance.

8.4 Conclusion

The UK Government was the first to publish its national strategy for sustainable development in 1994. The HE sector is an important contributor to achieving the country's targets and goals stated in the country's sustainability development strategies as HE sector buildings occupy 26.6 million m^2 and produce three million tons of CO_2 emission per year as mentioned in recent reports. Therefore, modernizing the ageing HE stock is important moreover, universities have committed to building sustainably through their estate department strategies in both the short and long term, as each HE institution has produced carbon management and implementation plans to achieve absolute CO_2 emission reduction, including capital projects and actions to embed carbon management within the institution. However, JM Consulting Ltd., which was consulting on behalf of HEFCE, reported recently a lack of sustainable performance in HE stock.

Despite the widely recognized benefits of implementing BIM on the buildings' sustainable performance, BIM implementation is still limited especially in HE estate departments. In the UK, the pressure of adopting BIM started with a governmental pressure where the Government required a fully collaborative 3D BIM in its Construction Strategy 2011 to be implemented on government projects as a minimum by 2016.

Procurement routes have been seen as one of the most critical steps/actions to fully implement BIM. BIM can be implemented in combination with any procurement route. However, BIM's greatest impact can occur when used in projects where stakeholders collaborate in an open and integrated way through the building life cycle.

References

AIA California Council. (2007). Integrated project delivery: A working definition. Retrieved from http://www.ipd-ca.net/images/Integrated%20Project%20Delivery%20Definition.pdf

AUDE. (2014). Higher education estates statistics report. Retrieved from http://www.aude.ac.uk/about-us/ems-report/ems-report-2014/

AUDE, (2008). *The Legacy of 1960s University Buildings*, (March).

Azhar, S., & Brown, J. (2009). BIM for sustainability analyses. *International Journal of Construction Education and Research, 5*(4), 276–292. doi:10.1080/15578770903355657.

Brett, G. (2012). *Does sustainable design lead to operational efficiency? [online]*. Retrieved from http://www.ukgbc.org/resources/blog/does-sustainable-design-lead-operational-efficiency

Burgan, B. A., & Sansom, M. R. (2006). Sustainable steel construction. *Journal of Constructional Steel Research, 62*, 1178–1183.

Bynum, P., Issa, R. R. A., & Olbina, S. (2013). Building information modeling in support of sustainable design and construction. *Journal of Construction Engineering and Management, 139*(1), 24–34 Retrieved from http://cedb.asce.org/cgi/www.display.cgi?299911.

Cabinet Office. (2011). Government construction strategy. *Construction, 96*, 43 .Retrieved from http://www.bimtaskgroup.org/reports/

Carbon Trust. (2008). Low carbon refurbishment of buildings. *Work*, 30p.

Carbon Trust. (2013). *Accelerating the move to a sustainable, low carbon economy.*

Carter, N. (2007). *The politics of the environment: Ideas, activism, and policy* (2nd ed.). Cambridge: Cambridge University Press.

Crowe, P. (2013). *Collaboration in English higher education estates.* DBEnv Thesis University of Salford School of the Built Environment.

Edwards, B. (1999). *Sustainable architecture: European directives and building design* (2nd ed.). Oxford: Butterworth-Heinemann Ltd.

Carbon Trust. (n.d.). *Energy efficiency in non-domestic buildings.* Retrieved August 27, 2015, from http://www.carbontrust.com/news/2013/09/energy-efficiency-in-non-domestic-buildings/.

Foulkes, J. (2012). *Design and build procurement in the context of bim and the government construction strategy.* Retrieved from: https://www.fgould.com/uk-europe/articles/design-and-build-procurement-context-bim-and-gover/

HEFCE, (2015). *Guide to funding 2015-16 How HEFCE allocates its funds.*

HEFCE. (n.d). *Sustainable development in higher education, update to strategic statement an action plan.*

Kapogiannis, G., Gaterell, M., & Oulasoglou, E. (2015). *ScienceDirect, 00.*

Laishram, B. (n.d.). *Building information modeling in public private partnership projects—Perspectives and hurdles.*

Latham, Sir M. (1994). *Constructing the team. Her Majesty's Stationery Service (HMSO).* London.

Learning and Skills & Council, L. and S. (2008, August). *From here to sustainability—The Learning and Skills Council's Strategy for Sustainable Development progress report.*

Mair, L. (2013). *London universities hesitate on BIM due to cost concerns.* Retrieved from http://www.constructionnews.co.uk/bestpractice/bim/london-universities-hesitate-on-bim-due-to-cost-concerns/8655716.fullarticle

Morrell, P. (2010). *Low carbon construction.* 231.

McGraw Hill. (2010). *SmartMarket Report.* Retrieved from http://construction.com/market_research/FreeReport/GreenBIM/MHC_GreenBIM_SmartM

Painting, N. J., Piroozfar, P. A. E., and Farr, E. R. P. (2014). *Refurbishment of higher education premises: Stakeholder engagement in the process and product,* (September), 955–964.

Paper, W. (2003). Building information modeling for sustainable design. *Building Design, 20,* 1–13.

Publication, C. I. B. R., Team, E., Bourdeau, L., Huovila, P., Lanting, R., and Gilham, A. (n.d.). *LF t. Sustainable Development.*

Schneider Electric, & O'Mara, M. (2012). *Why invest in high-performance green buildings?* 24.

Sebastian, R. (2011). BIM in different methods of project delivery. *CIB WB 78-W102 International Conference,* 26–28.

Sev, A. (2009). How can the construction industry contribute to sustainable development? A conceptual framework. *Sustainable Development, 17*(3), 161–173.

Sheth, B. A. Z. (2011). A refurbishment framework with an emphasis on energy consumption of existing healthcare facilities.

Smith, A., Horsburgh, D., Cowburn, C., & Lyttle, B. (2011). *HEFCE performance in higher education estates.*

Smith, A., 2014. *Sustainable development in higher education: Consultation outcomes.* pp. 1–36.

Stirton, L., & Tree, J. (2015). *IPD and BIM: A new dimension to collaboration.*

Succar, B. (2009). Building information modelling framework: A research and delivery foundation for industry stakeholders. *Automation in Construction, 18*(3), 357–375. doi:10.1016/j.autcon.2008.10.003.

Redclift, M. (2005). *Sustainability: Critical concepts in the social sciences* (Vol. 4 vols). London: Routledge.

UK Parliament. (2005). *Securing the future delivering UK sustainable development strategy.* CM6467, London.

Volk, R., Stengel, J., and Schultmann, F. (2014). *Automation in construction building information modeling (BIM) for existing buildings—Literature review and future needs* **.**

World Commission on the Environment and Development (WCED). (1987). *Our common future, the Brundtland report.* Oxford: Oxford University Press.

Part III
Building Performance and Design

Chapter 9
QUB/e: A Novel Transient Experimental Method for in situ Measurements of the Thermal Performance of Building Fabrics

Johann Meulemans, Florent Alzetto, David Farmer, and Christopher Gorse

Abstract This paper presents a novel transient experimental method developed in order to perform in situ measurements of the thermal performance of building fabrics: the QUB/e method. In one night, a QUB/e test yields the whole house heat loss coefficient (HLC) and the local in situ U-values. A comprehensive set of in situ measurements were carried out in a circa 1900 solid wall end-terrace house located in an environmental chamber to evaluate the thermal performance of the building fabric and to validate the QUB/e method. The accuracy of the QUB/e method was assessed against steady-state measurements before and after a deep retrofit, both the HLC and U-values were used in the comparison. The measurement of the HLC using the QUB/e method for heating durations down to one hour yielded accurate results (i.e. the relative differences from the value estimated with the steady-state method were smaller than 10%) provided the α-criterion lay within the recommended range (i.e. between approximately 0.4 and 0.7). The U-values measured in situ with the QUB/e method were in good agreement with the steady-state (ISO 9869-1) values (i.e. the relative differences were within the uncertainty bound of the measurement methods). The QUB/e method was thus deemed validated by comparison with reference U-values measured in accordance with ISO 9869-1.

J. Meulemans (✉) • F. Alzetto
Saint-Gobain Recherche, 39 quai Lucien Lefranc B.P. 135, 93303 Aubervilliers, France
e-mail: johann.meulemans@saint-gobain.com

D. Farmer • C. Gorse
Leeds Sustainability Institute, Leeds Beckett University, School of the Built Environment and Engineering, Leeds LS2 9EN, UK

M. Dastbaz et al. (eds.), *Building Information Modelling, Building Performance, Design and Smart Construction*, DOI 10.1007/978-3-319-50346-2_9

9.1 Introduction

The assessment of the energy performance of a building often relies on theoretical calculations. When the output of these calculations is compared to in situ measurements, the actual energy performance of a building often shows some discrepancy with the theoretical predictions (i.e. the so-called performance gap, see Stafford et al. (2012), Johnston et al. (2015) and references therein). The mismatch between the "theoretical" and the real performances of a building can be due to material differences, ageing, thermal bridging, moisture, quality of construction (e.g. poor detailing and/or workmanship) and other factors.

In order to address the issues behind the "performance gap", the overall energy performance of a building is not enough, we need to know the local thermal performance of the building assembly, i.e. the contribution of each building element (walls, windows, roof and floor) to the whole heat loss coefficient (HLC) must be determined.

A novel dynamic experimental method, called QUB/e, which addresses most of the shortcomings of the available methods, was recently proposed (Meulemans and Alzetto 2016). The accuracy of the QUB/e method was assessed on a lightweight building in a real climate by comparison with calculations (i.e. in accordance with standards ISO 6946 (International Organization for Standardization 2007) and DIN EN 673 (German Institute for Standardization 2011) and quasi-static measurements (i.e. in accordance with standard ISO 9869-1 (International Organization for Standardization 2014). The agreement between calculated and measured U-values was good (i.e. the differences were within the uncertainty bound of the methods).

The QUB/e method is promising but its validation is incomplete, i.e. its accuracy (vs. quasi-static measurements) remains to be fully assessed. The main objectives of this research were two-fold:

1. Whole building thermal performance (heat loss coefficient, HLC): to improve the accuracy and reduce the duration of QUB tests (Mangematin et al. 2012; Pandraud and Fitton 2013; Pandraud et al. 2013, 2014; Alzetto et al. 2014; Bouchié et al. 2014; Alzetto et al. 2016a, b);
2. Local elements thermal performance (U-values): to validate the QUB/e method (Meulemans and Alzetto 2016) vs. steady-state measurements (i.e. in accordance with ISO 9869-1).

A comprehensive set of in situ measurements were carried out in a circa 1900 solid wall end-terrace house located in an environmental chamber to evaluate the thermal performance of the building fabric and to validate the QUB/e method. For both the whole house HLC and U-values, the accuracy of the QUB/e method was assessed by comparison with steady-state measurements before and after a full retrofit programme of the test house.

This chapter is organised as follows. The materials and methods used in this study are described in Sect. 9.2. The results obtained from in situ measurements are presented and discussed in Sect. 9.3. Concluding remarks can be found in Sect. 9.4.

9.2 Materials and Method

9.2.1 Salford Energy House

The Salford Energy House is a full-scale pre-1919 solid-wall Victorian end-terrace house constructed inside an environmentally controlled chamber at the University of Salford. The construction of the Salford Energy House Test Facility was achieved by using reclaimed materials and methods of the time. A guard house is also present so that the effects of a neighbouring property can be explored during experiments. For more details, the interested reader should refer to Ji et al. (2014), Farmer et al. (2015) and Alzetto et al. (2016b).

The baseline house was uninsulated and had single glazing. The full retrofit programme involved the application of mineral wool internal wall insulation (IWI) to the external walls of the test house. The IWI continued into the intermediate floor void and 400 mm along the party wall. The original windows were replaced by uPVC double glazed units (DGU). The loft was insulated with mineral wool quilt laid between and over the ceiling joists. PIR was installed between the joists of the suspended timber ground floor.

9.2.2 Monitoring Equipment

Electrical heaters (i.e. fan heaters and heating carpets) were placed within the test house in order to provide a uniform heating source. Air circulation fans were used during the steady-state measurements in order to further homogenise the air temperature (Johnston et al. 2013; Farmer et al. 2015). Heat flux plates (Hukseflux HFP01), type K thermocouples and PT100 resistance temperature detectors were used to monitor the heat flux densities on building elements and the air temperatures. A silicone paste was used to ensure a good thermal contact between the heat flux plates and the building elements. All sensors were connected to data loggers (Graphtek GL820 and dataTaker DT80). The data acquisition rate was set to one minute.

9.2.3 Static Measurements

9.2.3.1 Whole House Heat Loss Measurements (Heat Loss Coefficient)

A modified version of Leeds Beckett University's 2013 Whole House Heat Loss Test Method (Johnston et al. 2013) was used to obtain measurements of the test house HLC during each steady-state measurement period (Farmer et al. 2015; Alzetto et al. 2016b).

The test house and chamber were left undisturbed for a minimum period of 72 h during which data were collected at intervals of one minute. To ensure continuous heat flow through the building envelope to the test chamber, a temperature differential (ΔT) between the internal and external environment of 15 K was selected for the steady-state measurements of whole house heat loss and in situ U-values. The test chamber HVAC equipment was set to maintain an air temperature of 5 (± 0.5) °C.

The internal environment of the test house and the guard house were heated electrically using fan heaters. A mean internal air temperature of 20 °C was maintained within the test house using thermostatic air temperature controls connected to each heater. A relatively homogeneous air temperature throughout the test house was facilitated by the use of air circulation fans, ensuring a similar temperature difference throughout the building envelope. Considerable care was taken to ensure that heat flux plates (HFPs) were not unduly influenced by excessive air movement by positioning fans in such a way that air was not blown directly on to the HFPs.

A steady state was considered to be achieved if the heat flux density or the total power input differed by less than ± 5% from the value measured in the previous 24-hour period.

The HLC was calculated using the following equation:

$$HLC = \frac{\sum Q_j}{\sum {''}\, T_j}, \tag{9.1}$$

where Q_j and ΔT_j are, respectively, the electrical power input into the test house (in *W*) and the air temperature difference between the interior and exterior of the test house (in *K*) of the *j*th individual measurement. The calculation period was the last 24 h of the steady-state measurement period.

Q_j was obtained by measuring the electrical power input to the test house for the heaters, fans and logging equipment. ΔT_j was obtained by subtracting the arithmetic mean internal temperature of the house from the arithmetic mean chamber temperature.

The internal air temperature of the guard house was also maintained at 20 °C throughout each 72-h steady-state measurement period; this was to minimise inter-dwelling heat transfer across the party wall, as the party wall is considered a zero heat loss element in whole house heat loss calculations. HFPs were installed on the party wall to measure inter-dwelling heat transfer throughout each steady-state measurement period.

9.2.4 *In situ U-value Measurements*

In situ U-value measurements were undertaken in accordance with ISO 9869-1 (International Organization for Standardization 2014). In situ measurements of heat flux density, from which in situ U-values are derived, were taken at 49 locations on

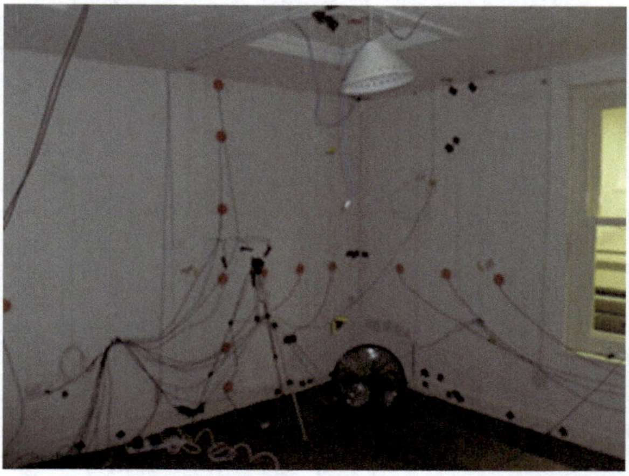

Fig. 9.1 Photograph of bedroom 1 in the test house showing HFPs (red discs) on the thermal elements

the thermal elements of the test house using heat flux plates (HFPs) as illustrated in Fig. 9.1. Only measurements of heat flux density obtained from those locations that were considered not to be significantly influenced by thermal bridging at junctions with neighbouring thermal elements (typically at distances greater than 500 mm from the junction) were used in the calculation of the in situ U-values. One HFP was also placed at the centre pane of the window in the first bedroom, i.e. only the U_g-value of this glazing could be derived from our measurements.

9.2.5 *Dynamic Measurements*

The QUB/e method (Meulemans and Alzetto 2016) was used to measure the whole house heat loss coefficient and local U-values. The QUB/e method is based on both the QUB and the heat flowmeter (HFM) (International Organization for Standardization 2014) methods. The principle of the QUB method is based on a single resistance and capacity model (Mangematin et al. 2012; Pandraud and Fitton 2013; Pandraud et al. 2013), and describes the temperature evolution as a single decaying exponential. By using two different constant powers in two different phases (respectively, noted 1 and 2), the whole heat loss coefficient (HLC) of a building can be evaluated in one night (Pandraud et al. 2014; Alzetto et al. 2016a).

With the QUB/e method, heat flux densities and nearby air temperatures for each building element of interest are monitored during a QUB test. The QUB analysis procedure is then used to derive the U-values of each building element:

$$U = \frac{a_2 q_1 - a_1 q_2}{a_2\text{''}\, T_1 - a_1\text{''}\, T_2}, \tag{9.2}$$

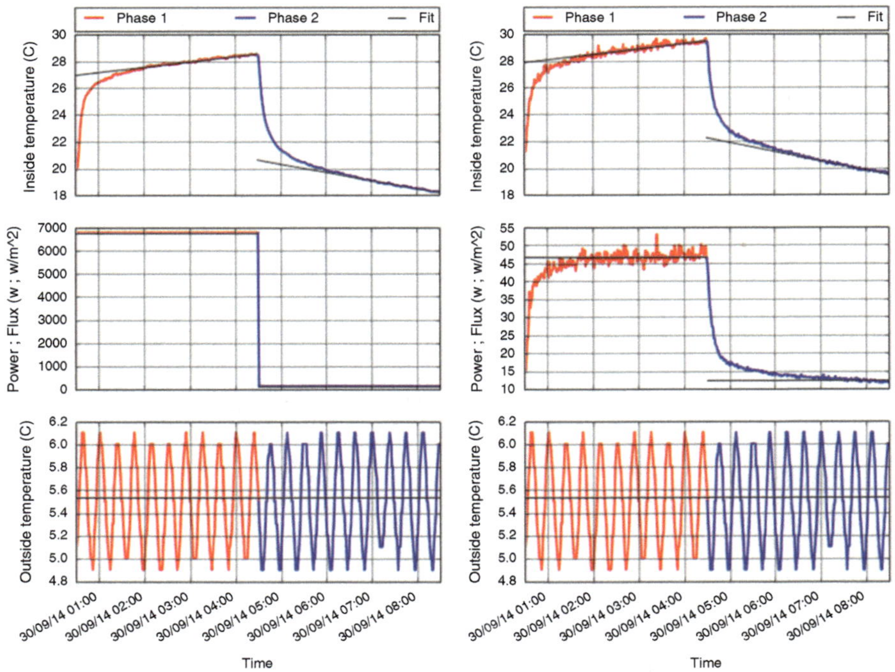

Fig. 9.2 Evolution of temperatures, heating power and heat flux density during a 4-hour QUB/e test—Baseline: whole house heat loss coefficient (left) and in situ U-value of an internal wall of bedroom 1 (right). The red, blue and black solid lines correspond to the heating phase, the free cooling phase and the linear regressions used to derive the quantities used in the QUB formula, respectively

where q_i, a_i and ΔT_i are, respectively, the mean heat flux density, the slope of the inside air temperature and the inside/outside air temperature difference at the 'end' of the ith phase (defined here as $t_i - \min(t_i/2, \tau)$ where t_i is the duration of the ith phase and $\tau = 2$ h). The evolution of the heating power, the air temperatures and the heat flux passing through a building element during a QUB/e test is illustrated in Fig. 9.2.

The error on the estimated HLC with the QUB method depends on a dimensionless parameter called the α-criterion (Pandraud et al. 2014; Alzetto et al. 2016a):

$$\alpha = 1 - \frac{HLC_{ref}'' \ T_0}{P_h}, \tag{9.3}$$

where HLC_{ref}, P_h and ΔT_0 are a reference heat loss coefficient (in $W.K^{-1}$), the heating power (in *W*) and the initial temperature difference (in *K*) between the internal and external environment (i.e. at the beginning of a QUB test), respectively. The HLC obtained from the steady-state measurements was used as a reference value here. HLC was studied for different heating duration: 0.5, 1 and 4 h. The total duration of each test corresponds to twice these values (i.e. total time = heating time + free cooling time).

While the influence of the α-criterion on the estimation of the HLC with the QUB method was already reported in a previous project undertaken at the Energy House (Pandraud et al. 2014; Alzetto et al. 2016a), shorter durations were investigated in this project.

Since the indoor air temperature was not kept constant during the QUB/e test, there was a temperature difference across the party wall separating the test house and the guard house. Since the heat fluxes were monitored during the QUB/e tests, the heat losses at the party wall could be accounted for. The HLCs were thus corrected in order to report only heat losses to the exterior environment and have a sound comparison with the static measurements (in that case, there were no heat losses at the party wall because both houses were kept at constant indoor air temperature of 20 °C).

The corrected HLC was computed with the following equation:

$$HLC_{corr} = HLC_{raw} - U_{part,\,eff} \times A_{part}, \tag{9.4}$$

where HLC_{corr}, HLC_{raw}, $U_{part,\,eff}$ and A_{part} are the corrected HLC (in $W.K^{-1}$), the raw HLC (in $W.K^{-1}$) obtained from the standard QUB analysis, the effective U-value of the party wall (in $W.m^{-2}.K^{-1}$) obtained from the QUB/e method and the area of the party wall (in m^2), respectively.

In order to obtain reliable values, we repeated QUB/e measurements and derived representative values based on a statistical analysis. The mean values and standard deviations were reported for both HLCs and U-values. The U-values estimated in situ with the QUB/e method were compared with measured U-values using the static method described earlier.

9.3 Results and Discussion

9.3.1 Whole Building Thermal Performance

Figure 9.3 shows the heat loss coefficients obtained with the QUB method for different heating durations. The HLCs are plotted against the α-criterion calculated with a reference value taken from the steady-state measurements: 238.8 (± 9.3) $W.K^{-1}$ and 59.3 (± 2.0) $W.K^{-1}$ for the baseline and retrofit stages, respectively. The experimental tests covered values of the α-criterion between 0.2 and 0.8.

For the baseline stage, comparison with static measurements suggested that the HLC derived from dynamic measurements could be slightly underestimated (around 10%). However, the use of air circulation fans during the static measurements in order to homogenise the inside air temperature led to higher convective losses. For uninsulated (or poorly insulated) buildings (or building elements), the R-value is very sensitive to internal surfaces resistances (both internal and external). The discrepancy between the mean R-values derived from static and "long" dynamic measurements (i.e. 4-h QUB tests) was approximately 0.07 $m^2.K.W^{-1}$. This value

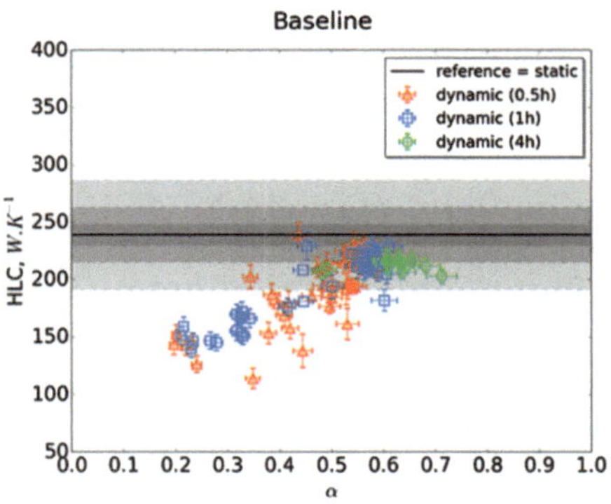

Fig. 9.3 Heat loss coefficient (HLC) vs. α-criterion: baseline (left) and retrofit (right). Symbols: QUB tests, solid line = reference value. Grey-shaded area corresponds to 10 and 20% around the reference value, respectively. The reference value was taken from the steady-state measurements

corresponds to a difference of air velocity of a few meters per second in the vicinity of the building elements (e.g. see International Organization for Standardization (2007) and references therein) consistent with the use of air circulation fans. The 'true' HLC of the building should therefore lie between both values.

Another source of discrepancy arose from the effective heat losses at the party wall. The guard house was heated at a constant air temperature of 20 °C during the tests. For the static measurement performed at 20 °C, there were no heat losses through the party wall. Since the temperature was not kept constant during the dynamic tests, effective heat losses or gains at the party wall should be accounted for in the calculation of the HLC value derived from QUB test. An effective U-value was derived. However, there was a great dispersion in the obtained value and only five locations were monitored in our analysis.

The effective heat losses attributed to the party wall during the QUB/e tests were prone to uncertainty reflecting the non-uniformity of the party wall (e.g. chimney breast vs. cavity wall).

For "short" durations (i.e. 0.5 and 1 h), the HLC values estimated with the QUB method depended on the value of the α-criterion (cf. Fig. 9.3). The HLC was underestimated (i.e. up to 50%) for "low" values of the α-criterion (i.e. smaller than 0.35). It should be noted that it was not possible to reach high values of the α-criterion (i.e. greater than 0.7) at the baseline stage due to constraints on the available heating power. The 4-hour QUB tests did not exhibit any influence of the α-criterion although the values tested remained around 0.5 (i.e. between 0.45 and 0.70).

For the retrofit stage, the observed overestimation of the HLC (up to 70%) for "short" durations (i.e. 0.5 and 1 h) was attributed to α-criterion values larger than 0.7 (cf. Fig. 9.3). The 4-h QUB tests did not exhibit any influence of the α-criterion although the values tested remained around 0.5 (i.e. between 0.45 and 0.65). Unlike the baseline stage, the static measurements were not impacted by the use of air circulation fans because the sensibility to surface thermal resistances was almost nil.

The α-criterion can be viewed as a confidence index regarding the accuracy of each individual QUB measurement. If we only consider tests with values around 0.5

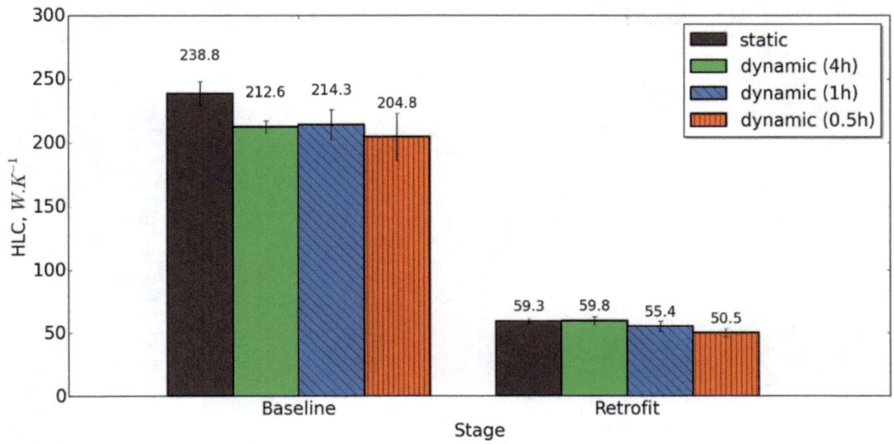

Fig. 9.4 Heat loss coefficient—dynamic vs. static measurements

(i.e. 0.45–0.75 and 0.4–0.6 for the baseline and retrofit stages, respectively), the associated distribution of HLCs can be considered Gaussian. The obtained mean values were compared with the values derived from the static measurements in Fig. 9.4. The agreement between static and dynamic measurements was relatively good for both stages. 1-h and 4-h QUB tests yielded relative differences smaller than 10%. For the (very) "short" heating durations (i.e. 0.5 h), we obtained relative differences of 15%.

The associated uncertainty in the estimated HLC values with (very) "short" heating durations (i.e. 0.5 h) could be large due to dispersion in the estimated HLC for single QUB tests (cf. Fig. 9.3). For a single (very) "short" QUB test (i.e. 0.5 h), the estimated HLC might exhibit a relative difference up to 40% (vs. the reference value) even for an α-criterion within the recommended range (i.e. between approximately 0.4 and 0.7 (Alzetto et al. 2016a)). This lack of robustness might be linked to the bias of the QUB method and should be further investigated.

For a single QUB test, the minimum heating duration needed to yield an accurate HLC value (i.e. relative difference with the value estimated with the steady-state method smaller than 10%) was 1 h, provided great care was taken for the choice of the α-criterion (i.e. values within the recommended range). This is consistent with previous findings (Pandraud et al. 2014; Alzetto et al. 2016a).

9.3.2 *In situ U-values*

The U-values measured in situ with the QUB/e and the steady-state (i.e. ISO 9869-1) methods are plotted against each other in Fig. 9.5. Each symbol corresponds to the in situ U-value of each HFP placed on the different elements of the building (walls, glazings, roof and floor) for different heating durations (i.e. 0.5, 1 or 4 h).

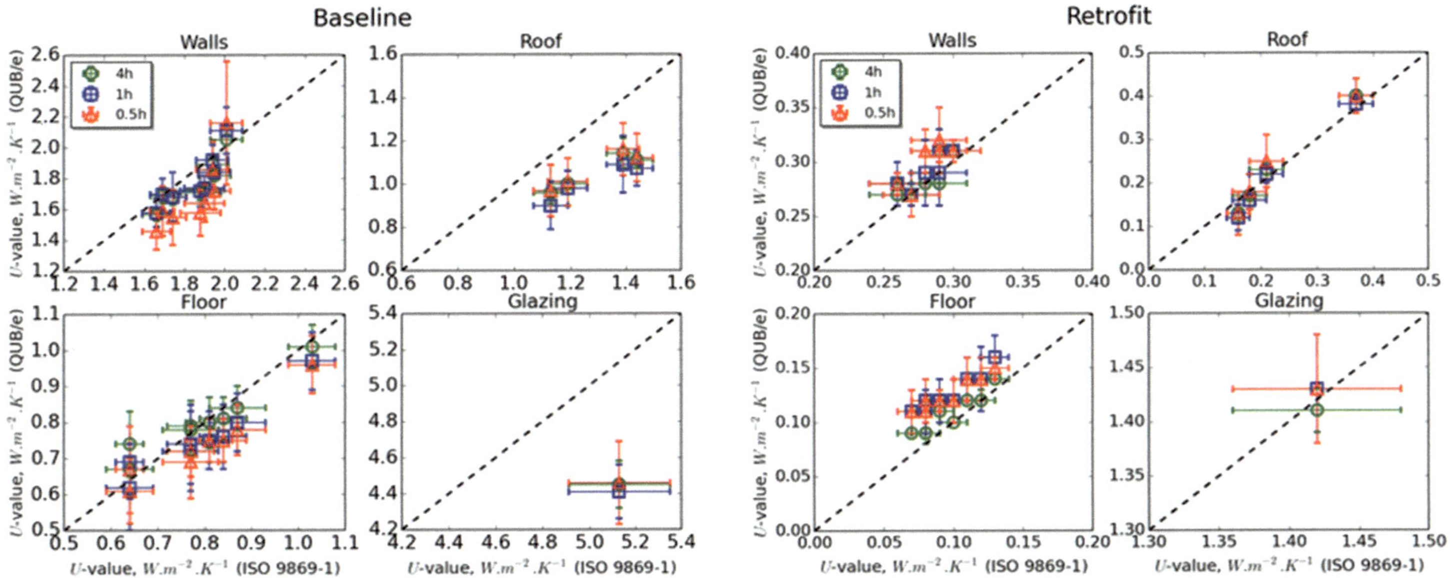

Fig. 9.5 Local U-values—dynamic vs. static measurements: baseline (left) and retrofit (right). For comparison, the $y = x$ reference curve is plotted for each case (dashed black lines)

The mean values and the associated standard deviations computed over the different QUB/e tests are given.

Figure 9.5 shows that there was little (or no) influence of the heating duration on the estimated U-values with the QUB/e method. The main difference lay in the associated uncertainties (i.e. standard deviations): the shorter the test, the higher the uncertainty. The main source of uncertainty in the QUB/e method arose from the determination of the slopes at the end of each phase (heating and free cooling) because the external air temperature was kept constant at 5.5 (± 0.5) °C during the tests. For (very) "short" durations (i.e. 0.5 h), the number of data points available to determine the slopes was much less hence the greater uncertainty in the reported values.

The U-values measured in situ with the QUB/e method were in good agreement with the steady-state (ISO 9869-1) values (i.e. the relative differences were within the uncertainty bound of the measurement methods) except for the glazing and the roof at the baseline stage. The QUB/e method was thus deemed validated by comparison with reference U-values measured in accordance with ISO 9869-1.

The observed difference for the single glazing can be attributed to different internal surface thermal resistances during the tests. As previously mentioned, air circulation fans were used for the steady-state measurements and the internal surface thermal resistance was thus greater than the one during the QUB/e tests (no air circulation fans were used). For uninsulated (or "poorly" insulated) elements, the U-value is highly sensitive to the surface resistances (both external and internal). The observed difference for the roof at the baseline stage should be further investigated.

9.4 Conclusions

A comprehensive set of in situ measurements were carried out in a circa 1900 solid wall end-terrace house located in an environmental chamber to evaluate the thermal performance of the building fabric and to further improve/validate the QUB/e method. The following conclusions were drawn from the analysis carried out:

1 The measurement of the HLC with the QUB method for heating durations down to one hour yielded accurate results (i.e. the relative differences with the value estimated with the steady-state method were smaller than 10%) provided the α-criterion lay within the recommended range (i.e. between approximately 0.4 and 0.7).
2 The measurement of the HLC with the QUB method for very "short" heating durations (i.e. down to half an hour) yielded highly dispersed HLC values (i.e. relative differences with the value estimated with the steady-state method up to 40% even for values of the α-criterion within the recommended range).
3 The QUB/e method was deemed validated by comparison with reference U-values measured in accordance with ISO 9869-1 (i.e. the relative differences were within the uncertainty bound of the measurement methods).

It should be stressed out that these findings apply to measurements performed in a controlled exterior environment (i.e. constant exterior air temperature, no wind, no rain, no solar gains) at the Salford Energy House Test Facility. For measurements in the field, variability in the exterior environment as well as the likely unavailability of reference (measured) values makes it more challenging to reach a similar level of accuracy with the same confidence level.

Acknowledgements The funding of the project was provided by Saint-Gobain. The project management was under the responsibility of Luke Smith (National Energy Foundation). Richard Fitton and Mo Benjaber (University of Salford) provided the technical support for the operation of the Energy Hub during the project. The thermal upgrade of the test house was carried out by Jacob Eco Energy under the supervision of Saint-Gobain Habitat UK. The installation/removal of the sensors was carried out with the help of Olivier Samin (Saint-Gobain Recherche) and the Centre for the Built Environment (CeBE) of the Leeds Sustainability Institute of Leeds Beckett University (LBU).

References

Alzetto, F., Gossard, D., & Pandraud, G. (2014). Mesure rapide de la perte thermique des bâtiments. In: *Ecobat Sciences & Techniques 2014 Congress*, 19–20 March 2014, Paris, France.

Alzetto, F., Pandraud, G., & Fitton, R. (2016a). The QUB method: A fast and reliable building envelope thermal diagnosis. *Energy and Buildings*, submitted.

Alzetto, F., Farmer, D., Fitton, R., Hughes, T., & Swan, W. (2016b). Comparison of whole house heat loss test methods under controlled conditions in six distinct retrofit scenarios. *Energy and Buildings*, submitted.

Bouchié, R., Alzetto, F., Brun, A., Boisson, P., & Thébault, S. (2014). Short methodologies for in-situ assessment of the intrinsic thermal performance of the building envelope. In: *Sustainable Places 2014*, 1–3 October 2014, Nice (France).

Farmer, D., Gorse, C., Miles-Shenton, D., Brooke-Peat, M., & Cuttle, C. (2015). Off-the-shelf solutions to the retrofit challenge: thermal performance. In: C. Gorse, & M. Dastbaz, (Eds.), *International SEEDS Conference 2015: Sustainable Ecological Engineering Design for Society*, 17–18 September 2015, Leeds (UK). Leeds (UK): Leeds Beckett University.

German Institute for Standardization. (2011). *DIN EN 673:2011 Glass in building—Determination of thermal transmittance (U-value)—Calculation method.* Berlin: DIN.

International Organization for Standardization. (2007). *ISO 6946:2007 Building components and building elements—Thermal resistance and thermal transmittance. Calculation method.* Geneva: ISO.

International Organization for Standardization. (2014). *ISO 9869–1:2014 Thermal insulation—building elements—in-situ measurement of thermal resistance and thermal transmittance—Part 1: Heat flow meter method.* Geneva: ISO.

Ji, Y., Fitton, R., Swan, W., & Webster, P. (2014). Assessing the overheating of the UK existing dwellings—A case study of replica Victorian end terrace house. *Building and Environment, 77*, 1–11.

Johnston, D., Miles-Shenton, D., & Farmer, D. (2015). Quantifying the domestic building fabric a 'performance gap'. *Building Services Engineering Research and Technology, 36*(5), 614–627.

Johnston, D., Miles-Shenton, D., Farmer, D., & Wingfield, J. (2013). *Whole house heat loss test method (Coheating).* Leeds: Leeds Metropolitan University.

Mangematin, E., Pandraud, G., & Roux, D. (2012). Quick measurements of energy efficiency of buildings. *Comptes Rendus Physique, 13*(4), 383–390.

Meulemans, J., & Alzetto, F. (2016). QUB/e—A novel dynamic experimental method for in situ measurements of the thermal transmission properties of building elements. *Energy and Buildings*, submitted.

Pandraud, G., & Fitton, R. (2013). QUB: Validation of a rapid energy diagnosis method for buildings. In: *IEA–EBC Annex 58, 4th Expert meeting, 8–10 April 2013, Holzkirchen (Germany).*

Pandraud, G., Gossard, D., & Alzetto, F. (2014). Experimental optimization of the QUB method. In: *IEA–EBC Annex 58, 6th Expert meeting, 14–16 April 2014, Ghent (Belgium).*

Pandraud, G., Mangematin, E., Roux, D., & Quentin, E. (2013). QUB: A new rapid building energy diagnosis method. In: K. Kabele, M. Urban, K. Suchý, & M. Lain, (Eds.), *Proceedings of CLIMA 2013, the 11th REHVA World Congress and the 8th International Conference on IAQVEC, 13–19 June 2013, Prague (Czech Republic).* Prague (Czech Republic): Society of Environmental Engineering (STP).

Stafford, A., Bell, M., & Gorse, C. (2012). *Building confidence—A working paper* (Report no: 008). Leeds: The Centre for Low Carbon Futures.

Chapter 10
Energy Efficiency in Residential Buildings in the Kingdom of Saudi Arabia

A. Al kanani, N. Dawood, and V. Vukovic

Abstract Due to a rapidly escalating population and a high level of economic growth, the Kingdom of Saudi Arabia is experiencing a vigorous infrastructure expansion, especially with respect to residential buildings. As a result, energy demand for residential buildings is of a very high level, whereby approximately 70% of electricity is consumed by air conditioning systems alone for interior cooling throughout the year due to the hot and humid Saudi climate. This high energy consumption sheds light on the size of the problem in Saudi Arabia. As a result, this indicates the urgent need to adopt a strategy to reduce the excessive use of energy in residential buildings. One such strategy can be based on passive architectural design principles relating to the materials being used for the construction of the building, as these offer the potential for a cost-effective solution for energy reduction and major savings in electricity use for cooling purposes in residential buildings. Hence, a comprehensive study of the energy consumption in the cities of Jeddah, Riyadh, and Dammam across the Kingdom of Saudi Arabia has been conducted using Sefaira energy simulation software. The investigated cities are located within the three distinct climatic regions representing the most populated areas of Saudi Arabia. Investigations were conducted in order to evaluate the impact of insulation material use on energy performance of a typical residential building. Using the Sefaira energy simulation software, many different types of building facades for external walls of residential buildings have been assessed in order to evaluate suitable construction materials for the Saudi hot weather. The results of the assessment show that adding a thermal insulation of polyurethane to external walls and adopting an appropriate construction type could reduce by over 30% energy consumption for cooling in residential buildings of Saudi Arabia. The presented analyses and energy simulations are likely to elucidate the potentials for sustainable development and costs for implementing energy-saving design measures in the construction sector of Saudi Arabia.

A. Al kanani (✉) • N. Dawood
School of Science and Engineering, University of Teesside, Middlesbrough TS1 3BA, UK
e-mail: A.Alkanani@tees.ac.uk

V. Vukovic
School of Science and Technology, University of Teesside, Middlesbrough TS1 3BA, UK

M. Dastbaz et al. (eds.), *Building Information Modelling, Building Performance, Design and Smart Construction*, DOI 10.1007/978-3-319-50346-2_10

10.1 Introduction

Energy efficiency of residential buildings has recently become a major issue because of growing concerns of CO_2 and other greenhouse gas emissions and scarcity of fossil fuels. Residential buildings worldwide account for a surprisingly high 40% of global energy consumption (Krarti 2012; Krarti 2011; Bribian et al. 2009). According to Olofsson et al. (2010) and Meier et al. (2002), an efficient residential building is one that applies energy-efficiency measures and technologies while operating as per design, supplies the amenities and features appropriate for that kind of residential building, and which can be operated in such a manner as to have a low energy use compared to other similar residential buildings.

Due to a rapidly escalating population and a high level of economic growth, the Kingdom of Saudi Arabia is experiencing a vigorous infrastructure expansion, especially with respect to residential buildings. Domestic energy demand is increasing (Elgendy 2011; Taleb and Sharples 2011; Almatawa et al. 2012). As a result of this, energy demand in the form of heating/cooling systems, ventilation, hot water, and lighting in Saudi Arabia is of a very high level and is constantly increasing. Saudi Arabia consumes about one third of its oil production (Elgendy 2011) and its buildings are consuming about 30% of its allocated domestic usage (Almatawa et al. 2012).

The energy-efficient design concept intervention is essential due to the scarcity of natural resources, and unfortunately, the issues of energy consumption, energy performance, and energy efficiency in particular are not generally given serious consideration regarding the designs of residential buildings in Saudi Arabia. Research shows Saudi residential buildings severely lack the means to ensure energy efficiency. As a result of poorly designed residential buildings in Saudi Arabia, nearly 70% of household energy is used for cooling purposes alone (Taleb and Sharples 2011; Akbari et al. 1996).

Domestic energy consumption in residential buildings in Saudi Arabia accounts for more than half of the country's energy demand (SEEC 2013). As almost all Saudi residential buildings depend on air conditioning for interior cooling throughout the year and due to the hot and humid local Saudi climate, these residential buildings are consuming over and above their allocated domestic usage of electricity alone (Aldossary et al. 2013).

Azhar et al. (2011) and Krygiel and Nies (2008) suggest that the use of BIM can assist in the following areas of building energy performance: Building orientation (selecting a good orientation can reduce energy costs), building massing (to analyse building form and optimize the building envelope), day-lighting analysis, water harvesting (reducing water needs in a building), energy modelling (reducing energy needs and analysing renewable energy options can contribute to low energy costs), sustainable construction materials (reducing material needs and using recycled materials), and site and logistics management (to reduce waste and carbon footprints). Design options for energy efficiency can be tracked and studied in a BIM model along with spatial data to geographically locate and import building site information to place it within context and to contribute to an understanding of issues relating to climate, surrounding systems, and resources. The building can then be adjusted and engineered using real coordinates to reduce the impact on and utilise the surrounding environment to reduce energy requirements, for example solar orientation (Hardin 2011).

With such high residential energy consumption in Saudi Arabia, there is an urgent need to adopt a strategy to reduce the excessive use of energy in residential buildings. The reduction of domestic energy consumption in Saudi Arabia can be based on such architectural principles as the building's construction materials being used, as these offer the potential for a cost-effective solution for energy reduction and major savings in electricity use. In view of that, this paper presents a description of the research undertaken with the use of BIM software as a method to expedite change in the practices of the use of construction materials in Saudi residential buildings, with the use of 3D BIM software for energy analysis to foster environmentally responsible energy consumption practices in Saudi Arabia.

10.2 Literature Review

Buildings are major contributor in energy use according to the Saudi Energy Efficiency Centre (SEEC 2013). In 2011, Saudi buildings consumed almost 80% of the total electricity generated, of which 51.2% is used by residential buildings, and energy consumed for air conditioning represents 70% of the total national electrical demand (SEEC 2013).

Moreover, it has been estimated that 2.32 million new residential buildings will be built by 2020, indicating an even more significant increase in electricity demand associated to residential buildings for the Kingdom of Saudi Arabia in the coming years (Alrashed and Asif 2012).

Building envelope components of residential buildings such as the walls, roof, floor, and windows has other functions than just structural or architectural elements. Indeed, building envelope components can be designed to maintain a safe and comfortable indoor environment. In particular, building envelope components affect the energy required for thermal comfort within buildings. For instance, the heat storage capability of some building envelope components, such as walls, can help in controlling the indoor temperatures without the need of mechanical systems. So there are sustainable approaches to achieve thermal comfort in buildings without utilizing significant amounts of energy, especially for cooling or heating (Alaidroos and Krarti 2015).

However, the lack of thermal insulation and the absence of sustainable standards in the Saudi construction industry have led to an assortment of low quality residential buildings in the nation's existing built environment. At the present time, there is no standard criterion for residential buildings to raise the level of quality and efficiency. This is due to the fact that energy efficiency performance of both residential and commercial buildings was initially neglected by the Saudi Building Code since energy consumption was low and no serious threat from peak loads was expected. As a result, over 60% of buildings in the Kingdom of Saudi Arabia are not thermally insulated (SEEC 2013; Alshenaif 2015).

According to Fang et al. (2013), BIM is a critical element in reducing industry waste including wasted energy, adding value to industry products, and decreasing environmental damage. Their research on the extent of the contribution that BIM can make to the energy performance of buildings is well worthwhile, at the same time increasing awareness of the main capacities of this new technology and its potential contribution towards energy efficiency in residential developments. Their research claims that BIM has helped to promote energy analysis and performance in their projects (Fung and Cheng 2011).

Through the use of highly energy-efficient construction materials and building operation optimisation technologies, the impacts to life cycle energy and emissions consumption from the operational phase can be shifted back to the material production and construction phase (Blanchard and Reppe 1998). Integration of LCA software and BIM energy analysis software to automate this process will not only allow for efficiencies in LCA assessment procedures, but also enable design changes to be made prior to construction and assist building management in the optimisation of a building's environmental footprint throughout its operation (Russell-Smith and Lepech 2012). Since BIM has great potential for promoting sustainable building design, it is inevitable that BIM should be more utilized in the Saudi construction industry. This is backed up by Azhar et al. (2011), who states that the use of BIM for sustainable design can broaden further than commercial buildings.

Consequently, the rapid increase in residential energy consumption requires a current and fundamental strategy to reduce this energy consumption. This is essentially in order to improve and set out minimum standards for conventional design, construction practices and standards, as well as reducing the overall impact energy use of buildings upon the environment (Geyer 2012; Kubba 2012; Kassim et al. 2013). So together with practical innovation, the residential buildings of Saudi Arabia could have a considerable impact upon the energy performance, whereby they will be more energy-efficient, costless to operate and promote the construction of high-energy performance buildings.

10.3 Research Methodology

In order to assess the optimal building design for residential buildings through the varying climates across the Kingdom of Saudi Arabia, a multiple case study approach was undertaken. The multiple case study involved analysis of a new two-storey residential building in the cities of Jeddah, Riyadh, and Dammam with the specific aim of assessing the current and potential improvements in terms of energy and consumption with regard to energy efficiency of the building.

10.3.1 Criteria for case study selection

Due to the differences in average temperature and climate, three cities have been selected in the geographically diverse kingdom of Saudi Arabia to conduct the case studies. The location of the first case study building is based in the city of Jeddah, which is located on the coast of the Red Sea in the west of Saudi Arabia. It experiences a subtropical hot arid desert **climate, having** extremely hot summer temperatures and warm winter temperatures. The second city selected is Riyadh, the capital city of Saudi Arabia, which is located in the middle of the kingdom. It experiences extremely hot and dry summers and cold dry winters. Thirdly, the city of Dammam has been selected. This is a port city on the east coast of the Persian Gulf. This city experiences extremely hot summers with high humidity levels and cold winters.

10.3.2 Case Study Description

The case study buildings will consist of a two-storey residential building in each of the selected cities, with an average of 185 m^2 on each floor of the building. The ground floor is used as the main living and entertaining space, whereas the first floor area is used as the bedroom space. The structure of the building consists of reinforced concrete for the floor and ceilings. The walls are made from concrete blocks without any insulation used in the walls and roof. The outer façade of the building is covered by a cement sand render.

The methodology involved an analysis of the properties' design and construction fabric and materials. Three-dimensional models were designed for each property based on the design, and then assessed using energy simulation software (1) without the use of insulation materials and also (2) with the use of insulation material.

10.3.3 Use of Simulation Software Tools

Energy simulation has been conducted using Sefaira energy simulation software in order to perform whole building energy analysis for the purpose of comprehensively studying the influence on energy consumption, daylight analysis and carbon emissions upon residential buildings in the cities of Jeddah, Riyadh, and Dammam in order to gain meaningful energy performance results. The philosophy of undertaking energy simulation upon residential buildings is to create a model where the user can specify in detail parameters that influence the building performance, with resulting performance predictions that are as close to reality as possible.

The simulated results provided real time climate data, energy consumption, daylighting metrics, as well as the building geometry. As a result, the analyses and the results from the energy simulations are likely to determine the optimal construction materials for both the maximum and minimum effect upon energy performance of residential buildings, as well as exemplify costs for energy savings in the residential building sector of Saudi Arabia.

10.3.4 Research Limitations

This research is limited to the impact of energy performance of residential buildings in the Kingdom of Saudi Arabia. Secondly, all simulations are carried out only by using Sefaira energy simulation software. Finally, it is limited to only presenting the research data from simulations with and without insulation on external walls of residential buildings owing to the fact that the literature review affirms that the lack of thermal insulation has led to an assortment of low quality residential buildings in Saudi Arabia's existing built environment. Therefore, these were the first set of variable factors to be assessed when conducting the case studies.

10.4 Analysis and Discussion

The energy consumption was simulated and analysed using Sefaira real time energy analysis software. The results of the simulations for the energy consumed in the residential building in each city were evaluated according to the design of the case study residential building, including the building fabric used. The simulations provided energy consumption analysis, daylight analysis, and information regarding energy demand for heating/cooling purposes.

At first, energy simulation was conducted without the use of insulation materials in the external walls of the residential building and then the details of the design were changed by adding insulation materials to the external walls. The building energy consumption and its associated daylight analysis were re-assessed in order to examine the potential improvements with the use of insulation materials in the external walls of the residential building, which was applied as an energy efficiency measure.

The energy consumption data were then analysed and the results compared with and without the use of insulation materials to the external walls of the case study residential building in each of the selected cities across the Kingdom of Saudi Arabia.

The results of the impact of wall insulation on the energy consumption of the case study residential buildings show a similar pattern for all three cities across the Kingdom of Saudi Arabia, with most of the optimized energy performance being achieved by adding the external wall insulation.

The simulation results show that by adding wall insulation to the case study residential building in Jeddah, it achieves 35% in energy savings, while the same insulation in Riyadh gives almost 35% savings in the total energy consumption. While in

Dammam, the same insulation applied to the same case study residential building has 33.5% savings in energy consumption. These findings are attributed to the fact that the lack of optimal architectural design and construction materials, especially insulating materials, has been the cause of the high-energy consumption and demand.

From a comparison of the diagrams shown in Figs. 10.1, 10.2, 10.3, 10.4, 10.5, 10.6 and 10.7, the results show an overall difference of between 30 and 40% by the addition of insulating materials to the external walls. Importantly, these results are significant in view of the fact that the case studies were conducted in three different weather conditions in three different cities across the Kingdom of Saudi Arabia to investigate how energy consumption varied due to the changes in weather from extreme heat to much colder temperatures.

According to Blom et al. (2011), "Energy consumption in dwellings contributes significantly to their total negative environmental impact". Thus, as a result, it is evident that according to the energy simulations, there is high-energy consumption in residential buildings in all three cities, with the biggest issue causing a high demand for energy is the high level of electricity consumption for cooling. Aldossary et al. (2013) state, among other energy conservation measures, a lack of optimal architectural design and construction materials has been the cause of this high energy demand in residential buildings in Saudi Arabia. Preceding this, Balaras et al. (2007) has earlier specified that, as an example, 33–60% of energy can be saved by using efficiently designed external walls.

Hence, the in-depth Sefaira analysis has allowed an understanding of what the energy was consumed for in each case. As illustrated, the case of the highest energy consumption is for cooling purposes due to the extremely hot and arid climate of Saudi Arabia.

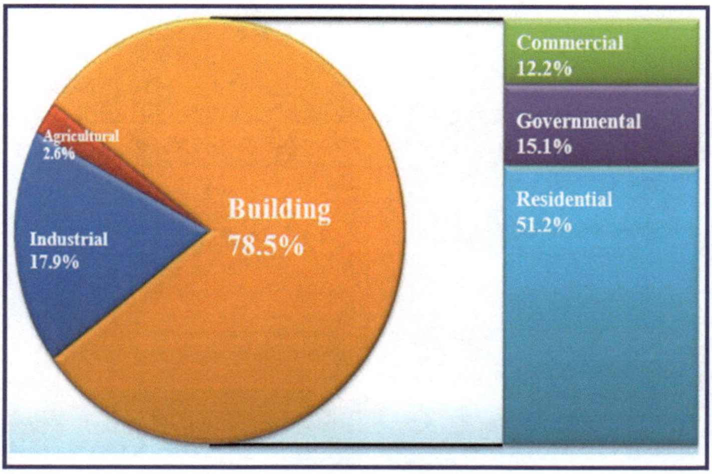

Fig. 10.1 Energy Consumption by Sector in Saudi Arabia Source: SEEC & K.A.CARE, (2013)

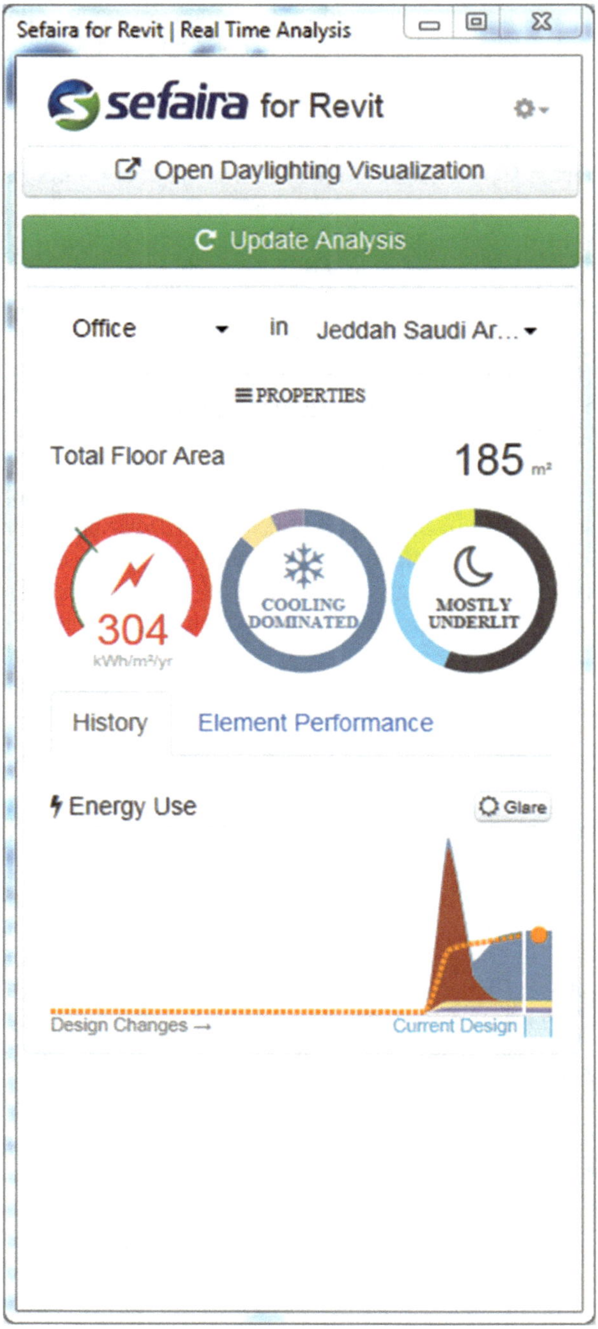

Fig. 10.2 Energy consumption analysis on the case study residential building in the city of Jeddah KSA—without the use of insulation materials on external walls

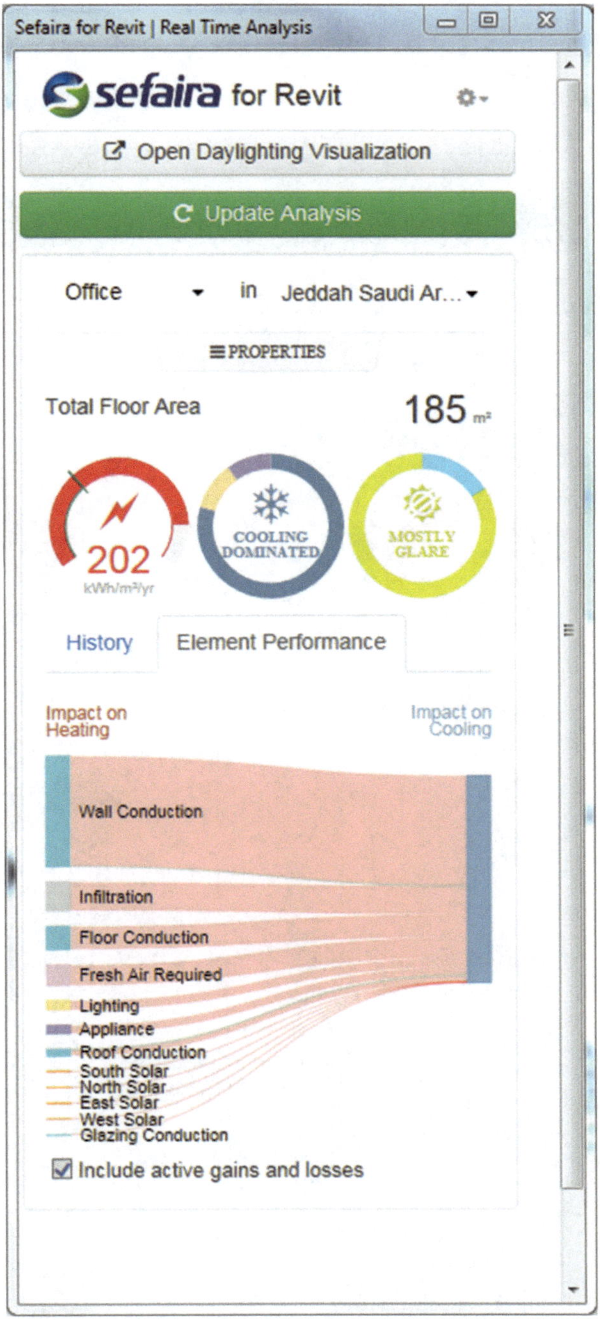

Fig. 10.3 Energy consumption analysis on the case study residential building in the city of Jeddah KSA—with the use of insulation materials on external walls

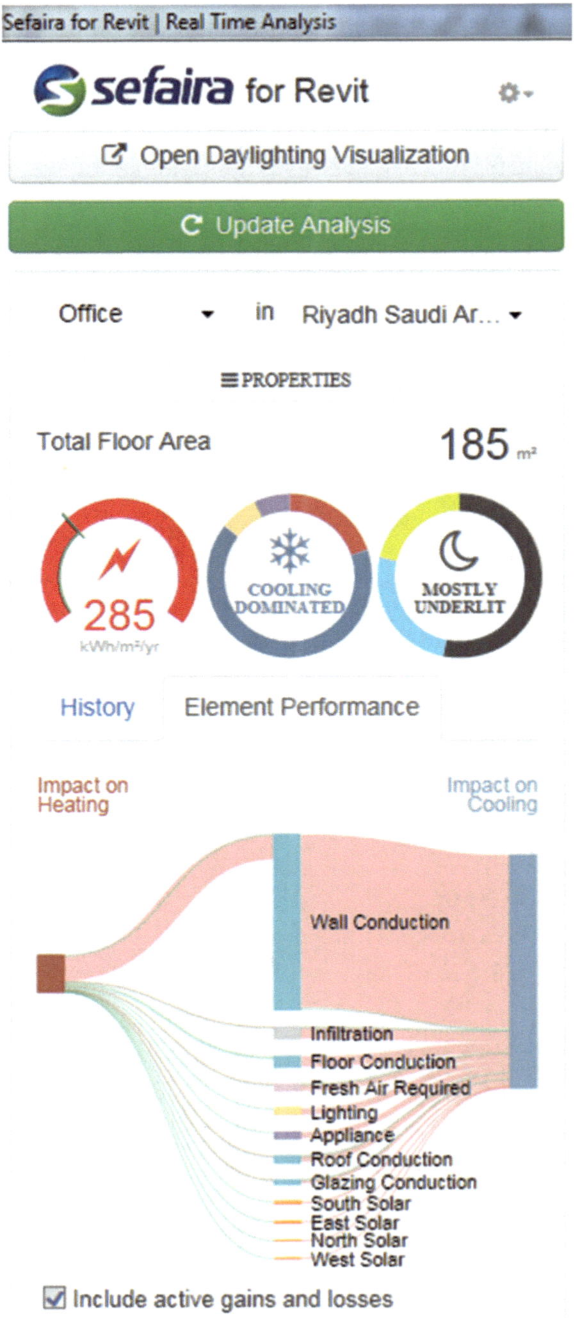

Fig. 10.4 Energy consumption analysis on the case study residential building in the city of Riyadh KSA—without the use of insulation materials on external walls

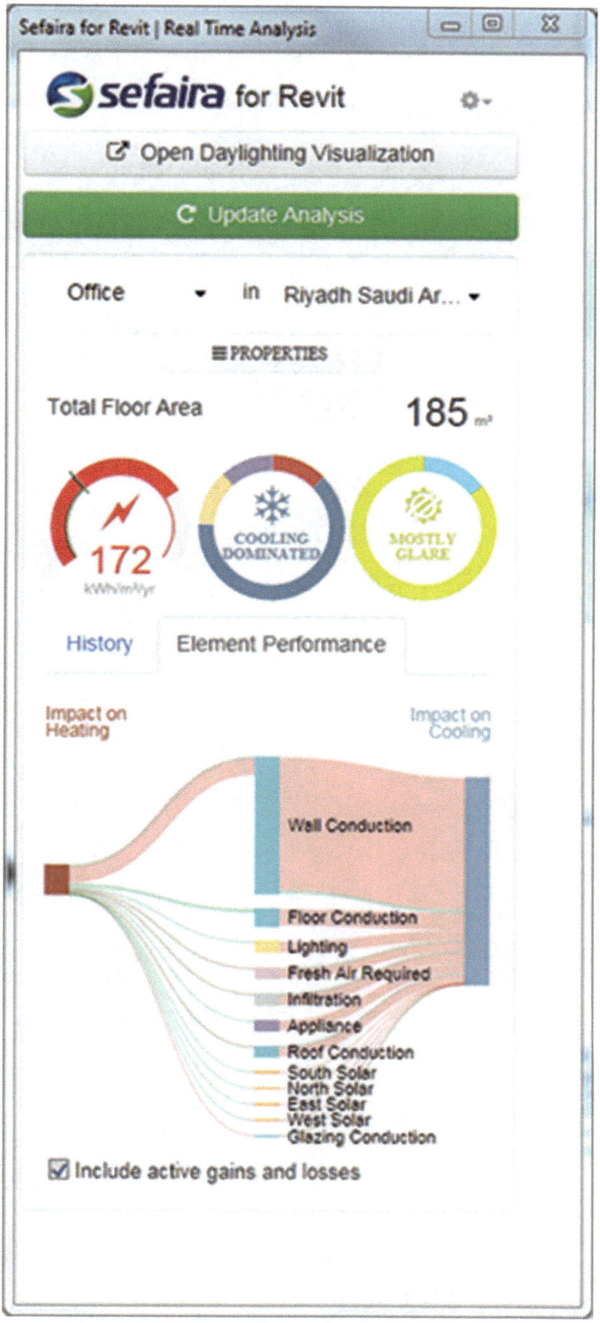

Fig. 10.5 Energy consumption analysis on the case study residential building in the city of Riyadh KSA—with the use of insulation materials on external walls

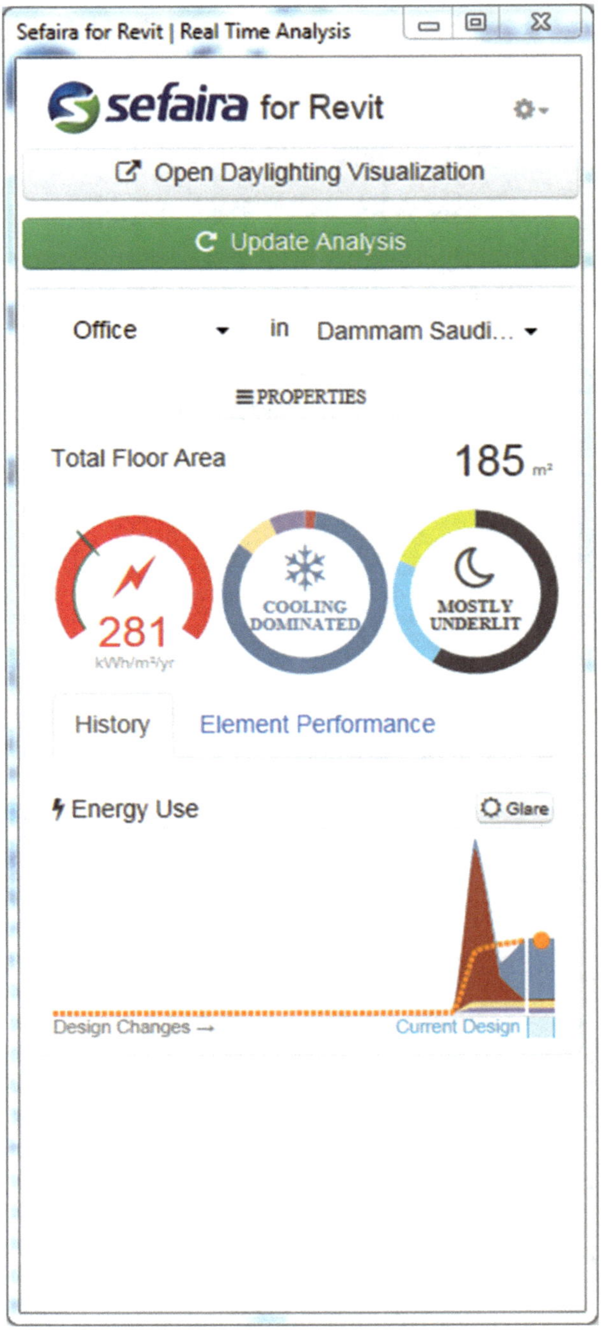

Fig. 10.6 Energy consumption analysis on the case study residential building in the city of Dammam KSA—without the use of insulation materials on external walls

Fig. 10.7 Energy consumption analysis on the case study residential building in the city of Dammam KSA—with the use of insulation materials on external walls

Subsequently, in order to reduce this high demand for domestic energy consumption and for residential buildings to become energy-efficient in the hot arid climate of Saudi Arabia, it is important to enhance the building envelope performance to reduce cooling thermal loads through thermal insulation. This illustrates the need to employ optimal insulation and architectural solutions.

10.5 Conclusion

The energy consumption of a residential building case study across three cities in Saudi Arabia has been investigated by examining data from energy simulations conducted using the Sefaira energy analysis software. The energy efficiency measure for the building envelope design considered in this study was wall insulation associated with exterior walls.

The results of the study demonstrated high-energy consumption, thus the energy optimization approach is to be based on energy performance to identify optimal and effective energy efficiency measures. These analyses and energy simulations offer the potential for a cost-effective solution for energy reduction and major savings in electricity use for cooling purposes, as well as elucidate the potentials for energy efficiency in the residential building sector of Saudi Arabia.

Finally, the reduction of domestic energy consumption can be based on such architectural principles as the building's construction and insulation materials being used, as these have proven how a few design changes could have a significant impact on the energy performance of residential buildings in the Kingdom of Saudi Arabia.

References

Akbari, H., Morsy, M. M., Al-Baharna, N. (1996). *Electricity saving potentials in the residential sector of Bahrain* (vol. 1, pp. 11–20). Lawrence Berkeley National Laboratory Publication.

Alaidroos, A., & Krarti, M. (2015). Optimal design of residential building envelope systems in the Kingdom of Saudi Arabia. *Energy and Buildings, 86*, 104–117.

Aldossary, N. A., Rezgui, Y., & Kwan, A. (2013). Domestic energy consumption patterns in a hot and humid climate: A multiple-case study analysis. *Applied Energy, 114*, 353–365.

Almatawa, M. S., Elmualim, A. A., & Essah, E. A. (2012). Passive and active hybrid approach to building design in Saudi Arabia. In C. A. Brebbia (Ed.), *Eco-architecture IV: Harmonisation between Architecture and Nature*. Southampton, UK: WIT Press.

Alrashed, F., & Asif, M. (2012). Prospects of renewable energy to promote zero-energy residential buildings in the KSA. *Energy Procedia, 18*, 1096–1105.

Alshenaif, M. A. (2015). High performance homes in Saudi Arabia Revised Passivhaus principles for hot and arid climates. M.Sc. Thesis, Philadelphia University.

Azhar, S., Carlton, W., Olsen, D., & Ahmad, I. (2011). Building information modelling for sustainable design and LEED® rating analysis. *Automation in Construction, 20*(2), 217–224.

Balaras, C. A., Gaglia, A. G., Georgopoulou, E., Mirasgedis, S., Sarafidis, Y., & Lalas, D. P. (2007). European residential buildings and empirical assessment of the Hellenic building stock, energy consumption, emissions and potential energy savings. *Building and Environment, 42*, 1298–1314.

Blanchard, S., P. Reppe. (1998). *Life cycle analysis of a residential home in Michigan*. Master's thesis, University of Michigan, Ann Arbor.

Blom, I., Itard, L., & Meijer, A. (2011). Environmental impact of building-related and user related energy consumption in dwellings. *Building and Environment, 46*(8), 1657e69.

Bribian, I. Z., Uson, A. A., & Scarpellini, S. (2009). Life cycle assessment in buildings: State-of-the-art and simplified LCA methodology as a complement for building certification. *Building and Environment, 44*, 2605–2612.

Elgendy, K. (2010). KAUST: A sustainable campus in Saudi Arabia. *Carboun Journal*. Retrieved February 22, 2016, from http://www.carboun.com/sustainable-design/kaust-a-sustainable-campus-by-the-red-sea/

Elgendy, K. (2011). Sustainability in the desert. *Detail Green Magazine*. Retrieved February 22, 2016, from http://www.carboun.com/wp-content/uploads/2011/09/1419 _background_elgendy_EN-copy.pdf

Fang, G., Park, H., & Wong, J. (2013). D3 City project—Economic impact of BIM-assisted design validation. *Automation in Construction, 22*, 577–586.

Fung, A., Cheng, I. (2011). How BIM & GIS help Project Planning and Utility Management with Safety & Sustainability in Mind—Housing Authority's Experience. *Proceedings of the 2nd International Conference on Utilities Management and Safety (ICMAS 2011)*.

Geyer, P. (2012). Systems modelling for sustainable building design. *Advanced Engineering Informatics, 26*(4), 656–668.

Hardin, B. (2011). BIM and construction management: Proven tools, methods and workflows. *Sybex*.

K.A.CARE. (2012). Building the Renewable Energy Sector in Saudi Arabia. *International Renewable Energy Agency*. Retrieved March 30, 2016, from http://www.irena.org/document-downloads/events/maltaseptember2012/5_ibrahim_babelli.pdf

K.A.CARE. (2013). *The Vision | King Abdullah City for Atomic and Renewable Energy*. Retrieved March 30, 2016, from http://www.kacare.gov.sa/en/?page_id=84

Kassim, T. et al. (2013). Automated sustainability compliance checking process proof of concept. In *Proceeding of CONVR 2013, 13th International Conference in Construction Applications of Virtual Reality*.

Krarti, M. (2011). *Auditing of building system* (2nd edn). Hard back.

Krarti, M.. (2012). *Weatherization and energy efficiency improvement for existing homes—An Engineering approach*. (Hard back).

Krygiel, E., & Nies, B. (2008). Green BIM: Successful sustainable design with building information modelling. *Sybex*.

Kubba, S. (2012). Handbook of green building design and construction. *LEED, BREEAM and GREEN GLOBES*.

Meier, A., Olofsson, T., & Lamberts, R. (2002). What is an energy efficient building? *9th National Meeting of Environmental Technology Built*, Foz do Iguacu Parana, Brazil.

Olofsson, T., Andersson, S., & Sjogren, J. U. (2010). Building energy parameter investigations based on multivariate analysis. *Energy and Buildings, 41*, 71–80.

Russell-Smith, S., & Lepech, M. (2012). Activity-based methodology for life cycle assessment of building construction. In: *CIBSE ASHRAE Technical Symposium*, Imperial College London, 1–13.

SEEC. (2013). Saudi Energy Efficiency Centre. Retrieved March 30, 2016, from http://www.seec.gov.sa/2013/03/27/مباني/?lang=en

Taleb, H. M., & Sharples, S. (2011). Developing sustainable residential buildings in Saudi Arabia: A case study. *Applied Energy, 88*(1), 383e91.

Chapter 11
Fossil Fuel Reliant Housing in Nigeria: Physio-climatic Regionalism as an Energy/Cost Efficient Perspective to Providing Thermal Comfort

Alolote Amadi and Anthony Higham

Abstract The study investigates trends in housing design in Nigeria and assesses the degree of adherence to thermal comfort requirement. It reveals that building designs mostly reflect uniformity despite the variety of climatic conditions in Nigeria. This is shown from the results of a survey and statistical analysis of the distribution of external thermal design features, including roofing, cladding, openings, shading, and layout and landscaping, for 1000 houses at sample locations in the three climatic regions of Nigeria. The analytical outcome shows the close similarity in housing designs between disparate climatic regions and the predominant typology of design features. The study further evaluates meteorological data, using the Oligyay Bioclimatic Chart developed in the 1950s, to approximate the average thermal comfort requirements for the regional climatic sequences of the hot-dry north, dry sub-humid middle belt and warm humid south of Nigeria, with Kano, Minna and Port Harcourt used as reference points of comparisons. The outcome of the analysis shows that a greater percentage of urban houses in Nigeria do not portray the regional comfort characteristics required for thermal comfort, which may account for the wasteful reliance on the use of fossil-fuelled generators in buildings for house cooling.

11.1 Introduction

Nigeria's physical regions are closely related with latitude, and as such environmental variables such as vegetation, soils and climate are strongly correlated with their north-south location (UNDP 2000). Nigeria's climate therefore follows this correlation, with hot-dry weather conditions being found in the north of the country, which progressively changes to warm humid conditions southwards towards the

---

A. Amadi (✉) • A. Higham
School of the Built Environment, University of Salford, Salford, UK
e-mail: A.I.Amadi@edu.salford.ac.uk

M. Dastbaz et al. (eds.), *Building Information Modelling, Building Performance, Design and Smart Construction*, DOI 10.1007/978-3-319-50346-2_11

Table 11.1 Application of the ET index in Nigeria

Location	Month	Effective temperature
Kano	January	Comfortable
	March	Warm, sometimes too warm at midday
	May	Too warm for most of the day
	July	Warm for most of the day
	September	Warm to comfortable
	November	Comfortable
Minna	January	Comfortable
	March	Comfortable
	May	Approaching warm at midday
	July	Comfortable
	September	Comfortable
	November	Comfortable
Port Harcourt	January	Warm
	March	Warm most of the day
	May	Warm
	July	Just approaching warm
	September	Just warm at midday
	November	Warm

Source: Ladell (1949)

coast (Ayaode 1973). Human comfort requirements will thus vary accordingly. The effective temperature (ET) index is the most widely used physio-climatic index, indicating the most appropriate thermal environment for human comfort (Ilyas et al. 1981). The ET index was developed by a research team for the American Society of Heating and Ventilating Engineers (ASHVE), by exposing the subjects to different degrees of atmospheric conditions of humidity, temperature and air movements and asking the subjects to rate their sensations of warmth and coolness (Giovanni 1994).

Ladell (1949) used the ET index to study several locations in Nigeria. Results from Ladell's study, for three locations in the north, middle belt and south of Nigeria, are presented in Table 11.1, showing the variations in the degree of thermal sensation induced by the climatic elements at various locations across the country. Various other indices for assessing the degree of sensation of the human body have also been developed. A significant amount of research has been done, focused on classifying climate with human physiology (Bob Manuel 1990). Many of these approaches use elements such as: temperature, humidity, radiation and wind. However, some researchers have applied the measures singly or in combination for the purposes of developing comfort indices. One of these physiologic indices applied specifically to Nigeria and also used for wider distributional studies in Africa is the relative strain index developed in 1965 (Griffith and Sidwell 1995). The index provides an indirect way of measuring strain, by relating the application of heat stress to the maximum stress that can be tolerated without breaking down. The index is written in the form: $R_s = E_r/E_m$ where E_r = evaporative cooling required and E_m = maximum evaporative cooling possible. Griffith and Sidwell (1995) indicate that such values range from less than 10% to 100% over the African continent. The Nigerian specific study provided a value of 50% in January and between 40 and 70% for July. The distribution across Nigeria in July suggested southern parts of the country have a value of circa 40%, the west about 50% and the northern parts 60–70%.

Griffith and Sidwell (1995) have also applied the 'Predicted Four Hour Sweat Rate Index' (P4SR) across the African continent. The P4SR index assumes that the rate at which man sweats is a good index of heat stress (Ladell 1949). A monogram was constructed empirically, which was then used to predict the amount of sweat measured in litres perspired over a 4 h period. The limiting P4SR is 4.5 litres, although nobody should be exposed to rates above 2.5 in 4 h. The P4SR is considered to be one of the most accurate of the existing indices. Griffith and Sidwell (1995) presented the distribution of the P4SR in both January and July. The work recorded relatively high levels of perspiration in West Africa, with Nigeria having values between 1.5 and 2.0 litres. More recent studies have been conducted in Nigeria by Ogunsote and Prucnal-Ogunsote (2002), Akande and Adebamowo (2009) and Omonij and Matzarakis (2011). Ogunsote and Prucnal-Ogunsote (2002) as well as Omonij and Matzarakis (2011) compared the predictive capacity of several thermal indices. The studies showed that the thermal analysis, which incorporates the effects of relative humidity, air movement, metabolic rate and clothing, based on the ET index developed by ASHVE, have a high degree of accuracy. Akande and Adebamowo (2009) adopted other bioclimatic indices based on the American Society of Heating, Refrigerating and Air-Conditioning Engineers (ASHRAE, 2007) seven-point scale (−3 to 3), which subjectively rates thermal sensation, and the three-point thermal preference (TP), for rating the thermal need. Using the ASHRAE scale, Akande and Adebamowo (2009) reported similar thermal discomfort sensations as recorded by Ladell (1949) for Kano in Northern Nigeria.

In sustainable building designs and cost analysis, thermal comfort and wellbeing based on these physio-climatic indices are now routinely considered due to the pronounced effect of climate on human physiology. As a result, comfort indices have been applied to building designs using various methods such as the Mahoney scale, the Evans scale and the bioclimatic chart, to define comfort requirements in buildings suitable for about 80% of normal healthy people (Giovanni 1994; Ogunsote and Prucnal-Ogunsote 2002). Investigating the perceived reliance on passive design parameters for public buildings in the warm humid conditions of Southern Nigeria using the Mahoney table, Lawal et al. (2012) noted that a greater percentage of the design parameters necessary for climate control did not conform. This is because the specific features of design and structural materials affect the response of a building structure to exposure of inherent climatic elements, and a lack of which in turn determines the energy demands and cost required to maintain thermal comfort within buildings. All other studies on passive housing in the literature are limited to specific zones of Nigeria. No comprehensive study has been identified in the literature with a wider coverage, across the three climatic zones.This study comparatively investigates the energy demands of housing, their design features across Nigeria and the level of adherence to their respective climatic requirements using the bioclimatic chart.

11.2 Literature Review

11.2.1 Energy Demand of Housing in Nigeria: Life Cycle Cost of Domestic Power Generator Use

Several studies have investigated the energy demands of residential buildings in different parts of Nigeria. (Lawal et al. 2012; Agajelu et al. 2013; Omoruyi and Idiata 2015). In the hot climate of Nigeria, it was shown that most buildings rely primarily on the use of domestic power generating sets to achieve thermal comfort indoors (Omoruyi and Idiata 2015). Ibidapo-Obe and Ajibola (2011) espouse that Nigeria has one of the lowest net electricity generation per capita rates in the world with the majority of the populace having no electricity supply. A point reinforced in a recent survey conducted by Omoruyi and Idiata (2015) which revealed the majority of respondents (66.7%) was reliant on the use of petrol- and diesel-powered generators to boost their economic activities.

Initial capital cost is habitually used as the only financial consideration in the procurement of buildings (Higham et al. 2015). Life cycle cost analysis however shows long-term financial commitments, implied by investment decisions in buildings with consideration for the time value of money (Agajelu et al. 2013). According to Loh et al. (2009:20–21), the '... Lack of attention at the early design phase ...has led to an unsustainable built environment. Choice of building materials at the early stage of the design process, obviously has a direct bearing on energy consumption, cost performance and greenhouse gas emissions'. Environmental comfort and energy-saving cost analysis should also constitute the primary considerations at the conceptual, construction and usage phases of buildings. Morphological adaptation of buildings to microclimate so they use less energy, thus mitigating the environmental impact of the building over its life cycle, is thus considered an optimal sustainable solution (Higham et al. 2015). Indeed, Loh et al. (2009) emphasised that for maximum effect and economy, the aim should be the integration of appropriate design and technology into the overall building form and not simply to apply technology as an afterthought or as *sustainable bling* to satisfy the demands of regulators (Higham and Thomson 2015). The more effective integration of sustainable features may cost more initially, but the long-term running costs would be lower, leading to overall cost savings (Loh et al. 2009).

In Nigeria, fossil-fuelled plants provide a major source of electricity generation for short periods (5–7 h per day), unfortunately Nigeria does not have a continuous supply of power, as such the country experiences power outages for the majority of the day/night. A recent survey of residents undertaken by Omoruyi and Idiata (2015) revealed the daily running cost of domestic generators in residential areas in the southwest of Nigeria (used for periods of power outages) was between N250 (10% of respondents) and N1250 (4% of respondents), with the vast majority of respondents (86%) spending N750 on average. These figures suggest the annual running costs for domestic power generation ranges from N91,250 to N456,250, with the majority of Nigerians spending N273,750 per annum on domestic power generation. Agajelu et al. (2013) carried out a detailed life cycle cost analysis of a diesel power generating system, for an off-grid residential building in the southeast of Nigeria, using both net present value (NPV) and internal rate of return (IRR) methodologies. Assuming a system lifespan of 25 years for a 2.5 KVA diesel generator, a real interest rate of 9% per annum and average hourly electrical load demand data obtained from the Power Holding Company of Nigeria (PHCN). The analysis revealed the life cycle cost of power generation was N7,098,192.00. On the basis of this study, Agajelu et al. (2013) concluded that despite the attractiveness of the low initial cost of diesel generators, the long-term cost was high, due to running costs associated with its fuel consumption, at N135/litre, and maintenance costs. Furthermore, the pump price of fossil fuels, regulated by the Nigerian National Petroleum Company (NNPC), is experiencing high levels of price inflation in Nigeria, due to scarce supplies and world demand. At the time of writing, diesel is currently sold at N250–300 per litre at filling stations and N400 per litre by the black market, often in adulterated forms.

11.2.2 Environmental, Health and Safety Impacts of Domestic Fossil Fuel Combustion in Buildings

The combustion of fossil fuels has severe environmental consequences. On a global scale, the environmental impact of the combustion of fossil fuels is linked to global warming and climate change, which has led to international treaties such as the Kyoto Protocol (Ibidapo-Obe and Ajibola 2011). Combustion of fossil fuels locally impacts on the environment, causing acid rain and air pollution in coastal communities. The World Health Organization (WHO) and other international agencies have expressed concerns about the health impact of burning fossil fuels. The global death toll due to pollution caused by fossil fuel burning for electricity generation has been estimated at 0.3 million people annually. Inhalation of emissions such as carbon monoxide, sulphur dioxide, nitrogen oxides, from small- and large-scale generators is rife in Nigeria (Omoruyi and Idiata 2015). However, in Nigeria, the total fossil fuel-based deaths are not known due to the absence of detailed mortality records. Omoruyi and Idiata (2015) revealed that the associated health and social hazards of combustion of fossil fuels are also major concerns to both building inhabitants and their neighbours in residential areas, as reported cases of impaired hearing, impaired visibility, deafness, sleeplessness, choking sensations and dizziness have been associated with the use of generators.

Adulteration of fossil fuels in Nigeria is another environmental, health and safety dimension to the dangers posed by domestic power generators' use in buildings (Kamil et al. 2008; Osueke and Ofondu 2011). The NNPC defined adulteration of fossil fuels as 'the illegal or unauthorized introduction of foreign substances into fuel with the result that the product does not conform to the requirements and specifications of the product' (Osueke and Ofondu 2011:32). The most common forms of adulteration of fossil fuels in Nigeria were listed as the introduction of lubricants into kerosene to serve as diesel, kerosene into petrol, kerosene into diesel and used lubricants into diesel. Increased profit and acute scarcity drive the trend of fuel adulteration by independent marketers and roadside black markets (Centre for Science and Environment 2002). It has been argued that malfunctioning generator engines together with safety concerns associated with the adulteration of fossil fuels have led to increased gaseous emissions. This hypothesis was scientifically tested by Kamil et al. (2008); their experiments showed increased emissions of hydrocarbons, carbon monoxide and toxic carcinogenic substances when adulterated fossil fuels were used. This finding has subsequently been reiterated in experiments conducted by Osueke and Ofondu (2011) which showed a 36% increase in SFC emissions when adulterated fossil fuels were used. In addition to the environmental impacts, the Department for Petroleum Resources (DPR 2016) has recently reported the serious safety risks presented by adulterated fuels such as petrol adulterated with highly inflammable kerosene. Asserting the highly explosive mixture has resulted significant loss of life due to explosions triggered by the fuel, which is widely used due to Nigeria's current fuel crisis.

The environmental and cost implications of fossil fuel-reliant housing thus reinforce the need for adequate thermal diagnosis of housing types situated in the heterogeneous climatic setting of Nigeria. The study thus carries out a comprehensive and spatially distributed survey of housing design features, to serve as a platform for inferring the degree of thermal adherence, which can offer explanation for this negative trend of fossil fuel reliance in domestic energy consumption.

11.3 Research Method

Nigeria has three major climatic zones: the dry subhumid climate in the Northern parts of the country, the moist subhumid climate in the middle belt and the humid climate in the Southern region (Ayaode 1973). A survey was carried out to assess the design characteristics of house types in these major climatic zones of Nigeria. Table 11.2 shows the states of Nigeria subdivided based on climatic region. Fifty locations across the climatic zones were cluster sampled for one-third of the states in each climatic zone, randomly selected as representative of the houses in Nigeria. Twenty houses were physically surveyed using direct observation at each location.

The residential locations were sampled from urban areas, for both lower- and higher-income neighbourhoods, to ensure adequate representativeness of the population of urban houses. A total of 1000 houses were surveyed, exclusive of rural areas. Data analyses involved assessing and comparing physical characteristics of the houses for similarity in the percentage distribution of design features using the statistical tool of analysis of variance (ANOVA).

Further to this preliminary investigation, bioclimatic evaluation is used to define the typical thermal design requirements of houses in the climatic zones based on selected stations. The choice of climate samples also offers the possibility of comparing diverse environmental settings and their effects on housing design in Nigeria. The method of Oligyay, based on the building bioclimatic chart (BBC) developed in 1953, is used. The BBC was the first attempt to develop a systematic procedure for

Table 11.2 Nigeria subdivided based on climatic regions

Warm humid		Dry subhumid		Dry hot	
1	Oyo	1	Kwara	1	Sokoto
2	Osun	2	Niger	2	Kebbi
3	Ondo	3	Abuja	3	Katsina
4	Ogun	4	Kaduna	4	Jigawa
5	Lagos	5	Bauchi	5	Yobe
6	Delta	6	Gombe	6	Borno
7	Enugu	7	Benue	7	Adamawa
8	Imo	8	Kogi	8	Zamfara
9	Akwa Ibom	9	Nasarawa	9	Kano
10	Cross River	10	Taraba		
11	Abia	11	Plateau		
12	Rivers				
13	Ebonyi				
14	Anambra				
15	Ekiti				
16	Bayelsa				
17	Edo				

adapting the design of buildings to the human comfort and microclimatic requirements of a given region (Giovanni 1994, 1998). Although several other adaptations and methods of designing for thermal comfort in buildings have been proposed, the Oligyay method remains the most widely used and referenced. The method of analysis as used in this study proceeds according to the following steps:

- Compilation of local climatic data for the selected stations, i.e. Kano, Minna and Port Harcourt.
- Application of regional evaluations to the bioclimatic chart by plotting of the combined monthly temperature and relative humidity values at regular intervals for each region. This was done for average conditions, i.e. climatic situation of typical average day of each month.
- Summary of architectural interpretation in terms of the required design elements and housing layouts in each region.
- Diagnosis of the relative adherence of the various elements of the houses surveyed in each region for degree of thermal comfort offered.

11.4 Data Analysis and Discussion

11.4.1 Survey Results

Table 11.3 shows the building types surveyed across the climatic zones. The highest percentage of houses sampled was terraced single rooms, typical of lower-income neighbourhoods. Detached bungalows and terraced block of flats constitute 21.9%

Table 11.3 Housing types surveyed

Housing type	Humid climate	Moist subhumid	Dry subhumid	% of type
Terraced single rented rooms	141	68	25	23.4
Detached bungalows	93	66	60	21.9
Terraced bungalow flats	121	44	38	20.3
Terraced duplex	26	43	89	15.8
Multi-storey block of flats	22	61	19	10.2
Duplex	47	23	14	8.4
Total	450	305	245	

and 20.3% of the houses sampled and were mostly in middle-class neighbourhoods. Multi-storey blocks of flats and duplexes were typical of the higher-income neighbourhoods surveyed.

Tables 11.4 and 11.5 show the results of the assessment and statistical analysis of the distribution of design features for the houses surveyed. From the analysis, it is shown that sandcrete block is the major material utilised for cladding in Nigeria and that the percentage use of various cladding alternatives amongst the three climatic zones does not vary significantly. Similar outcomes were noted for other design features including colour of finishing, which shows the predominant use of dark colours for walls and roof despite the hot climate of Nigeria. Roofing materials, window size, hoods, glazing and height above ground level also do not significantly vary. The general building outlays including the use of courtyards and landscaping, orientation of houses and building forms are similar. Building forms were noted as being mostly compact. This was noted particularly in lower-class neighbourhoods, where the houses occupy all of the plot space without provision for circulation spaces and adjacent buildings overly at the roof overhangs.

The only significant difference in the trend of housing designs based on the ANOVA results is in terms of the roof pitch, overhang and elevation of buildings above ground level. This is likely due to differences in the levels of precipitation recorded between the zones. The similarity in design elements indicates a climatically unresponsive approach to housing design in Nigeria, which has produced a typology of buildings without clearly discernible regional characteristics.

11.4.2 Comparative Bioclimatic Analysis

The bioclimatic chart, developed using the effective temperature scale (ET), defines comfort requirements for people at rest and normally clothed, shown as the comfort zone in the centre. Outside the comfort zone, there are indications of the different sensations. The 'difficult environment' and 'impossible environment' (93^0 – 96 °F) and (95 – 97 °F) ET curves are also shown. The chart is built up with temperatures in degrees Fahrenheit as ordinates and relative humidity as abscissa. Any climatic condition determined by temperature and relative humidity can be plotted on the

Table 11.4 Housing design features showing predominant trends

Adaptive mechanism	Zone Humid Design features	No	Zone Moist subhumid Design features	No	Zone Dry subhumid Design features	No	(%)
Cladding							
• Material	Sandcrete blocks	324	Sandcrete blocks	200	Sandcrete blocks	136	**66**
	Bricks	86	Bricks	78	Bricks	59	22.3
	Concrete walls	40	Concrete walls	27	Concrete walls	20	8.7
	Others	0	Others	0	Others	30	3
• Finishing	Light colour	123	Light colour	113	White colour	38	27.4
	Dark colour	200	Dark colour	100	Dark colours	145	**44.5**
	None	127	None	92	None	62	28.1
Roof							
• Material	Light reflective	141	Light reflective	124	Light reflective	44	30.9
	Dark reflective	135	Dark reflective	180	Dark reflective	100	**41.5**
	Heavy weight	90	Heavy weight	25	Heavy weight	81	19.6
	Others	29	Others	31	Others	20	8
• Pitch	High> 30^0	234	High> 30^0	50	High> 30^0	21	30.5
	Low< 30^0	180	Low< 30^0	212	Low< 30^0	82	**47.4**
	Flat	36	Flat	43	Flat	142	22.1
• Overhang	Wide	143	Wide	89	Wide	33	26.5
	Narrow	135	Narrow	156	Narrow	67	**35.8**
	None	137	None	35	None	78	25
	Parapet	45	Parapet	25	Parapet	67	13.7
Windows							
• Type	Casement	55	Casement	36	Casement	44	13.5
	Louvre	207	Louvre	100	Louvre	87	**39.4**
	Sliding	168	Sliding	149	Sliding	32	**34.9**
	Wooden	22	Wooden	18	Wooden	82	12.2
• Sizing	Large	170	Large	60	Large	55	28.5
	Medium	123	Medium	100	Medium	112	33.5
	Small	257	Small	145	Small	78	**48**
• Glazing	Single reflective	56	Single reflective	37	Single reflective	43	13.6
	Double reflective	32	Double reflective	41	Double reflective	32	10.5
	Transparent	362	Transparent	227	Transparent	170	**75.9**
• Shading	Hoods	122	Hoods	76	Hoods	67	26.5
	None	280	None	229	None	178	**68.7**
• Height above GL	0.5–1.5 m	138	0.5–1.5 m	123	0.5–1.5 m	200	46.1
	Above 1.5 m	312	Above 1.5 m	182	Above 1.5 m	45	**53.9**

(continued)

Table 11.4 (continued)

Adaptive mechanism	Zone Humid Design features	No	Zone Moist subhumid Design features	No	Zone Dry subhumid Design features	No	(%)
Externals/layout							
• Building form	Elongated	239	Elongated	142	Elongated	149	**53**
	Compact	211	Compact	163	Compact	96	47
• Orientation	East–west	78	East–west	45	East–west	43	16.6
	North–south	42	North–south	71	North–south	34	14.7
	Deviated	330	Deviated	189	Deviated	168	**68.7**
• Elevation	High	145	High	188	High	122	45.5
	Low	305	Low	117	Low	123	**54.5**
• Courtyard	External	94	External	49	External	34	17.7
	Internal	52	Internal	23	Internal	17	9.2
	None	304	None	233	None	194	**73.1**
• Fencing	Low	79	Low	69	Low	37	18.5
	High	256	High	76	High	32	36.4
	None	115	None	160	None	176	**45.1**
• Windbreaks	NA	–	NA	–	Trees	32	13.1
					none	213	**86.9**
• Landscaping	Present	89	Present	87	Present	57	23.3
	None	361	None	218	None	188	**76.7**

Table 11.5 One-way ANOVA results for similarity of building designs across the climatic zones

Design element	Specification	Significance at 0.05
Cladding	Material	0.101 (not significant)
	Finishing	0.062 (not significant)
Roof	Material	0.067 (not significant)
	Pitch	0.007 (significant)
	Overhang	0.005 (significant)
Window	Type	0.090 (not significant)
	Sizing	0.116 (not significant)
	Glazing	0.061 (not significant)
	Shading	0.147 (not significant
	Height above GL	0.118 (not significant)
Layout	Building form	0.462 (not significant)
	Orientation	0.080 (not significant)
	Elevation	0.030 (significant)
	Courtyard	0.12 (not significant)
	Fencing	0.230 (not significant)
	Landscaping	0.625 (not significant)

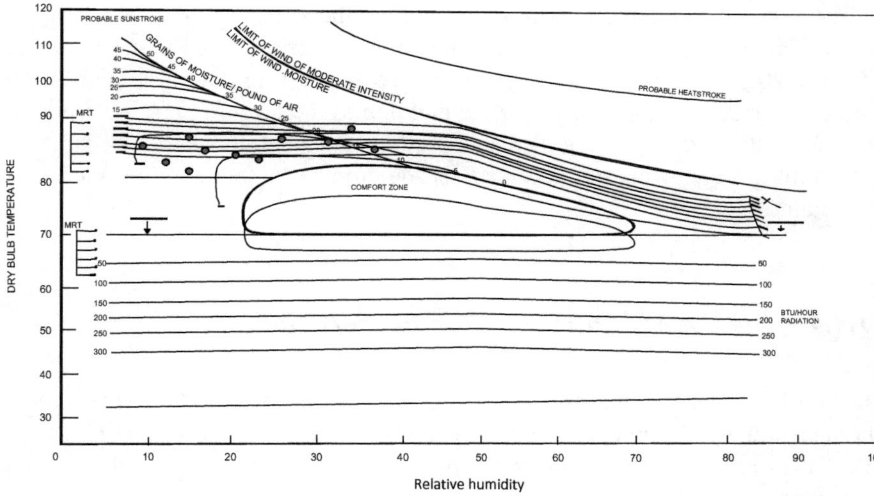

Fig. 11.1 Bioclimatic chart for Kano

chart. If the plotted point falls in the comfort zone according to the season, we feel comfortable in shade, and if the point falls outside the comfort zone, corrective measures are needed (Giovanni 1998). If the point is higher than the upper perimeter of the comfort zone, winds are needed to offset high temperatures and is calibrated with nearly parallel lines spaced 100 fpm apart and reaching a maximum of 700 fpm following the upper limit of the comfort zone perimeter, indicating the needed wind velocities under shade. The dotted lines indicate the grains of moisture per pound of air needed to reduce temperatures to the level of the upper comfort perimeter. At the lower perimeter of the comfort zone is the zone where solar radiation is needed. Port Harcourt, Minna and Kano have been used as reference points for interpretations of climatic data, to indicate and specify thermal requirements of direct geographic applicability. The plotted climatic data are all based on those available at the Nigerian Meteorological services (NIMET).

11.4.3 Bioclimatic Interpretation for Kano

From the chart plotted in Fig. 11.1, it can be seen that the majority of the plotted points fall outside the comfort zone. It is thus obvious that corrective measures are needed to restore adequate comfort sensations. All the plotted points are above the shading line implying a need for shade protection for buildings located in this area. The values of the plotted points show that most of the points are of high temperatures and low humidity indicating that the inherent thermal sensation in this climatic zone is that of dryness and hotness. Winds are thus of little help here as the recorded wind velocities in this area are already on the high side and will further require windbreaks.

The required corrective measures are therefore those that would induce evaporative cooling, which is the tool required to fight high temperatures and low humidity. About 5–20 gr/1 b of moisture is therefore needed to bring down the temperature within the inhabited living space of the building to the level of comfort. Building designs in this climatic region therefore need to promote the retention of moisture in buildings along with the necessary shade from solar radiation at such high temperatures.

11.4.4 Bioclimatic Interpretation for Port Harcourt

The Port Harcourt station falls into the warm humid zone of Ayaode (1973) climatic classification. All the points fall outside the comfort zone and are above the shading line. There is thus need for corrective measures to be incorporated into building designs in this region.

Most of the points plotted on the chart shown in Fig. 11.2 are of high humidity and high temperature. Evaporative cooling is the tool with which to fight high temperatures and high humidity. The moist sweaty feeling associated with these climatic characteristics can be alleviated by winds/air movements of adequate velocities, between 100 and 700 fpm, to counteract the high humidity levels.

The bioclimatic requirement for this region is thus to ensure air movements via regulation in the use of large openings for ventilation, depending on the time of the year and season. The rainy season months with lower temperatures lie closer to the comfort zone and would require lower wind velocities to restore comfort, as can be seen from the charted points for this season. Conversely the dry season months with

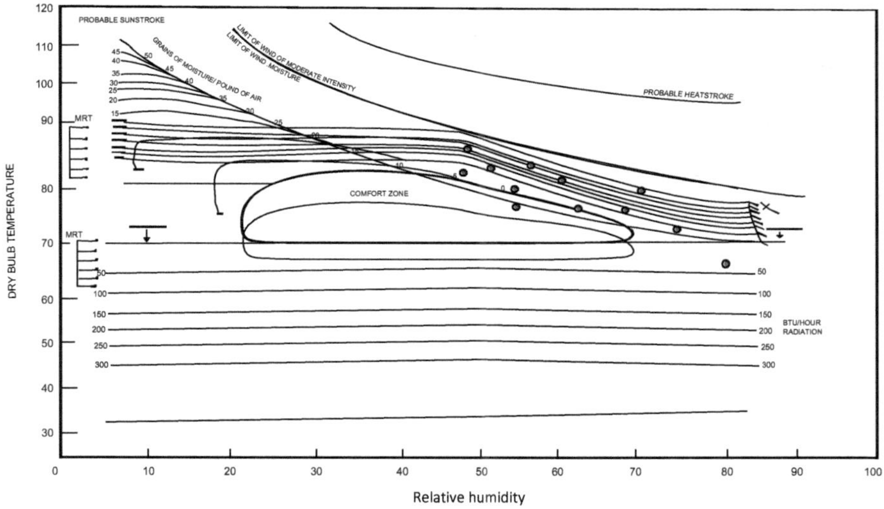

Fig. 11.2 Bioclimatic chart for Port Harcourt

higher temperatures are further away from the comfort zone and would require higher wind velocities. Building designs in this region should also strive to provide shading requirements.

11.4.5 Bioclimatic Interpretation for Minna

Minna displays a wider range of spread in the positions of the climatic points plotted in Fig. 11.3, with some points falling into either of the upper extremes of the climatic types, while others are dispersed in between. However, all the points are located above the shading line and mostly outside the comfort zone, indicating the need for corrective measures. The zone thus offers more flexibility in the mode of corrective measures that can be adopted to restore comfort. The bioclimatic needs for this region can therefore be tackled in one of three ways, depending on the need that is paramount at that particular time:

- By addition of the required air movement, by this mode winds ranging from 100 to 500 fpm can help restore comfort during months of high temperature and high humidity
- Through evaporative cooling, by adding 5–10 gr moisture of air during months of high temperature and low humidity as are usually associated with the dry season months
- By a combination of both measures as may be necessary

Housing design requirements for the three zones, summarised in Table 11.6, are thus specified in relation to this thermal diagnosis.

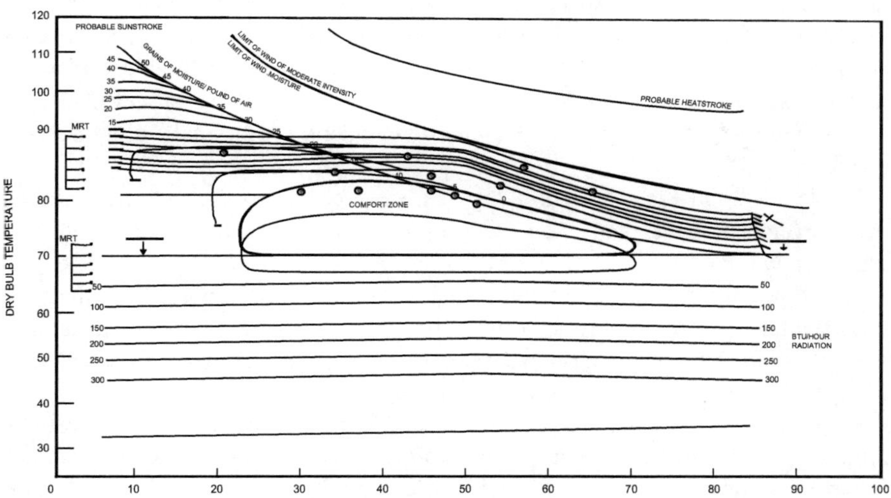

Fig. 11.3 Bioclimatic chart for Minna

Table 11.6 Housing design requirement for the climatic zones

Adaptive mechanism	Design requirement Warm humid	Design requirement Dry subhumid	Design requirement Dry hot
Cladding			
• Material	Blocks	Blocks/bricks	Bricks/concrete walls
• Finishing	Light colour	Light colour	White colour
Roof			
• Material	Light reflective	Light reflective	Heavy weight
• Pitch	High> 30^0	Low< 30^0	Flat/curved
• Overhang	Wide	Wide	Parapet
Windows			
• Type	Pivot/louvre	Pivot/casement	Sliding/casement
• Sizing	Large	Large	Small
• Glazing	Single reflective	Single reflective	Double reflective
• Shading	Hoods	Hoods	Hoods
• Height above GL	0.5–1.5 m	0.5–1.5 m	Above 1.5 m
Externals/layout			
• Building form	Elongated	Elongated	Compact
• Orientation	East–west	East–west	North–south
• Elevation	High	Low	Low
• Courtyard	External	External	Internal
• Fencing	Low	Low	High
• Windbreaks	NA	NA	Trees
• Landscaping	Needed	Needed	Needed

The available local meteorological data on the climatic elements has thus been used for the purpose of bioclimatic design. The difference between observation height used for data collection by the meteorological agency and living level is small enough to be disregarded, although other localised site features will play out in more specific designs. A further post hoc evaluation of the surveyed houses shows their respective degree of adherence to the identified thermal requirements in each zone.

Table 11.7 and Fig. 11.4 show that a greater percentage of external building design elements, which can be adapted to fulfil indoor thermal comfort, is not bioclimatically aligned.

The types of materials used for cladding and the roof pitch appear to be the most consistent design factors across the zones that display adherence to climate control, with a visible congruence of the radar chart for these features. The composite level of bioclimatic adherence to passive design features amongst the regions varies, with the least level of regional character evidenced in the dry-hot climate of Northern Nigeria. The middle belt with a more flexibility in climatic considerations is shown to have the highest level of design adherence, with 39% of passive design features present in the sampled houses.

Table 11.7 Post hoc evaluation of surveyed houses in relation to zonal thermal requirements

Adaptive mechanism	Humid		Dry subhumid		Dry hot	
	Design requirement	(%)	Design requirement	(%)	Design requirement	(%)
Cladding						
• Material	Sandcrete blocks	72	Sandcrete blocks	51.1	Bricks	19.9
			Bricks	40	Concrete	12.3
• Finishing	Light colour	27.3	Light colour	37	White colour	15.5
Roof						
• Material	Light reflective	31.3	Light reflective	40.7	Heavy weight	33.1
• Pitch	High > 30^0	52	Low < 30^0	69.5	Flat	57.9
• Overhang	Wide	31.8	Wide	51.1	Parapet	27.3
Windows						
• Type	Pivot	29	Pivot	39	Sliding	15
	Louvre	26	Casement	5.6	Casement	3
• Sizing	Large	42.9	Medium	52.5	Small	31.8
• Glazing	Single reflective	12.4	Single reflective	12.1	Double reflective	13.1
• Shading	Hoods	27.1	Hoods	24.9	Hoods	27.3
• HAGL	0.5 to 1.5 m	30.7	0.5 to 1.5 m	40.3	Above 1.5 m	18.4
Externals/layout						
• Building form	Elongated	53.1	Elongated	46.6	Compact	39.2
• Orientation	East–west	I7.0	East–west	14.8	North–south	13.9
• Elevation	High	32.2	Low	38.4	Low	50.2
• Courtyard	External	52	Internal	7.5	Internal	6.9
• Fencing	Low	17.6	Low	22.6	High	47.8
• Windbreaks	NA	–	NA	–	Trees	13.1
• Landscaping	Needed	19.8	Needed	28.5	Needed	23.3
Regional trend		32.4		39		27.6

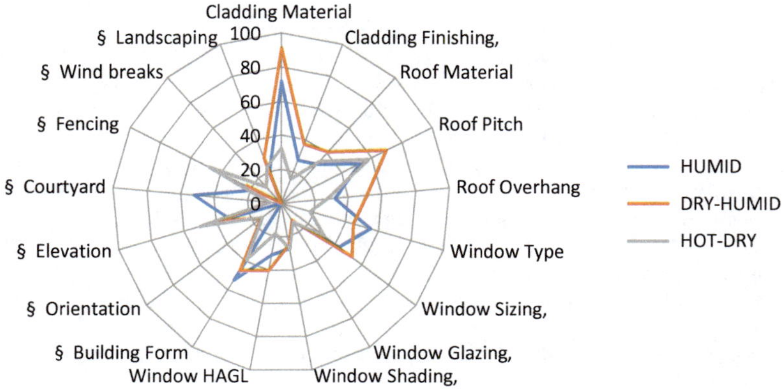

Fig. 11.4 Regional passive design trend for external housing features in Nigeria

11.5 Conclusion

The study outlines the predominant trends in housing design features and the discernible lack of differentiated forms of housing between the climatic zones, despite their varying thermal comfort requirements. The study submits that this trend is a needless waste of resources, which could be saved at the conceptual phase of building designs with the financial benefits more strongly evidenced during the early phases of cost planning. This is considering the long-term financial commitments due to the ever increasing cost, acute scarcity and adulteration of conventional fuel sources in Nigeria and the cost of purchasing and maintaining mechanical power generators, coupled with the complimentary health and safety implications of constantly inhaling poisonous gaseous emissions. The post hoc evaluations have been used to approximate the regional level of adherence of design features to the thermal requirements within each climatic setting. This may thus account for the trend of relying primarily on mechanical power generator sets in residential buildings. Based on the outcome of the bioclimatic analysis, the thermal comfort requirements of the three major climatic zones, the humid, moist subhumid and the dry subhumid, vary according to the needs of each region. The bioclimatic regional evaluations of these differing climatic situations have indicated the comfort needs of evaporative cooling, shading and wind effects required for each climatic environment. It can thus be inferred from the details of the evaluations that all applicable design techniques should be called into play to promote environmentally sustainable designs that interact positively with the various elements of their local environment. Bioclimatic evaluation is therefore the starting point for any housing design in the climatic zones of Nigeria, aiming at energy cost-efficiency in houses, by keeping the use of mechanical aids for climate control to a minimum. Indoor thermal environment should thus be one of the most influential factors on architectural expression and cost considerations in the different climatic settings of Nigeria.

References

Agajelu, O., Ekwueme, O. G., Obuka, N. S. P., & Gracefield, O. R. (2013). Life cycle cost analysis of a diesel/photovoltaic hybrid power generating system. *Industrial Engineering Letters, 3*(1), 9–30. Retrieved April 13, 2016, from www.iiste.org

Akande, O. K., & Adebamowo, M. A. (2009). Indoor thermal comfort for residential buildings in hot-dry climate of Nigeria. In *Proceedings of Conference: Adapting to Change: New Thinking on Comfort* (pp. 133–144). Cumberland Lodge, Windsor, UK.

ASHRAE. (2007). *ASHRAE standard 90.1*. Atlanta, GA: American Society of Heating, Refrigerating and Air-Conditioning Engineers.

Ayaode, J. O. (1973). *The Seasonal incidence of rainfall and weather patterns in Nigeria*. Lagos: Omotara Printing House.

Bob Manuel, H. (1990). Climatic index and conformability status of the humid tropics. *Journal of Architectural Focus, 40*(2), 31–33.

Centre for Science and Environment. (2002). A report on the independent inspection of fuel quality at fuel dispensing stations, oil tanks and tank Lorries [Online]. Retrieved April 11, 2016, from http://www.cseindia.org/html/cmp/air/fnladul.pdf

Department for Petroleum Resources (DPR). (2016). Fuel Scarcity and Adulteration [online]. Retrieved April 11, 2016, from www.dailytrust.com.ng/news

Giovanni, B. (1994). *Passive and low energy cooling of buildings*. New York: Van Nostrand Reinhold.

Giovanni, B. (1998). *Climate considerations in building and urban design*. New York: Van Nostrand.

Griffith, A., & Sidwell, A. C. (1995). *Constructability in building and engineering projects*. Basingstoke: Macmillan Press Ltd.

Higham, A., & Thomson, C. (2015). An evaluation of construction professionals' sustainability literacy in North West England. In A. Raiden & E. Aboagye-Nimo (Eds.), *Proceedings 31st Annual ARCOM Conference*, 7–9 September 2015, Lincoln, UK. Association of Researchers in Construction Management, 417–426.

Higham, A. P., James, H., & Fortune, C. (2015). Life cycle costing: Evaluating its use in UK practice. *Structural Survey, 33*(1), 73–87.

Ibidapo-Obe, O., & Ajibola, O. (2011). Towards a renewable energy development for rural power sufficiency. In *Proceedings of International Conference on Innovations in Engineering and Technology*.

Ilyas, M., Pang, C. Y., & Wah, C. A. (1981). Effective temperature comfort indices for some Malaysian Towns. *Singapore Journal of Tropical Geography, 2*(1), 27–31.

Kamil, M., Sardar, N., & Ansari, Y. (2008). Experimental study on adulterated gasoline and diesel fuels India Chem. *Engineering Journal, 89*(1), 23–22.

Ladell, W. (1949). *Nigeria; Water balance and heat load measurements*. Ibadan: Raymond House.

Lawal, A., Akinbami, J., & Akinpade, J. (2012). Assessing effectiveness of passive design parameters on active energy consumption in warm-humid climate. *Civil Engineering and Construction Technology, 3*(4), 140–147. Retrieved May 10, 2016, from http://www.academicjournals.org/JCECT

Loh, E., Dawood, N., Dean, J., & Crosbie, T. (2009). Development of trade-off algorithm with AHP for building lifecycle cost and Environment Impact Assessment. In *Proceeding of the 2nd International Conference on Whole Life Urban Sustainability and its Assessment*, Loughborough, UK.

Ogunsote, O. O., & Prucnal-Ogunsote, B. (2002). Choice of a thermal index for architectural design with climate in Nigeria. *Habitat International, 26*(1), 1–19.

Omonij, A. G., & Matzarakis, A. (2011). Climate and bioclimate analysis of Ondo State, Nigeria. *Meterologische Zeitschuit, 20*(5), 531–539.

Omoruyi, S. O., & Idiata, D. J. (2015). The environmental and cost implication of fossil fuel generators. *Nigeria International Journal of Emerging Technology and Advanced Engineering, 5*(2), 25–29.

Osueke, C. O., & Ofondu, I. O. (2011). Fuel adulteration in Nigeria and its consequences. *International Journal of Mechanical & Mechatronics Engineering IJMME-IJENS, 11*(4), 34–40.

UNDP. (2000). Climate Change Country Profiles Nigeria. Tyndall Centre for Climate Change Research. Retrieved May 7, 2016, from http://country-profiles.geog.ox.ac.uk

Chapter 12
Predicting Future Overheating in a Passivhaus Dwelling Using Calibrated Dynamic Thermal Simulation Models

James Parker, Martin Fletcher, and David Johnston

Abstract Energy used for space heating accounts for the majority of anthropogenic greenhouse gas emissions from the built environment in the UK. As the fabric performance of new build dwellings improves, as part of the UK's response to reducing national CO_2 emissions, the potential for excessive overheating also increases. This can be particularly pertinent in very airtight low-energy dwellings with high levels of insulation and low overall heat loss, such as Passivhaus dwellings. The work described in this paper uses calibrated dynamic thermal simulation models of an as-built Certified Passivhaus dwelling to evaluate the potential for natural ventilation to avoid excessive summertime overheating. The fabric performance of the Passivhaus model was calibrated against whole dwelling heat loss coefficient measurements derived from coheating tests. Model accuracy was further refined by comparing predicted internal summer temperatures against in-use monitoring data from the actual dwelling. The calibrated model has been used to evaluate the impact that user-controlled natural ventilation can have on regulating internal summer temperatures. Thermal performance has been examined using simulation weather files for existing climatic conditions and for predicted future climate scenarios. The extent of overheating has been quantified using absolute and adaptive comfort metrics, which exceed the relatively restricted measures used for regulatory compliance of dwellings in the UK. The results suggest that extended periods of window opening can help to avoid overheating in this type of low-energy dwelling and that this is true under both existing and future climatic conditions.

12.1 Introduction

Scientific consensus documented by the Intergovernmental Panel on Climate Change (IPCC) states that anthropogenic greenhouse gas emissions are changing the world's climate as described in the third synthesis report (Stocker et al. 2013).

J. Parker (✉) • M. Fletcher • D. Johnston
Centre for the Built Environment, Leeds Sustainability Institute, Leeds Beckett University, BPA223, City Campus, Leeds LS2 9EN, UK
e-mail: j.m.parker@leedsbeckett.ac.uk

M. Dastbaz et al. (eds.), *Building Information Modelling, Building Performance, Design and Smart Construction*, DOI 10.1007/978-3-319-50346-2_12

It is estimated that the built environment accounts for approximately 34% of these emissions world-wide (Yamamoto and Graham 2009) and 45% in the UK (The Carbon Trust 2009). The reinforced understanding that the consumption of fossil fuels is damaging the earth's atmosphere, along with fears over fuel cost and security dating back to the 1970s, has led to extensive research in the field of low-energy buildings, with a particular focus on reducing the amount of Carbon Dioxide (CO_2) emitted (Khasreen et al. 2009). Energy consumed through the conditioning of internal spaces remains the greatest source of these emissions (Pérez-Lombard et al. 2008) and climatic conditions in the UK and Northern Europe dictate that the largest proportion of this is used to provide space heating. Logically, this has led to a significant amount of academic and industry-led research designed to minimise the energy consumption associated with domestic space heating.

Despite space heating demands accounting for the greatest proportion of conditioning energy in the UK, overheating in dwellings is steadily becoming seen as a considerable problem and is predicted to become worse in the future aligned with a global rise in temperatures (Jentsch et al. 2014). Although they are likely to avoid the most severe impacts of climate change, countries with temperate climates, like many European nations, are predicted to experience more regular and intense heat waves in the future (Meehl and Tebaldi 2004). This has obvious implications for thermal comfort conditions, but also has potentially more serious repercussions for the health of occupants (Vardoulakis et al. 2015). An unintended consequence of reducing heat losses in low-energy dwellings is that the potential for overheating can be exacerbated (Gupta and Kapsali 2016; Mavrogianni et al. 2009). The Passivhaus standard is an established and validated technological solution to minimise heat losses from buildings. However, dwellings built to this standard have the potential to experience excessive overheating, particularly in a warmer future climate (Mcleod et al. 2013; Tabatabaei Sameni et al. 2015).

The contents of this paper present the results of fabric testing and in-use monitoring data from an occupied Certified Passivhaus dwelling in the UK. This measured data has been used to help calibrate a dynamic thermal simulation (DTS) model which has, in conjunction with information relating to user behaviour, been used to understand overheating over the first year of occupancy. The calibrated model has then been used to predict the extent of overheating in future climate scenarios and examine the potential to mitigate this overheating using natural ventilation.

12.2 Literature Review

A growing body of evidence supports the notion that overheating is becoming a significant problem in UK dwellings (Beizaee et al. 2013; Coley and Kershaw 2010; Gupta and Kapsali 2016; Pretlove and Kade 2016; Tabatabaei Sameni et al. 2015). The risk of overheating is not necessarily localised, but it is widely accepted that this is exaggerated in dense urban environments and there is strong evidence to support this in the UK (Mavrogianni et al. 2010, 2011; Gartland 2012; Oikonomou et al. 2012). Excessive overheating under existing climatic conditions has already been verified in the literature. A group of reports published by the Zero Carbon Hub were produced with the aim of increasing understanding of domestic overheating in England and Wales. Through working with government and industry partners, the publications provide practical advice and help to quantify the extent of the problem (Zero Carbon Hub 2015c).

Two large scale academic studies are cited in Zero Carbon Hub reports, both of which monitored over two hundred unheated properties during summer months. The first of these studies collected over forty-one summer days during 2007 (Beizaee et al. 2013). This study found that 21% of bedrooms exceeded 26 °C for more than 1% of night-time hours and 47% exceeded 24 °C for more than 5% of night-time hours. The second of these studies was undertaken in the Summer of 2009 (Lomas and Kane 2013). This study found that 27% of living rooms exceeded 28 °C for more than 1% of occupied hours (assumed) and that approximately 20% of bedrooms exceeded 24 °C during night-time hours for 30% of the monitoring period. In addition, the results obtained from a group of case studies using Housing Association properties have also been reported by the Zero Carbon Hub. Analysis of the data collected through these case studies found that issues relating to the summer bypass in Mechanical Ventilation and Heat Recovery (MVHR) units, large proportions of glazing, and insufficient ventilation all contributed to overheating in the sample dwellings (Zero Carbon Hub 2015b). One of the case studies focused on a Passivhaus development and found that a larger percentage of dwellings were considered to overheat when using an adaptive comfort criterion designed to rate conditions for vulnerable occupants. This has similarities with the case study dwelling described in this paper. The alternative means of assessing overheating are also discussed in the methodology section.

An academic paper which uses data from the same Passivhaus development described in the Zero Carbon Hub report provides further detail on the thermal comfort in these dwellings (Tabatabaei Sameni et al. 2015). Twenty-five flats built to the Passivhaus standard were monitored over three summers (cooling seasons) and more than two thirds of these dwellings were considered to overheat when using the Passivhaus assessment criteria. As mentioned above, conclusions noted that the overheating was considered to be more excessive when using adaptive comfort criteria for vulnerable occupants. It is important to note that analysis of the data suggested that the overheating was largely due to occupant behaviour rather than the construction of the dwellings; in many cases, residents had not activated summer bypass for the MVHR systems and did not increase ventilation by opening windows (Tabatabaei Sameni et al. 2015).

The extent of overheating in a range of Passivhaus dwellings has been evaluated in previous academic work (Mcleod et al. 2013). This research used a similar research methodology to that described later in this paper, utilising similar morphed simulation weather files. The main finding of this work was that excessive overheating can be avoided through the optimisation of a relatively small group of design parameters, including the ratio of glazing on specific facades and external shading devices. In addition to the Passivhaus study, there is a collection of published work that predicts the impact of climate change on future domestic overheating in the UK (Porritt et al. 2012; Gul et al. 2012; Jenkins et al. 2013). As with the Passivhaus example, the methodologies used by all of these researchers are fundamentally very similar; they all use DTS models in combination with morphed simulation weather files. Results from all of this work indicate that the example naturally ventilated buildings are likely to experience excessive overheating in the future based upon their existing designs. The methodology used here differs in that it is using measured fabric performance and monitored temperature data to refine the baseline model.

The potential to mitigate excessive overheating is relatively well-understood in the literature. There are various mitigation measures that can be integrated into the fabric of a building to help avoid thermal discomfort including: internal and external solar shading; increased natural ventilation (either through larger openings or longer opening periods); night-time purge ventilation (a form of natural ventilation coupled with thermal mass); and additional mechanical ventilation and/or air conditioning (Butcher 2014; Porritt et al. 2013). Obviously, the final options listed here are not passive and will result in additional energy consumption. Research conducted by Porritt et al. (2012) found that external shading, in particular, is very effective in reducing solar gains, but also found that treating exposed external surfaces with solar reflective paint and external wall insulation can also help to mitigate overheating. It is also worth noting that low-zero cost measures such as 'rules' for window opening and drawing curtains can also play an important role in avoiding heat gain, but it was suggested that night-time purge ventilation would be best managed through automated openings which would result in some additional energy consumption. This work also found that the extended occupancy in living spaces occupied by older occupants is an important consideration for modelling inputs in this type of analysis (Porritt et al. 2012).

12.3 Methodology

Current UK Building Regulations require overheating to be considered using a relatively simplistic modelling methodology as part of the Standard Assessment Procedure (HM Government 2013, 2014) and the need to evaluate the potential for overheating using a more sophisticated approach has been acknowledged at a policy level (Zero Carbon Hub 2015a). This work uses the adaptive comfort criteria developed by the Chartered Institute of Building Services Engineers (CIBSE) which take account of peoples' increased tolerance of warmer internal temperatures during extended periods of warm weather, placing an emphasis on the running mean temperature (CIBSE 2013). This metric can also be used to assess overheating for vulnerable occupants, which is pertinent to the case study dwelling. There are three separate criteria, with a 'pass' being dependent upon any two of the three criteria being met. The criteria include: threshold temperature exceeded $\not>3\%$ of occupied hours per year; daily weighted exceedance (degree hours) $\not>6$; and a temperature $\not>$

upper limit. An absolute threshold of no more than 1% of occupied hours exceeding 28 °C has also been used in this work; this has historically been defined as a suitable metric by CIBSE (CIBSE 2006).

Multiple environmental factors including building geometry, surrounding structures, building orientation, building fabric, solar gains, air tightness, internal heat gains, solar radiation, and wind have a direct impact on internal thermal conditions (Taylor et al. 2014). The complex interaction between these variables means that DTS software is an effective tool for evaluating potential overheating and natural ventilation strategies. Models used in this work were produced using IES Virtual Environment software, which is approved for UK Building Regulations compliance calculations for non-domestic buildings (IES 2014). It is not approved for any domestic regulatory compliance calculations, but offers a much more sophisticated dynamic calculation of thermal performance than the steady state models approved for regulatory use.

Morphed simulation weather files have been used in this work to predict the impact of future climate scenarios on the performance of the case study dwelling. The Prometheus project uses predictions made in the UK Climate Impact Projections 2009 (UKCIP09) to morph simulation weather files that can be used in this type of analysis (Eames et al. 2011). The Prometheus files reflect the change in climate under medium and high emission scenarios and are probabilistic in nature, creating files for both emission scenarios that include the 10th (unlikely to be more than), 33rd, 50th, 67th, and 90th (unlikely to be less than) percentiles for periods covering the 2020s (2010–2039), 2050s (2040–2069), and 2080s (2070–2099). For the purposes of this work, the 10th, 50th, and 90th percentile files for both emission scenarios for each time period have been used for comparison. The case study dwelling is in the North-East of England, and as such, the weather files for the Newcastle region have been used in the simulation models.

12.3.1 Case Study Building and Baseline Model

The case study dwelling is located at the east end of a terraced block of seven dwellings and is south-facing to maximise passive solar heat gain. It is a single storey building with two bedrooms, a bathroom, a hallway, three small storage cupboards, and an open-plan living and kitchen area and has a total conditioned floor area of approximately 66 m^2. The dwelling is approximately 7.8 m deep across the open-plan living area which allows for cross-flow ventilation. There is also a mezzanine-level plant room situated above the bathroom, hallway, and both bedrooms that houses the MVHR system and hot water storage tank which is only accessible via a loft hatch. The dwelling is neighboured by another house to the west and a small boiler house to the east, both of which have been included as adiabatic spaces in the model. There are also boundary walls to both the front and rear elevations that have been included in the geometry, as they provide some localised shading in addition to the roof overhangs. The geometry and layout of the DTS model can be seen in Fig. 12.1.

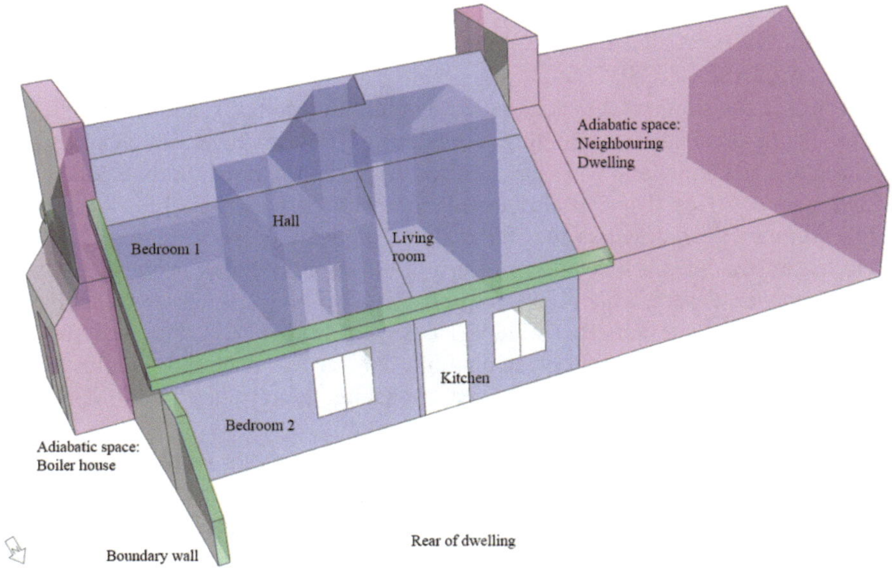

Fig. 12.1 Case study Passivhaus dwelling model geometry

A three-stage calibration process was undertaken on the baseline model. The first stage involved calibrating the fabric performance of the model based upon the *in situ* measurements obtained from an identically sized dwelling located at the opposite end of the same terrace as the case study dwelling. This method of calibration has been described in previous work (Parker et al. 2015). An initial model is created and then iteratively updated using a calculated Y-value, measured air change rates, and *in-situ* measured U-values. The results of this calibration exercise are shown in Fig. 12.2. The measured result is shown in bold italic text and the final value predicted by the model is shown below that in italic text. A very close match was achieved using this process with the modelled value of 46.65 W/K being within 0.04 W/K of the measured value. Examples of updates in this process include the measured wall U-value when adjusted with a calculated Y-value of 0.149 W/m^2 K and the air change rate per hour was 0.023 (measured when pressure equalised in the adjoining dwelling); these differ from the design values of 0.104 W/m^2 K and 0.03 air changes per hour, respectively.

The second stage of calibration process involved comparing the predictions made by the fabric-calibrated model under occupied conditions with metered data from the actual dwelling. The metered gas consumption from 2014 was compared with the value predicted by the model. Error between monthly values has been measured using the Mean Biased Error (MBE) and Cumulative Variation of Root Mean Square Error (CVRMSE) using industry standard error margins for monthly data (ASHRAE 2002). To be considered calibrated, the predicted monthly consumption must be within 5% for the MBE and within 15% for the CVRMSE (ASHRAE 2002). It is inevitable that there will be some error between the predicted and the

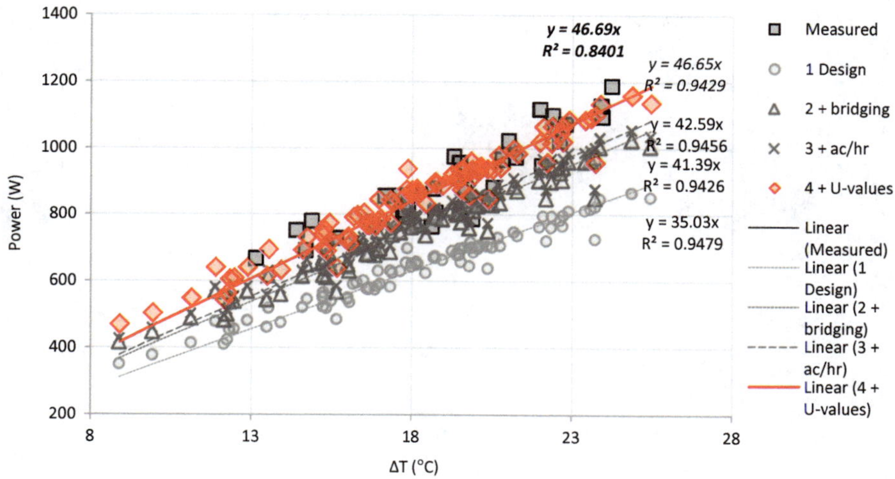

Fig. 12.2 Results from in-situ coheating test and fabric-performance-calibrated models

metered values over a set period of time if actual weather data from the same period is not used in the simulation weather file. A simulation weather file based upon site data from 2014 was not available in this instance. Therefore, for this stage of calibration, a comparison was made between the external temperature data from the available simulation weather files with the measured external air temperature to identify the most appropriate baseline weather file. Simulation weather files for the Newcastle area produced through the Prometheus research project (Eames et al. 2011) and by CIBSE for regulatory compliance calculations (CIBSE 2006) were available to use in the baseline simulations. Test Reference Year (TRY) and DSY files were available from both sources. When compared with the daily average temperatures from 2014, it was the CIBSE DSY file that produced the closest match. Daily average temperatures for the Prometheus TRY file, the CIBSE DSY file, and those measured on site are compared in Fig. 12.3. The annual average temperature from the 2014 site data was 10.4 °C. This compares most closely with the average from the CIBSE DSY file of 10.1 °C. The Prometheus file averages were 9.1 °C and 9.3 °C for the TRY and DSY files, respectively, and the CIBSE TRY average was 9.6 °C. The CIBSE DSY file was therefore selected for this stage of calibration.

Occupant density was calculated based upon actual floor areas and anecdotal evidence of the occupants' behaviour. There are two elderly residents within the case study dwelling, one leaves the house during the daytime to attend work and the other is retired and remains in the dwelling most days. Occupancy profiles reflect this, with an assumed 100% occupancy rate in living areas between 07:00 and 09:00, which reduces to 50% between 09:00 and 17:00 and returns to 100% until 22:00. An input of 3.30 W/m^2/100 lux was used for the lighting heat gains and consumption and the equipment heat gains in the living areas are based upon default NCM values for this zone type (HM Government 2013). For both lighting and equipment, the usage patterns were extended from the default NCM profiles to match the described in-use occupancy patterns.

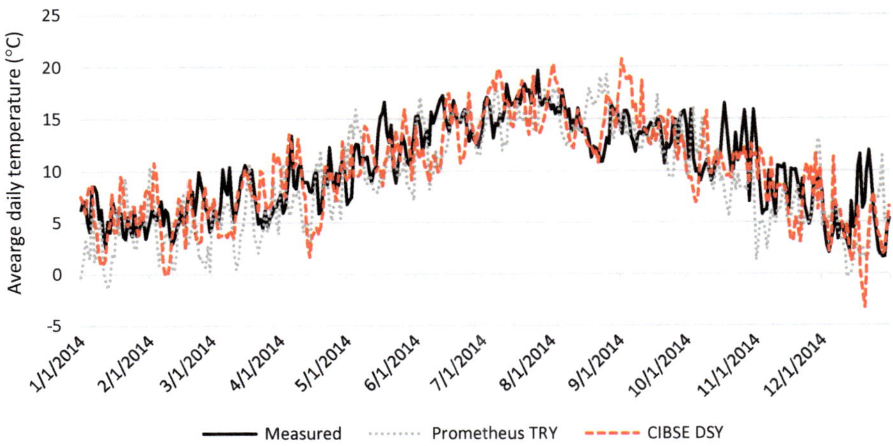

Fig. 12.3 Daily average temperature from measured 2014 data and simulation weather files

The dwelling is conditioned using an MVHR system with integral heater battery. The MVHR system is included in the model with a heat recovery efficiency of 88% and provides 0.47 air changes per hour. Additional space heating is generated through a small radiator housed within an airing cupboard at the centre of the dwelling and a towel radiator in the bathroom. Heat for space heating and domestic hot water is provided via a wet centralised heating system, fuelled by a small gas-fired condensing boiler serving the entire terrace. A roof-mounted solar-thermal water heater, with a total area of 3 m^2, is also used for hot water. Analysis of the in-use monitoring data suggests that the space heating set point used in the dwelling is 23 °C as the internal temperatures very rarely drop below this value. This is considerably higher than the default values used in the NCM thermal templates.

When compared with monthly gas consumption data from 2014, consumption predicted by the model had an MBE of 1.24% and a CVRMSE of 4.30%, both of which are well within the respective thresholds of 5% and 15% for these error measures. This version of the model used a fixed (scheduled) infiltration rate, but there is however an additional step required to produce a model that can be used to more accurately assess the impact of natural ventilation using opening windows. For the purposes of this research, it was necessary to use the bulk air movement application (MacroFlo) of the IES software. This application links air movement driven by wind speed, direction, and buoyancy to the thermal simulation engine in the DTS software. In this version of the model, infiltration is calculated using the external weather condition parameters and the crack flow coefficient of the openings. To ensure that the predicted performance remained calibrated to the actual data, it was necessary to use an input of 0.09 $l/s^{-1}{\cdot}m^{-1}$ $Pa^{-0.6}$ for the crack flow coefficient of the external openings; this value provided the closest match to the metered data. This resulted in an error of −0.16% for the MBE and 2.10% for the CVRMSE when predicted monthly gas consumption is compared with the metered data from 2014. A comparison of the gas consumption for 2014 and that predicted by the models

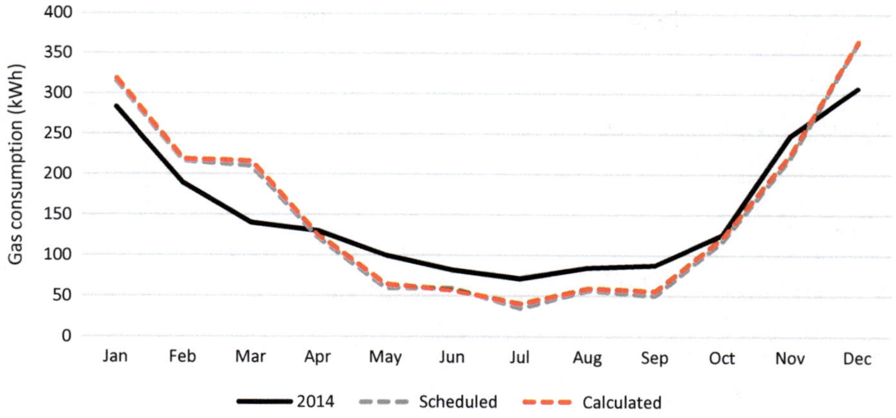

Fig. 12.4 Comparison of metered gas consumption with modelled consumption using scheduled and calculated infiltration

including scheduled and calculated ventilation is shown in Fig. 12.4. Hot water generated through the solar thermal system and a demand of 2.04 L per person per hour have been accounted for in this modelled estimate.

The final stage of calibration involved comparing modelled internal temperatures with those measured during 2013 and 2014. This was achieved by plotting the measured and modelled internal temperatures against the measured and modelled external temperatures. As the purpose of this research was to understand potential overheating in the dwelling, it was important that the predicted internal temperatures were consistent with those measured on site. The data collected on site indicated that there was significant overheating in the dwelling and anecdotal evidence suggested that this was due to the occupants not opening any windows (as per their instructions relating to heat retention), coupled with them not operating the MVHR summer bypass feature. Internal blinds were used to provide some shading on the southern façade during summer months. In anecdotal evidence, the occupants reported not opening any windows during 2013, but introduced some window opening in 2014 under very hot conditions. Figure 12.5 illustrates the relationship between external temperatures and internal temperatures. Included in Fig. 12.5 are measured data from 2013 and 2014 and simulated data from two versions of the model. The first includes no natural ventilation at all; the second version assumes that windows were opened when internal temperatures reached 30 °C. It was the second version that was used as the final baseline model against which all alternative operational and climate scenarios have been compared, as it demonstrates the most consistency with the performance of the in-use dwelling.

The building design incorporates an extended roof overhanging on the south-facing front façade which was intended to provide some shading in the summer months. This extends by 500 mm from the front wall of the dwelling and is included in the model geometry as local shading. All window units are triple-glazed and have an overall U-value of 0.828 W/m^2 K. The g-value (a measure of solar energy

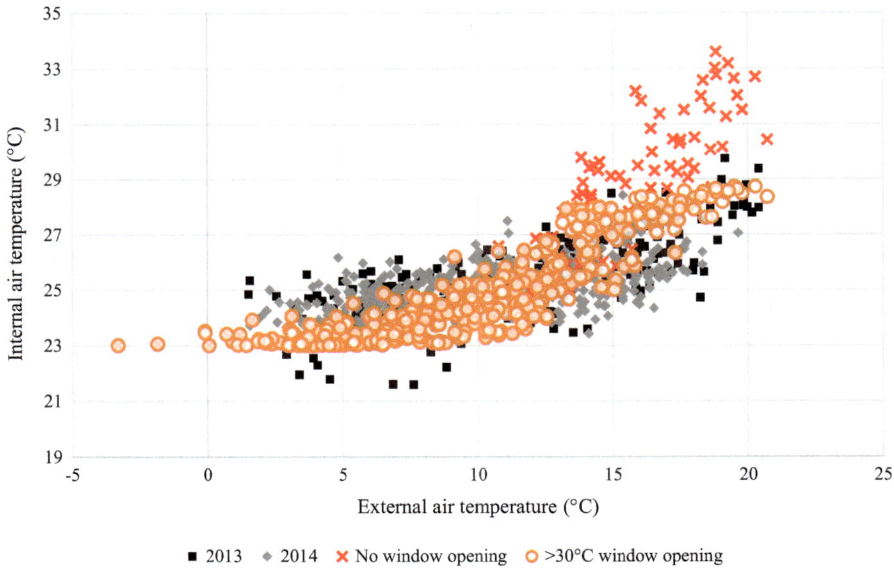

Fig. 12.5 Comparison of measured and modelled external and internal air temperature

transmittance with a value of 0 indicating no transmittance) of the glazing is 0.53 and blinds are assumed to be in operation during summer months and are lowered when incident radiation reaches 200 W/m^2.

12.4 Results and Discussion of the Overheating Analysis

Analysis of the extent of overheating can be divided into three sections. The first briefly evaluates the extent of overheating recorded by the measured data and reviewed as part of the post-occupancy evaluation work. The second section uses the calibrated baseline model to evaluate whether operational changes can either mitigate or completely avoid excessive overheating. The third section considers performance in future climate scenarios and assesses the potential for simple operational changes to avoid excessive overheating. It is important to note that all of this analysis focuses on overheating in the open-plan living/kitchen space only and does not include analysis of the circulation or bedroom areas which will be the subject of further work.

12.4.1 Measured Internal Temperatures

The case study dwelling was monitored in-use for a period of 24 months throughout 2013 and 2014. As part of this monitoring, local external air temperature and internal air temperature in the open plan lounge/kitchen area were measured at 10 min

Table 12.1 Measured overheating

	Category I (young/infirm)					Category 2 (new build)			
Description	%>28 °C	C1	C2	C3	Criteria failed	C1	C2	C3	Criteria failed
2013 Monitored Data	8.3	53	44.8	4	1 and 2	24.5	29.8	3	1 and 2
2014 Monitored Data	1.6	33.3	41.3	4	1 and 2	12.3	26.3	3	1 and 2

intervals. Due to the lightweight nature of the structure and absence of large sources of radiant heat, air temperature has been assumed to be equal to mean radiant temperature when determining operative temperature. External temperature measurements have been used to generate the exponentially weighted running mean daily temperature and applied to the methodology defined in CIBSE TM52 (CIBSE 2013) with the results presented in Table 12.1 below. In addition to TM52, the percentage of occupied hours exceeding 28 °C has been considered. For all analysis of measured data, occupied hours are the same as described for the modelling phase (07:00–22:00 for the combined living space) and data is for the duration 1st May–31st September of each year. In Table 12.1, and all subsequent presentations of the results, the three criteria defined in TM52 have been abbreviated to C1, C2, and C3. Results for the adaptive comfort criteria are presented for category I (young/infirm) and for category II (new build). Category I accounts for the reduced capacity of the young/infirm to tolerate and physiologically respond to higher temperatures.

As can be seen from the results, the dwelling fails on both C1 and C2 of the TM52 assessment during both years, although there is an observed improvement in 2014. This is supported by the absolute temperature threshold criteria which, although above the 1% limit for both years, is considerably reduced in 2014. It is known that during 2014 residents were encouraged to increase the use of natural ventilation which it is assumed accounts for the decrease in overheating.

12.4.2 Modelled Internal Temperatures in Baseline Scenario

Different operational scenarios have been used to evaluate the potential for overheating to be mitigated in the baseline model. As mentioned previously, the MVHR system incorporates a bypass mechanism. As previously mentioned, opening windows are both bottom and side hung and can be either tilted open to an angle of 20°, or side opened to an angle of 90°. The MVHR bypass mechanism and the opening windows form the basis of the different operating scenarios examined using the baseline model. The opening of the windows (both tilted and side opening) was evaluated at different opening threshold temperatures, along with the potential for night-purge ventilation. The operating scenarios and results from this analysis are noted below in Table 12.2. In all scenarios that include additional window opening, it is assumed that the MVHR bypass mechanism is also in operation.

Table 12.2 Predicted overheating for the baseline climate scenario

			Category I (young/infirm)				Category II (new build)			
Ref:	Description	%>28 °C	C1	C2	C3	Criteria failed	C1	C2	C3	Criteria failed
1.0	Baseline	19.9%	81.0	117.0	10.0	1 and 2 and 3	65.6	103.0	9.0	1 and 2 and 3
1.1	MVHR bypass (25 °C set point)	7.6%	56.7	59.0	6.0	1 and 2 and 3	33.1	46.0	5.0	1 and 2 and 3
1.2	MVHR bypass (24 °C set point)	6.7%	47.1	59.0	6.0	1 and 2 and 3	27.7	46.0	5.0	1 and 2 and 3
1.3	Bottom hung >28 °C set point	0.1%	54.6	30.0	3.0	1 and 2	24.6	19.0	3.0	1 and 2
1.4	B. hung night purge (>28 °C)	0.1%	54.6	30.0	3.0	1 and 2	21.7	18.0	2.0	1 and 2
1.5	Side hung >28 °C set point	0.0%	54.4	30.0	3.0	1 and 2	20.9	18.0	2.0	1 and 2
1.6	Side hung night purge (>28 °C)	0.0%	54.4	30.0	3.0	1 and 2	20.9	18.0	2.0	1 and 2
1.7	Bottom hung >25 °C set point	0.0%	1.3	8.0	1.0	2	0.0	0.0	0.0	–
1.8	Bottom hung night purge (25 °C)	0.0%	0.3	2.0	1.0	–	0.0	0.0	0.0	–
1.9	Side hung >25 °C set point	0.0%	0.2	2.0	1.0	–	0.0	0.0	0.0	–
1.10	Side hung night purge (25 °C)	0.0%	0.2	2.0	1.0	–	0.0	0.0	0.0	–

It can be seen from the results that the MVHR summer bypass mechanism alone cannot provide sufficient ventilation to avoid excessive overheating using either a 25 °C or 24 °C operating set point. It passes neither the absolute nor adaptive comfort assessments. It does, however, help to considerably reduce the extent of overheating when compared to the baseline scenario. These results do demonstrate that, in theory, excessive overheating can be avoided without the need for any physical changes to the building fabric or conditioning system. When assessed against the absolute metric of no more than 1% of occupied hours exceeding 28 °C, all of the scenarios using opening windows fall under this threshold. Using the category I adaptive comfort assessment, it is not until an opening threshold of 25 °C is introduced that overheating can be mitigated and at least two from the three criteria are met.

It is important to note that these modelled scenarios assume a perfect operating scenario where the MVHR bypass is operated and windows are opened at the exact time the set point and threshold temperatures are reached. In practice, it is highly unlikely that an occupant could respond in this way, especially without any prompt generated by internal air temperature sensors. These scenarios therefore represent behaviour that is arguably more aligned with an automated system. This will be discussed further in the conclusions of this paper along with issues related to perceived human comfort.

It is worth noting that the night purge has little impact on the results in the baseline scenario. It is important to note that the focus of this research is in the living space and the metrics used to assess overheating are considered in the context of occupied hours. The night purge of this space therefore has little impact on these results and this is exacerbated by the lightweight thermal mass of the dwelling.

12.4.3 Modelled Internal Temperatures in Future Climate Scenarios

Following the analysis completed in the previous section, two different air temperature thresholds for opening windows were selected to evaluate performance in future climate scenarios. The most obvious opening threshold temperature to evaluate is 25 °C, as this avoids overheating in all of the baseline scenarios in which it was tested. The opening threshold temperature of 28 °C has also been examined, as the in-use data suggests this is closer to the temperature at which occupants are opening windows. Both opening threshold temperatures have been evaluated for bottom hung window opening during the daytime, side hung opening during the daytime, and night purge versions of both opening types. Results from the future climate scenarios are shown in Table 12.3 (2020s), Table 12.4 (2050s), and Table 12.5 (2080s). All future weather files are the 50th percentile prediction from each given scenario.

As may be expected in the context of the baseline scenario results, in the 2020s scenario, it is not until windows are opened at the 25 °C set point that conditions in the living space meet the adaptive comfort criteria. All scenarios with opening windows avoid exceeding 1% of occupied hours above 28 °C, with one exception, the bottom hung windows with daytime opening at 28 °C. Using night purge ventilation does start to have a slightly more significant impact than in the baseline scenario and improves performance enough for the opening threshold temperature of 28 °C with night purge ventilation to avoid exceeding 1% of occupied hours above 28 °C. It also allows the 25 °C opening threshold to meet all three criteria in the high emissions scenario, although the version with no night purge only fails one of the criteria and would therefore be considered comfortable. All results for the 2020s scenarios

Table 12.3 Predicted overheating under 2020s medium and high emission scenario 50th percentile weather conditions

			Category I (young/infirm)				Category II (new build)			
Ref:	Description	%>28 °C	C1	C2	C3	Criteria failed	C1	C2	C3	Criteria failed
2020s medium emissions scenario										
2.1	MVHR bypass (25 °C)	14.9%	66.6	46.0	5.0	1 and 2 and 3	45.9	33.0	4.0	1 and 2
2.2	Btm hung >28 °C set point	1.3%	61.3	30.0	3.0	1 and 2	24.0	18.0	2.0	1 and 2
2.3	Btm hung night purge (>28 °C)	0.7%	59.3	28.0	3.0	1 and 2	20.8	16.0	2.0	1 and 2
2.4	Side hung >28 °C set point	0.7%	57.9	29.0	3.0	1 and 2	19.7	17.0	2.0	1 and 2
2.5	Side hung night purge (>28 °C)	0.7%	57.9	29.0	3.0	1 and 2	19.7	17.0	2.0	1 and 2
2.6	Btm hung >25 °C set point	0.5%	0.6	14.0	2.0	2	0.3	6.0	1.0	-
2.7	Btm hung night purge (>25 °C)	0.5%	0.5	14.0	2.0	2	0.3	6.0	1.0	-
2.8	Side hung >25 °C set point	0.5%	0.5	14.0	2.0	2	0.3	6.0	1.0	-
2.9	Side hung night purge (>25 °C)	0.5%	0.5	14.0	2.0	2	0.3	6.0	1.0	-
2020s high emission scenario										
3.1	MVHR bypass (25 °C)	16.8%	66.9	60.0	7.0	1 and 2 and 3	46.9	48.0	6.0	1 and 2 and 3
3.2	Btm hung >28 °C set point	0.9%	58.4	27.0	3.0	1 and 2	17.4	16.0	2.0	1 and 2
3.3	Btm hung night purge (>28 °C)	0.9%	58.4	27.0	3.0	1 and 2	17.4	16.0	2.0	1 and 2
3.4	Side hung >28 °C set point	0.8%	58.1	28.0	3.0	1 and 2	16.5	16.0	2.0	1 and 2
3.5	Side hung night purge (>28 °C)	0.8%	58.1	28.0	3.0	1 and 2	16.5	16.0	2.0	1 and 2

(continued)

Table 12.3 (continued)

			Category I (young/infirm)				Category II (new build)			
Ref:	Description	%>28 °C	C1	C2	C3	Criteria failed	C1	C2	C3	Criteria failed
3.6	Btm hung >25 °C set point	0.8%	0.7	8.0	2.0	2	0.0	1.0	1.0	–
3.7	Btm hung night purge (>25 °C)	0.8%	0.1	3.0	1.0	–	0.0	0.0	0.0	–
3.8	Side hung >25 °C set point	0.8%	0.0	0.0	0.0	–	0.0	0.0	0.0	–
3.9	Side hung night purge (>25 °C)	0.8%	0.0	0.0	0.0	–	0.0	0.0	0.0	–

are presented in Table 12.3. Although the category II results are significantly lower for each of the assessment criteria, all scenarios are deemed to fail the assessment apart from when the 25 °C opening threshold is introduced.

In keeping with both the baseline and 2020s scenarios, all of the models using a 28 °C opening threshold fail to pass the adaptive comfort assessment in all of the 2050s and 2080s scenarios. Contrary to the baseline and 2020s scenarios, the opening of windows at 28 °C is not sufficient to avoid exceeding 1% of occupied hours above 28 °C in all but two cases. In the medium emission scenario for the 2020s, the larger aperture, side hung windows meet but avoid exceeding the threshold 1%. The 1% threshold is also exceeded by all versions of the model in the 2080s high emission scenario, although the adaptive criteria assessment is passed in the majority of cases.

All of the 2050s and 2080s scenarios include a version of the model that fails the adaptive comfort assessment while using a 25 °C opening threshold. This only occurs when using the bottom hung opening option during the daytime only. When night purge ventilation is added to this operation, the living space conditions again pass the adaptive comfort assessment. This suggests that night time cooling could become more important in the future. The side hung window opening options avoid failing the adaptive comfort criteria assessment completely in all scenarios. With the exception of the 2080s medium emissions scenario, the bottom hung openings using the 25 °C set point fail the adaptive comfort assessment under category I, but pass under category II.

12.4.4 Limitations and Further Work

It is important to note that there are some limitations to this work. Occupant behaviour was anecdotal and it would be useful for window opening activity to be monitored in future work. The length of the paper also limited the inclusion of work examining building performance in multiple probabilistic weather scenarios and the

Table 12.4 Predicted overheating under 2050s medium and high emission scenario 50th percentile weather conditions

			Category I (young/infirm)				Category II (new build)			
Ref:	Description	%>28 °C	C1	C2	C3	Criteria failed	C1	C2	C3	Criteria failed
2050s medium emissions scenario										
4.1	MVHR bypass (25 °C)	18.9%	69.1	54.0	6.0	1 and 2 and 3	51.2	42.0	5.0	1 and 2 and 3
4.2	Btm hung >28 °C set point	1.2%	56.3	41.0	4.0	1 and 2	22.2	27.0	3.0	1 and 2
4.3	Btm hung night purge (>28 °C)	1.2%	56.7	41.0	4.0	1 and 2	22.2	27.0	3.0	1 and 2
4.4	Side hung >28 °C set point	1.0%	56.0	41.0	4.0	1 and 2	21.0	27.0	3.0	1 and 2
4.5	Side hung night purge (>28 °C)	1.0%	55.9	41.0	4.0	1 and 2	21.1	27.0	3.0	1 and 2
4.6	Btm hung >25 °C set point	0.8%	3.3	39.0	5.0	1 and 2 and 3	1.2	28.0	4.0	2
4.7	Btm hung night purge (>25 °C)	0.7%	2.0	35.0	4.0	2	1.0	24.0	3.0	2
4.8	Side hung >25 °C set point	0.6%	1.8	35.0	4.0	2	1.0	24.0	3.0	2
4.9	Side hung night purge (>25 °C)	0.6%	1.8	35.0	4.0	2	1.0	24.0	3.0	2
2050s high emission scenario										
5.1	MVHR bypass (25 °C)	22.3%	70.9	54.0	6.0	1 and 2 and 3	53.3	42.0	5.0	1 and 2 and 3
5.2	Btm hung >28 °C set point	2.0%	57.0	24.0	3.0	1 and 2	19.9	13.0	2.0	1 and 2
5.3	Btm hung night purge (>28 °C)	2.0%	57.0	24.0	3.0	1 and 2	20.0	13.0	2.0	1 and 2
5.4	Side hung >28 °C set point	1.7%	56.2	24.0	3.0	1 and 2	18.4	13.0	2.0	1 and 2
5.5	Side hung night purge (>28 °C)	1.7%	56.3	24.0	3.0	1 and 2	18.5	13.0	2.0	1 and 2

(continued)

Table 12.4 (continued)

			Category I (young/infirm)				Category II (new build)			
Ref:	Description	%>28 °C	C1	C2	C3	Criteria failed	C1	C2	C3	Criteria failed
5.6	Btm hung >25 °C set point	1.3%	4.0	22.0	4.0	1 and 2	1.3	13.0	3.0	2
5.7	Btm hung night purge (>25 °C)	0.9%	2.0	17.0	3.0	2	0.7	8.0	2.0	2
5.8	Side hung >25 °C set point	0.9%	1.6	15.0	3.0	2	0.6	7.0	2.0	2
5.9	Side hung night purge (>25 °C)	0.9%	1.6	15.0	3.0	2	0.6	7.0	2.0	2

potential for increases in heating consumption when the described control strategies are introduced. However, this was considered in the modelling analysis and more sophisticated opening schedules; using higher opening threshold temperatures during the shoulder seasons can help to minimise this. Further work will consider the impact that using calibrated and non-calibrated models can have on this type of analysis, with a particular focus on conductive heat transfer, the performance of this building type in other UK locations, and thermal comfort in the other zones of the building.

Despite many of the modelling inputs being based upon either as-built or measured data, there are still a number of assumptions that have had to have been made for model inputs. The values used for lighting and equipment gains are based upon NCM default values for dwellings. It may be possible to refine these inputs in future work based upon metered electricity consumption. The operation of blinds is also only based upon anecdotal evidence. All of these values will have some impact on potential overheating and further work will aim to refine these inputs. Another potential source of heat gain that is not accounted for in this version of the model are the heat gains associated with the hot water storage tank that is fed by the roof-mounted solar thermal collector. The storage tank is housed in the separate loft-space plant room above the bathroom and hallway and heat gain into the living space is therefore likely to be negligible but should be accounted for in future work.

Finally, it is possible to procure weather data for specific time periods from relatively local weather stations. There is, however, a cost associated with this and this resource was unfortunately not available for this piece of work. Any further work that is designed specifically to consider the impact of calibrated models on overheating assessment should aim to acquire actual simulation weather files for the period during which in-use data is collected.

Table 12.5 Predicted overheating under 2080s medium and high emission scenario 50th percentile weather conditions

			Category I (young/infirm)				Category II (new build)			
Ref:	Description	%>28 °C	C1	C2	C3	Criteria failed	C1	C2	C3	Criteria failed
2080s medium emissions scenario										
6.1	MVHR bypass (25 °C)	24.5%	73.4	57.0	7.0	1 and 2 and 3	55.0	45.0	6.0	1 and 2 and 3
6.2	Btm hung >28°C set point	2.2%	50.4	25.0	3.0	1 and 2	13.4	12.0	2.0	1 and 2
6.3	Btm hung night purge (>28 °C)	2.2%	50.2	25.0	3.0	1 and 2	13.4	12.0	2.0	1 and 2
6.4	Side hung >28 °C set point	1.7%	47.9	24.0	3.0	1 and 2	12.5	11.0	2.0	1 and 2
6.5	Side hung night purge (>28 °C)	1.7%	48.2	24.0	3.0	1 and 2	12.5	11.0	2.0	1 and 2
6.6	Btm hung >25 °C set point	0.9%	3.5	14.0	2.0	1 and 2	3.5	14.0	2.0	1 and 2
6.7	Btm hung night purge (>25 °C)	0.4%	1.6	7.0	2.0	2	0.0	1.0	1.0	–
6.8	Side hung >25 °C set point	0.3%	1.4	6.0	1.0	–	0.0	0.0	0.0	–
6.9	Side hung night purge (>25 °C)	0.3%	1.4	6.0	1.0	–	0.0	0.0	0.0	–
2080s high emission scenario										
7.1	MVHR bypass (25 °C)	28.3%	73.0	82.0	8.0	1 and 2 and 3	55.6	68.0	7.0	1 and 2 and 3
7.2	Btm hung >28 °C set point	2.9%	41.5	32.0	4.0	1 and 2	10.9	20.0	3.0	1 and 2
7.3	Btm hung night purge (>28 °C)	3.0%	41.5	32.0	4.0	1 and 2	10.9	20.0	3.0	1 and 2
7.4	Side hung >28 °C set point	2.6%	39.7	32.0	4.0	1 and 2	11.0	20.0	3.0	1 and 2
7.5	Side hung night purge (>28 °C)	2.6%	39.6	32.0	4.0	1 and 2	11.0	20.0	3.0	1 and 2

(continued)

Table 12.5 (continued)

			Category I (young/infirm)				Category II (new build)			
7.6	Btm hung >25 °C set point	2.2%	3.8	37.0	5.0	1 and 2 and 3	1.5	26.0	4.0	2
7.7	Btm hung night purge (>25 °C)	1.6%	2.1	31.0	4.0	2	1.1	20.0	3.0	2
7.8	Side hung >25 °C set point	1.4%	1.9	31.0	4.0	2	1.1	20.0	3.0	2
7.9	Side hung night purge (>25 °C)	1.4%	1.9	31.0	4.0	2	1.1	20.0	3.0	2

12.5 Conclusions

Results from the in-use monitoring revealed overheating during two consecutive years, although there was an observable improvement when mean external temperatures exceeded 16 °C during the second summer when cooling strategies were employed. Despite this, the dwelling failed to pass C1 and C2 of the CIBSE TM52 overheating assessment in either summer, supporting the assertion that summertime overheating is not an unusual phenomenon in Passivhaus Certified dwellings in the UK (Tabatabaei Sameni et al. 2015). Feedback from occupants via Building Use Study (BUS) surveys in 21 similar dwellings on the same development (including the case study dwelling) suggested uncomfortable temperatures during summer (Siddall et al. 2014). This was exacerbated by security concerns around leaving windows open for purging overnight, misinformation about MVHR operation, and an unfamiliarity with the summer bypass function.

Simulation model outputs show that this type of compact Passivhaus dwelling, in this region of the UK, can avoid excessive overheating in living spaces through the use of natural ventilation alone. This is, however, dependent upon windows being opened when internal temperatures reach a set point temperature of 25 °C. If windows are opened at this set point, then the case study dwelling would avoid excessive overheating in all the medium or high emissions scenarios examined here, although night purge ventilation would also need to be employed if windows were only tilted open to 20° during the day. In reality, it is unlikely that occupants will strictly open all windows in the dwelling when internal temperatures reach the 25 °C set point temperature and it may be necessary to automate the MVHR summer bypass controls and window openings if potentially dangerous overheating levels are to be avoided in the future. If full automation is not considered to be practical, then some occupant alerts may be considered to prompt the introduction of additional ventilation.

Ultimately, the results presented in this paper indicate that this type of low-energy Passivhaus dwelling can avoid excessive overheating in current and future climate scenarios if control strategies for additional natural ventilation are clearly defined. This is important in the context of future UK housing policy as the potential for this type of dwelling to significantly reduce emissions is well understood. However, there is some concern that comfort cannot be maintained in all seasons which could limit the widespread implementation of this low-energy solution.

References

ASHRAE. (2002). *ASHRAE Guideline 14-2002, measurement of energy and demand*. Atlanta: Savings.
Beizaee, A., Lomas, K. J., & Firth, S. K. (2013). National survey of summertime temperatures and overheating risk in English homes. *Building and Environment, 65*, 1–17.
Butcher, K. (2014). *TM55: Design for future climate: Case studies*. London: Connelly-Manton.
CIBSE. (2006). *Guide A: Environmental design*. London: Author.
CIBSE. (2013). *TM52: The limits of thermal comfort: Avoiding overheating in European buildings*. London: Author.
Coley, D., & Kershaw, T. (2010). Changes in internal temperatures within the built environment as a response to a changing climate. *Building and Environment, 45*, 89–93.
Eames, M., Kershaw, T., & Coley, D. (2011). On the creation of future probabilistic design weather years from UKCP09. *Building Services Engineering Research and Technology, 32*, 127–142.
Gartland, L. 2012. Heat Islands: Understanding and mitigating heat in urban areas.
Gul, M., Jenkins, D. P., Patidar, S., Banfill, P. F. G., Menzies, G., & Gibson, G. (2012). Tailoring a future overheating risk tool for existing building design practice in domestic and non-domestic sectors. *Building Services Engineering Research and Technology, 33*, 105–117.
Gupta, R., & Kapsali, M. (2016). Empirical assessment of indoor air quality and overheating in low-carbon social housing dwellings in England, UK. *Advances in Building Energy Research, 10*(1), 46–68.
HM Government. (2013). *National Calculation Methodology (NCM) modelling guide (for building other than dwellings in England and Wales)*. London: BRE.
HM Government. (2014). *UK Building regulations. Part L1A: Conservation of fuel and power in new dwellings*. London: RIBA Publishing.
IES. (2014). *Virtual environment*. 2014.2.1.0 ed.
Jenkins, D. P., Ingram, V., Simpson, S. A., & Patidar, S. (2013). Methods for assessing domestic overheating for future building regulation compliance. *Energy Policy, 56*, 684–692.
Jentsch, M. F., Levermore, G. J., Parkinson, J. B., & Eames, M. E. (2014). Limitations of the CIBSE design summer year approach for delivering representative near-extreme summer weather conditions. *Building Services Engineering Research and Technology, 35*, 155–169.
Khasreen, M. M., Banfill, P. F. G., & Menzies, G. F. (2009). Life-cycle assessment and the environmental impact of buildings: A review. *Sustainability, 1*, 674–701.
Lomas, K., & Kane, T. (2013). Summertime temperatures and thermal comfort in UK homes. *Building Research and Information, 41*(3), 259–280.
Mavrogianni, A., Davies, M., Batty, M., Belcher, S. E., Bohnenstengel, S. I., Carruthers, D., et al. (2011). The comfort, energy and health implications of London's urban heat island. *Building Services Engineering Research and Technology, 32*, 35–52.
Mavrogianni, A., Davies, M., Chalabi, Z., Wilkinson, P., Kolokotroni, M., & Milner, J. (2009). Space heating demand and heatwave vulnerability: London domestic stock. *Building Research and Information, 37*, 583–597.

Mavrogianni, A., Davies, M., Wilkinson, P., & Pathan, A. (2010). London housing and climate change: Impact on comfort and health—preliminary results of a summer overheating study. *Open House International, 35*, 49–59.

Mcleod, R. S., Hopfe, C. J., & Kwan, A. (2013). An investigation into future performance and overheating risks in Passivhaus dwellings. *Building and Environment, 70*, 189–209.

Meehl, G. A., & Tebaldi, C. (2004). More intense, more frequent, and longer lasting heat waves in the 21st century. *Science, 305*, 994–997.

Oikonomou, E., Davies, M., Mavrogianni, A., Biddulph, P., Wilkinson, P., & Kolokotroni, M. (2012). Modelling the relative importance of the urban heat island and the thermal quality of dwellings for overheating in London. *Building and Environment, 57*, 223–238.

Parker, J. M., Farmer, D. & Fletcher, M. (2015, December 10–12). Calibrating whole house thermal models against a coheating test. In *System Simulation in Buildings 2014 Proceedings of the Ninth International Conference*. Liege: Atelier des Presses, pp. 211–219.

Pérez-Lombard, L., Ortiz, J., & Pout, C. (2008). A review on buildings energy consumption information. *Energy and Buildings, 40*, 394–398.

Porritt, S. M., Cropper, P. C., Shao, L., & Goodier, C. I. (2012). Ranking of interventions to reduce dwelling overheating during heat waves. *Energy and Buildings, 55*, 16–27.

Porritt, S. M., Cropper, P. C., Shao, L., & Goodier, C. I. (2013). Heat wave adaptations for UK dwellings and development of a retrofit toolkit. *International Journal of Disaster Resilience in the Built Environment, 4*, 269–286.

Pretlove, S., & Kade, S. (2016). Post occupancy evaluation of social housing designed and built to Code for Sustainable Homes levels 3, 4 and 5. *Energy and Buildings, 110*, 120–134.

Siddall, M., Johnston, D. & Fletcher, M. (2014). *Occupant satisfaction in UK Passivhaus dwellings* (pp. 491–496). 18th International Passive House Conference 2014 Aachen, Germany.

Stocker, T. F., Qin, D., Plattner, G. K., Tignor, M., Allen, S. K., Boschung, J., et al. (2013). *IPCC, 2013: Climate change 2013: The physical science basis*. Contribution of Working Group I to the Fifth Assessment Report of the Intergovernmental Panel on Climate Change. Cambridge: Cambridge University Press.

Tabatabaei Sameni, S. M., Gaterell, M., Montazami, A., & Ahmed, A. (2015). Overheating investigation in UK social housing flats built to the Passivhaus standard. *Building and Environment, 92*, 222–235.

Taylor, J., Davies, M., Mavrogianni, A., Chalabi, Z., Biddulph, P., Oikonomou, E., et al. (2014). The relative importance of input weather data for indoor overheating risk assessment in dwellings. *Building and Environment, 76*, 81–91.

The Carbon Trust. (2009). *Building the future today: Transforming the economic and carbon performance of the buildings we work in*. London: Author.

Vardoulakis, S., Dimitroulopoulou, C., Thornes, J., Lai, K. M., Taylor, J., Myers, I., et al. (2015). Impact of climate change on the domestic indoor environment and associated health risks in the UK. *Environment International, 85*, 299–313.

Yamamoto, J., & Graham, P. (2009). *United Nations environment programme: Buildings and climate change*. Paris: United Nations.

Zero Carbon Hub. (2015a). *Assessing oveheating risk: Evidence review*. London: Author.

Zero Carbon Hub. (2015b). *Monitoring overheating: Housing association case studies*. London: Author.

Zero Carbon Hub. (2015c). *Overheating in homes: The big picture*. London: Author.

Chapter 13
A Method for Visualising Embodied and Whole Life Carbon of Buildings

Francesco Pomponi and Alice Moncaster

Abstract Embodied and whole life carbon of buildings are increasingly gaining attention. However, embodied carbon calculation is still far from being common practice for sustainability assessment of buildings. Some of its greatest difficulties lie with the long life span of buildings which implies a great unpredictability of future scenarios and high uncertainty of data. To help understand which life cycle stages should get the most attention when considering a building project, this chapter proposes a new visualisation method based on Sankey diagrams for whole life carbon that allows one to cluster the carbon emitted in each of the life cycle stages as identified in current BS 15978 standards. With the proposed method, the carbon figures can be further broken down to account for building assemblies and components. Additionally, the method is equally suitable to account for physical quantities of what is embedded in buildings and their components. As such it can supplement some units of existing assessment methods (e.g., metal depletion measured in mass units of Fe_{eq}) and turn it into mass units of embodied steel. With such new metric, a life cycle assessment would include knowledge on flows as well as quantities. Such information could then be linked to the building permanently and smartly to be updated when necessary as the building evolves, changes, and gets upgraded, building on the theoretical foundations of the shearing layers of buildings. As such, this information could be embedded within BIM which is fully suitable to store parametric details for each building component.

13.1 Introduction

Embodied carbon is a significant part of whole life carbon emissions of buildings and with operational energy (and therefore carbon) being continuously reduced, embodied carbon will represent the totality of carbon figures in Zero Energy Buildings

---

F. Pomponi (✉) • A. Moncaster
Centre for Sustainable Development, Department of Engineering, University of Cambridge, Trumpington Street, Cambridge CB2 1PZ, UK
e-mail: fp327@cam.ac.uk

M. Dastbaz et al. (eds.), *Building Information Modelling, Building Performance, Design and Smart Construction*, DOI 10.1007/978-3-319-50346-2_13

(ZEBs). However, both practitioners and academics lament several issues in embodied carbon calculations, as emerged in a research symposium and focus groups on the topic held at the University of Cambridge in April 2016 (CUBES 2016). Some of the issues that emerged are:

- Lack of uniform and standardised methodologies
- Lack of available data
- Complexity of the calculations
- Difficulty to predict plausible scenarios for future uses and end-of-life stages of buildings

Whilst some of these issues are certainly technical and require several and plural approaches to be addressed, during the focus groups it seemed that sometimes complexity was perceived even where there was not. To help in such respect, and after evaluating available possibilities, this short chapter suggests a new visualisation method for embodied and whole life carbon of buildings that allows one to cluster the carbon emitted at each of the life cycle stages as identified in current BS 15978 standards.

13.2 Visualising Embodied Carbon

Sankey diagrams are widely used to show flows and are based on the simple but extremely effective idea that the width of the arrows is proportional to the quantity of the flow. They are frequently used in Material Flow Analysis research (Haas et al. 2015) or to track worldwide flows of a specific element (Allwood and Cullen 2012). Sankey diagrams in building's research are however unusual although they could also help towards embedding circular economy thinking in the built environment. However, this particular aspect is outside the scope of this short chapter. For the purpose of showing the visualisation method and discussing the benefits and challenges that go with it, we use numerical results from previous research (Moncaster and Symons 2013). The objective of this representation is to present embodied carbon figures, and potentially also other environmental impact categories, in an innovative way which plots the life cycle stages according to existing standards (BSI 2011) for the whole life of the building (Fig. 13.1).

From the diagram in Fig. 13.1 it is immediately noticeable which life cycle stages account for the highest shares of embodied carbon, and which instead are barely noticeable. This representation does not suggest that a certain life cycle stage should be minimised prior to others. Rather it wants to help see where the greatest opportunities for reductions lie. Also, our proposed method includes a time element on the horizontal axis, which helps identify which activities span over a significant time horizon and, as such, might be affected by a lot of uncertainty about what happens over many years (such as the B2 stage of Fig. 13.1). Similarly, it helps visualise uncertainty and variability of what happens distant in the future such as the C stage (end of life of the building). In the latter case, the uncertainty is not related to a long

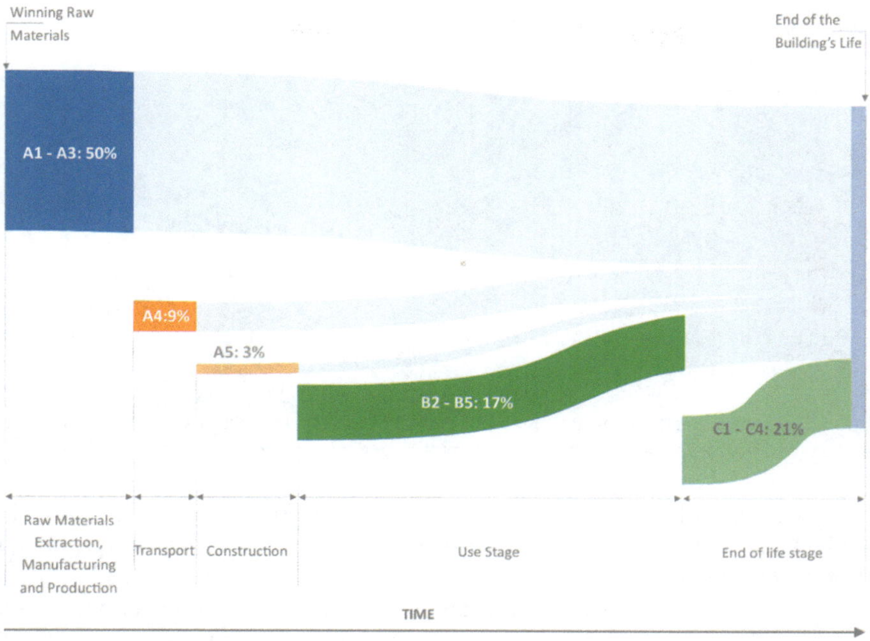

Fig. 13.1 Sankey diagram of whole life embodied carbon (coding of life cycle stages according to BS EN 15978:2011—The numbers refer to a specific case and are only used for illustrative reasons here)

time span of the specific activity but rather to the extreme uncertainty of what will happen after decades or centuries if one imagines to use this visualisation tool at the design stage of a new building. In both B and C stages, uncertainty analysis should play an important role in the assessment to ensure that the numbers produced have some meaningfulness—and the diagram in Fig. 13.1 may help to flag this aspect.

With the proposed method, the carbon figures can be further broken down to account for building assemblies and components. In a software environment, this could be done—for instance—by double clicking on each stage which would open up a sub-Sankey related to the components and assemblies of that specific stage. This approach could go further down on a tier-by-tier basis and allow to group or detail the level information according to the necessity. A BIM environment seems particularly suitable to do so, due to its parametric approach which goes well with the bill of quantities regularly used in embodied carbon assessment.

Furthermore, the method is equally suitable to account for physical quantities of what is embedded in buildings and their components to overcome one of the shortcomings of embodied carbon as a single metric, i.e. the risk of neglecting that environmental impacts might just be shifted from one impact category to the another (Pomponi et al. 2016). One example is to enrich some existing units of more comprehensive life cycle assessment (e.g., metal depletion measured in mass units of Fe_{eq}) and further it to become mass units of embodied steel. To keep both pieces of

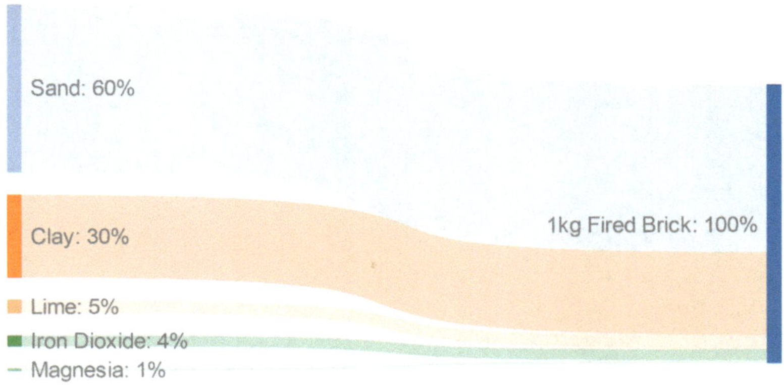

Fig. 13.2 Proposed metric to enrich embodied carbon as a measure to circularity in the built environment—numbers are solely for explanatory reasons and taken from Punmia et al. (2003)

information, these diagrams could be used to show the total amount of Fe_{eq} used in a building or one of its components and also how that equivalency figure is split into different metal sources and end-uses.

An example of this new form of metric is given in Fig. 13.2 for a mass unit of a fired brick. The Sankey diagram could of course carry on both ways to reach virgin raw materials on one end and the whole building on the other. With such new metric, a life cycle assessment would include knowledge on flows as well as quantities. Such information could then be linked to the building permanently and smartly to be updated when necessary as the building evolves, changes, and gets upgraded, building on the theoretical foundations of the shearing layers of buildings. Even in this second example, such information could be embedded within BIM which is fully suitable to store parametric details for each building component.

13.3 Conclusions

This short chapter has discussed the idea of visualising embodied carbon as a means to simplify the understanding and use of embodied carbon assessment in buildings and the built environment. In previous research, we had indeed realised that both practitioners and academics seemed to ask for simpler and easier ways of communicating embodied carbon results and for visualisation tools that would be richer than a simple pie chart or bar graph. The Sankey-based diagrams that we have proposed include an element of time which helps understand, or at least remember, elements characterised by high uncertainty either because of a long time span or due to happening in a very distant future. The Sankey chart also quickly allows to identify the life cycle stages which account for the most, thus pointing at where the greatest opportunities for reduction lie. These diagrams could also be developed further to include more comprehensive information (e.g., materials quantities and physical status) for a richer life cycle assessment or to start embed elements of circular economy thinking in buildings and the built environment. As such, collaboration on both initiatives is particularly welcomed.

References

Allwood, J. M., & Cullen, J. M. (2012). *Sustainable materials. With both eyes open.* Cambridge: UIT Cambridge.

BSI 2011 BS EN 15978:2011. (2011). *Sustainability of construction works—Assessment of environmental performance of buildings—Calculation method.*

CUBES. (2016). Research Symposium and Focus Groups. Cambridge University Built Environment Sustainability (CUBES) Embodied Carbon Symposium 2016. 11 April.

Haas, W., Krausmann, F., Wiedenhofer, D., & Heinz, M. (2015). How circular is the global economy?: An assessment of material flows, waste production, and recycling in the European Union and the world in 2005. *Journal of Industrial Ecology, 19*, 765–777.

Moncaster, A. M., & Symons, K. E. (2013). A method and tool for 'cradle to grave' embodied carbon and energy impacts of UK buildings in compliance with the new TC350 standards. *Energy and Buildings, 66*, 514–523.

Pomponi, F., Piroozfar, P. A. E., & Farr, E. R. P. (2016). An investigation into GHG and non-GHG impacts of double skin façades in office refurbishments. *Journal of Industrial Ecology, 20*, 234–248.

Punmia, B., Jain, A. K., & Jain, A. K. (2003). Basic civil engineering. *Firewall Media.*

Chapter 14
Models for Sustainable Electricity Provision in Rural Areas Using Renewable Energy Technologies - Nigeria Case Study

Abdulhakeem Garba, Mohammed Kishk, and David R. Moore

Abstract Sustainable electricity generation and supply in Nigeria has been a perennial challenge even though the country is one of the world's leading exporters of oil and a member of organisation of petroleum exporting countries (OPEC). The reasons for this problem include persistent vandalism of energy infrastructure, high cost of gridline network and weak transmission, and distribution facilities. Existing capacity only provides electricity to 34% and 10% of urban centres and rural areas, respectively. Decentralised renewable energy technologies (RETs) may be a sustainable and economical alternative for meeting electricity demands of the rural communities representing two-thirds of the total country's population. This research thus investigates alternative RETs that may provide sustainable electricity to Nigerian rural areas. Interview method was used. The findings reveal that the most suitable RETs in order of priority are biomass, solar PV, small hydropower, solar thermal, and wind energy systems. In addition, biomass energy systems (BES) being the most selected, has been subjected to further investigation; unlike the national energy policy under representation of BES, 77% of the interviewees agreed that BES utilisation in the country's rural areas are suitable and desirable. Also, for implementation of BES, all the identified drivers and enablers should be taken into consideration. However, some identified constraints to adoption and development of BES include supply chain limitation, substantial land, and water requirements for set-up and processing. Thus, this study recommends that the existing rural areas energy policies be reviewed.

A. Garba (✉) • D.R. Moore
The Scott Sutherland School, The Robert Gordon University, Aberdeen AB10 7QB, UK
e-mail: hakeemgarba@gmail.com

M. Kishk
Aberdeen Business School, The Robert Gordon University, Aberdeen AB10 7QE, UK

M. Dastbaz et al. (eds.), *Building Information Modelling, Building Performance, Design and Smart Construction*, DOI 10.1007/978-3-319-50346-2_14

14.1 Introduction

Electricity generation and supply sustainability in Nigeria has been a recurrent challenge despite the country being one of the world's leading exporters of oil and gas and a member of organisation of petroleum exporting countries (OPEC) (Energy Commission of Nigeria 2005). Existing capacity in the country only provides electricity to 34% and 10% of urban centres and rural areas, respectively (Sambo 2009; Garba and Kishk 2014). The reasons for this problem include high cost of gridline network, investment imbalance and weak transmission, and distribution facilities. The periods between March and April 2016 have witnessed many cases of inability to generate and supply electricity representing up to 80% of the total grid supply loss as a result of vandalism of energy infrastructure (Nnadim 2016). The country has never experienced over 5000 mega-watt (MW) capacity grid supply of electricity for a population of approximately 170 million, and after over a decade of energy sector privatisation (Garba and Kishk 2015). Similarly, there have been frequencies of undelivered generated electricity particularly for far reaching locations, due to weakness of the transmission and distribution network constituting up to 40% losses (World Bank 2005). Also, utility companies lack interest in delivering electricity to rural areas due to high cost of grid network extension in relation to their low energy consumption (Sambo 2009).

Ohunakin et al. (2011) reported that natural gas, which is the major source of electricity generation in the country, is also experiencing supply shortages with only one-third of the required 1.2 billion cubic feet/day being supplied to the thermal plants in the country. Likewise, large hydro sources which have significantly contributed to the national grid over the past four decades have suffered from the effect of climate change and inadequate maintenance of their turbines leading to a reduction of their contribution to the national grid (Garba and Kishk 2015). Sambo (2009) reported that even on normal days, the average "rural" price of fossil fuel products are over 200% of the cost in cities, hence, making them unaffordable and unattractive to rural communities.

These problems have significantly affected rural communities, resulting in unemployment, lack of development of local businesses and industries, endemic rural to urban migration and affected growth of local economy (Ikeme and Ebohon 2005). Fuel wood has become the major source of energy for these communities with consumption representing approximately one-third of the country's total primary energy (Sambo 2009).

Numerous energy policies targeting rural communities have been developed in the past, including rural electrification fund and consumer assistance fund among others (Ikeme and Ebohon 2005) with a view to enhancing rural electricity supplies. However, reality reveals that Nigerian rural areas electricity problems perhaps are connected with the centralised electricity supply system using fossil fuel sources. Also, the high cost of grid extensions to poor low consumption communities (living below US$ 1.25 per day (UNICEF 2011)) largely based on agriculture, act as a constraint. Thus, a sustainable means of electricity provision that is not reliant on grid extension systems and fossil fuel sources has to be employed. Decentralised renewable energy technologies (RETs) may be suitable for mitigating the problem of electricity in Nigerian rural areas. Thus, this study aims to investigate alternative RETs that may be employed in providing sustainable electricity to Nigerian rural areas.

14.1.1 Previous Studies (Related Work)

Several researchers have considered application of decentralised RETs in provision of sustainable electricity to rural areas. Garba and Kishk (2014) evaluated six major RETs (solar PV, wind, small hydropower, biomass, geothermal, and ocean energy systems) using a systematic review method, and strengths, weaknesses, opportunities and threats, (SWOT) analysis for each RET in order to assess their individual sustainability indicators. The findings by order of priority reveal that biomass, solar PV, small hydropower, and wind are the best means of providing sustainable electricity for Nigerian rural areas. Mahapatra and Dasappa (2012) reported the economic evaluation of biomass, solar PV, and grid extension systems. The study concluded that biomass is the most suitable and economical means of providing sustainable electricity to Indian rural areas. They further argued that a biomass energy system (BES) has significant advantages over solar PV system; "the increase in operation hours in biomass gasification system will only increase the fuel requirements. However, the increase in load demand does not require increase in the gasifier rating, as the gasifier turndown ratio is quite high"; while in the case of solar PV "as the operation hours increase, the system size also increases and consequently, the capital cost of the system". Dasappa (2011) also reported that biomass is among the optimal alternatives energy sources for sustainable electricity provision in Sub-Saharan Africa (SSA) rural areas given the universal availability of the resources. "Efficient use of biomass in Africa can meet both cooking and electricity generation needs. Using a small fraction (~30%) of the existing agricultural and forest residues, distributed power generation potential of about 15,000 to 20,000 MW is possible". Demirbas (2001) argued that a biomass energy system is cost competitive with fossil fuel sources; while Evans et al. (2010) disagrees, and argued that BES is more expensive than grid extension system but cheaper than solar PV (Evans et al. 2009).

However, Owen et al. (2013) stated that the bulk of SSA countries, Nigeria inclusive, have relegated biomass as an energy system of the past, as inefficient and symbolising poverty (based on their national energy policies (NEP)) despite its universal availability in their countries. These governments deliberately refuse to take advantage of contemporary realities in respect of technological opportunities connected to BES. Instead they continuously focus on fossil fuel energy sources to meeting their energy demand (Owen et al. 2013). For example, by the end of 2030, the number of SSA citizens depending on biomass consumption will increase by 60% (IEA 2010), but the NEP in these countries contradicts this reality.

This is because the NEPs have been based on a wrong assumption that biomass utilisation can be substituted with petroleum products and electricity. Meanwhile, there is a significant shift across developed countries back to low carbon renewable energy, particularly biomass based, with a view to achieving a sustainable and low-carbon energy strategy (Owen et al. 2013).

Given all of the above, this study argues that sustainable utilisation of BES for electricity provision in the Nigerian rural areas is perhaps the way forward as it has advantages of: suitability for electricity generation, employment creation opportunities (such as energy plantation and waste management), climate change mitigation, and energy security (particularly in Nigerian rural areas where fossil fuel products are sold at 200% over the regulated prices, and perennial shortages represent the new normal (Sambo 2009). Also, by the end of 2014, there was substantial electricity generation globally from BES representing around 93GW, showing the development of biomass conversion technologies (REN21 2015).

14.2 Methodology

The aim of this study is to investigate alternative RETs that are feasible in providing sustainable electricity to Nigerian rural areas. In achieving this aim, interview method was used. This is because the problem under study needs detailed investigation that requires knowing what, how, and why Nigeria's rural areas has only 10% electricity accessibility despite its massive energy resources (Naoum 2007). Gray (2004) argued that interview method is more desirable than questionnaires where questions are complex, and there is opportunity for probing further where necessary. The use of a questionnaire method is unsuitable in this context, as a large sample cannot be drawn following the outcomes of exploratory studies conducted at the beginning of the research work, which showed that the RETs industry in Nigeria is full of quack practitioners. In addition, care has been exercised in selecting appropriate sample at the structured interview stage. Structured interview method was selected because it is suitable for descriptive studies for the purpose of generalisation. This method can commence with open-ended questions then move to closed-ended questions (Naoum 2007).

The interviewees were chosen purposefully in this study. Interviewees' selection was based on criteria such as place of work, qualifications, and contributions related to RETs. Also, through review of literature, where some of the participants were identified; their addresses and names were obtained and later contacted via emails and telephones.

Given the fact that the research under study has no sufficient variables that are reported or tested empirically in the literature using questionnaire or other methods, content analysis was used to analyse the interview sessions. This is because it is more of a deductive approach which can lead to generalisation of the result outcomes (Gray 2004) and measure evidence in positivistic way (Fellows and Liu 2008). Also, it objectively and systematically identifies distinctive features among the data with a view of making inference (Gray 2004).

14.3 Data Collection

Data was collected in two phases which includes two levels of interview sessions. Exploratory interview was first conducted. The interview questions at this stage seeks to identify suitable RETs for providing sustainable electricity to rural areas, appropriate incentives for successful adoption, and the way forward (but only section on the most suitable RETs was reported in this study). This was followed by structured interview, where questions on the strategy for adoption and details on the most selected RET in term of drivers, enablers, and constraints in the context of Nigerian rural areas were asked. During these two levels of data collection especially structured interview, real RETs practitioners were consulted and their background checks were conducted by asking their details from their colleagues, consulting their human resources department where opportunity present itself, their places of work and also based on their contribution to RETs. Hence, this assisted in validating the collected data. Patton (1990) suggested strategies for conducting interviews was followed whereby exact wordings and sequence of questions are determined in advance, all the interviewees are asked the same basic questions and questions are worded in a completely open-ended format. All the interview sessions were conducted face-to-face and conducted in the interviewees' office premises. Both exploratory and structured interview sessions lasted between 21–28 and 30–46 min, respectively. A letter of expression of interest in respect of participation and anonymity was first sent to all participants, anonymity concern was also highlighted during the interview sessions by the researcher as a way of reassuring the interviewees. Other ethical issues were also observed, and this helps to improving the level of co-operation from the interviewees.

Initially 20 participants were contacted, 4 persons declined because of their schedules in their offices, while 3 persons did not respond to the emails and called made to them. 13 persons participated and have been considered suitable for this study. Although this may appear to be an insufficient sample for generalisation, but Kothari (2009) opined that a small sample is considered appropriate for technical survey. See Table 14.1 for details of the interviewees.

14.4 Data Analysis

The data analysis was carried out using content analysis for both exploratory and structured interviews. Content analysis approach comprises of the preparation (transcribing), organising (coding, themes/categories development), and reporting (Vaismoradi et al. 2013). The analysis commenced by identifying key points (coding) from the transcribed interview paragraphs, and a combination of related codings developed into concepts. Then, clusters of concepts with identical features were grouped together to form themes such as identification of the most suitable RET in rural areas. This iteration process continues by constant comparison of the

Table 14.1 Details of interviewees

Interviewees	Establishment	Qualification	Year of experience
1	Academic	PhD	18
2	Research Centre	PhD	28
3	Academic	PhD	16
4	Regulator	Msc	14
5	Research Centre	PhD	31
6	Research Centre	PhD	27
7	Research Centre	PhD	20
8	Research Centre	PhD	23
9	Research Centre	Msc	8
10	Practitioner	PhD	35
11	Regulator	Msc	28
12	Research Centre	Msc	11
13	Regulator	PhD	27

sentences and paragraphs from the transcribed interviews until new concepts and themes could no longer emerge; hence, data saturation is achieved. One theme emerged for exploratory study and three themes emerged for the structured interviews. These are presented in the findings.

14.5 Findings

14.5.1 Exploratory Study (Phase 1): RETs for Providing Sustainable Electricity to Rural Areas

The outcomes of the exploratory interview analysis reveals that five major RETs have been selected by the interviewees as a means for delivering sustainable electricity to Nigerian rural areas. See Fig. 14.1 for details. Biomass was identified as the leading energy system, followed by solar PV and the least is wind energy system. The summary of the reasons for the selection of these RETs includes resource availability, level of development of the systems (RETs systems maturity in Nigeria), cost competitiveness, and policy support.

14.5.2 Phase 2: Structured Interview Outcomes

Following the outcome of the exploratory interviews, biomass energy system (BES) was identified as the most suitable means of providing sustainable electricity to the rural areas. Thus, BES is further subjected to investigation among the interviewees using structured interview approach.

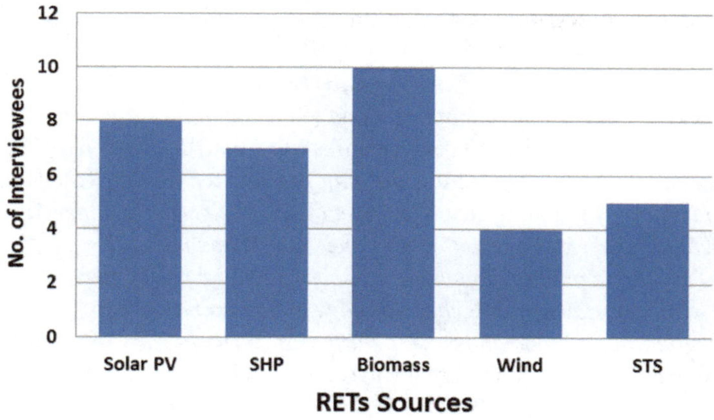

Fig. 14.1 RETs suggested for Nigerian rural areas electricity provision

The responses of interviewees in respect of BES provision of electricity in rural areas indicates that 10 out of 13 interviewees support the energy system, representing 77% of the participants, while 3 interviewees were against its application. The next sections present structured interview findings, which are classified under the following: drivers, enablers, and constraints of BES utilisation in Nigerian rural areas.

14.5.3 Theme 1: Drivers of BES Selection in Rural Areas

The following represents motives of some of the interviewees in this respect. Interviewee 3 was of the view that the drivers of the utilisation of BES is its inclusion in the national energy policy and biomass resource availability. Interviewee 11 was of the view that the drivers include reduction in CO_2 footprint, rising energy demand and conflict neutral energy source.

14.5.3.1 Rising Energy Demand

Following the rise in energy demand occasioned by the growth in population, particularly in the country's rural communities in relation to long gestation period of most carbon-based power plants in the country, is indicative that existing practice of centralised grid using fossil fuel energy system may not meet the immediate energy demand of these communities. Interviewee 3 says that "*To install and test run similar capacity of Egbin (gas) thermal station, we need about 36 months; while similar capacity to Mambilla hydropower station, may require six years or more*". Interviewee 8 opined that "*Due to the developmental period, we need something of immediate outcome such as RETs to meet the rising energy demand in the country over the coming years, looking at the rate our population is growing*".

14.5.3.2 Biomass Resources Availability

According to interviewee 6 "*Biomass has always been rural areas friendly, because that is where you find the raw materials and the technologies are not so complex to manage*". Interviewee 5 also expressed similar view that "*It is very feasible, because we have a lot of resources and that is the major one. Once you have the fuel, the next stage is technology*"; adding that "*There is biomass electricity generators already developed globally*". Interviewee 13 observes that "*The driving force is the biomass resources available in these communities and the energy policy that encourages the generation of electricity from such technology in a sustainable way*". "*We have a lot of waste from animal husbandry, in addition to agricultural waste*" according to interviewee 8. These findings are in agreement with (Mohammed et al. 2013), (Garba and Kishk 2014) and (ECN 2005).

14.5.3.3 Conflict Neutral Energy Source

Interviewee 3 says that renewables are conflict free energy sources "*once you have them, nobody can shut the atmosphere from sun radiation and biomass plantation photosynthesis*". This is in agreement with Owen et al. (2013) "Domestically-sourced biomass can help diversify domestic energy supply, leading to increased energy security and independence from imports". Similarly, BES can enable other Nigerian regions to have access to electricity through other sources given the unabated youth restiveness in the Niger delta region.

14.5.3.4 Climate Change Mitigation

Given that Nigeria is the second largest gas flaring country in the world (Oseni 2012), the adoption of BES by these communities will help in mitigating climate change effects considering their enormous electricity needs. Interviewee 11 opined that "*Biomass system utilization for rural communities will curb greenhouse gas emission in the country*". This finding also agrees with Owen et al. (2013) "Biomass is potentially carbon-neutral and can replace fossil fuels sources especially in power generation".

Disagreement with Biomass Application in Nigerian Rural Areas

However, 3 out of 13 interviewees disagree with BES utilisation in Nigerian rural areas. Their reasons for rejecting BES includes lack of biomass technology in the country, deficiency in local know-how, location peculiarity, and policy issues. Although, their disagreement with it utilisation was not far-fetched.

Interviewee 1 indicates that "*Nigerian rural areas are not mature enough for biomass electricity generation. Although, the potentials exist but the maturity is not*". Interviewee 10 added that "*Anything that needs monitoring in Nigerian rural areas poses some challenges and even the basic investment that is required to have kerosene stove, let alone RETs ownership*". Interviewee 7 argues that "*Biogas for electricity generation is not viable at the moment; the yield for the gas generation is not much*". This latest response may not be unconnected with the existing practice in the country, where biogas is mainly used as heating gas for school laboratories and cooking gas in the prison yards.

Interviewee 10 claims that "national renewable energy and energy efficiency policy (NREEEP) *is not promoting the use of biomass system especially when it requires cutting down of trees or forest in order to feed*". In any case, renewable policy doesn't de-emphasise the use of biomass as long as it is through the use of waste and energy plantation. In fact, NREEEP (2015) stresses the utilisation of BES: "To promote efficient use of agricultural residues, municipal wastes, animal and human wastes and energy crops as bioenergy sources". Furthermore, interviewee 1 stressed that "*If such an investment is to be located in rural areas, then it will require monitoring; hence, it will required people from these communities, to manage it. Do they have the technology, its know-how and even awareness*? No". He added that: "*You will find that, it (biomass) will be very expensive and abandoned in the long run and subsequently go back to wood burning*". Based on the existing practice in Nigeria, this problem is not only peculiar to BES; it is a general problem to RETs. Hence, there is the need for the practice and experience to be gained with a view to develop the RETs (BES).

Despite reservations of critics of BES utilisation as expressed above, they still agreed that biomass is good for rural communities. For instance, interviewee 1 agreed that "*Biomass utilisation is a very good idea but, there are sustainability questions to be answered particularly in terms of cost competitiveness with fossil fuel (FF) and environmental benign*". Thus, biomass is now cost competitive with fossil fuel based electricity generation particularly in the developing countries. This is in agreement with Mahapatra and Dasappa (2012) and Garba and Kishk (2015). Evans et al. (2010), however, disagree with this assertion.

Interviewee 7 also suggested that "*If there is a proper organisation, biomass is good but using the dwindling forestry product- no*". "*Biomass should only be used in most suitable locations*" interviewee 10 added. All of the above concerns have already been covered by (NREEEP 2015) despite relegation of biomass by NEP among other RETs and FF sources in Nigeria.

The benefits of using BES as opined by interviewee 6 includes "*Biomass has tripod advantages which include sanitising the environment, produce gas for cooking and electricity generation and the wastes are used as organic fertiliser*". "*You can create a business case by growing grass, corn and any visible waste can be bought; thus, you are creating a chain of business for people*" interviewee 12 added. Interviewee 10 says that "*Bio-digester can be used to solve waste problems that arise from bush burning, plant and animal waste often disposed in our open abattoirs and farmlands*".

Furthermore, interviewee 11 advocated that "*The critics of biomass that, it is not totally renewable should be ignored. This is only because they want to sell their oil*". He then asked: "*Have you been informed* about *what they have gone through before they can get oil up to this level*"*? Interviewee 13 added that "Based on whole life cycle assessment, solar and wind still have elements of pollution*". This finding is in agreement with Manish et al. (2006).

14.5.4 Theme 2: Enablers of BES in Rural Areas

For successful development of biomass energy system (BES) in Nigerian rural areas, there are certain things that need to be taken into consideration. Interviewee 5 suggested that certain prerequisites need to be considered for BES to work in rural communities, these include "*adequate water supply, the need to train local people to handle the facilities and appropriate siting of biomass plant based on availability of biomass resources*". Interviewee 12 added "*It depends on the policy in place and how one wants to implement it*".

14.5.4.1 Water Availability

It is very necessary to build BES plant where there is adequate water resource. According to interviewee 5 "*If you build a BES plant in a village where only a hand dug well* is available for *feeding their animals and communities utilisation, there might be problem of water shortage and eventually could lead to abandonment*". Interviewee 6 commented "*Areas and locations with good water level or close to water sources, and have the biomass resources can have the technology implemented*". Interviewee 8 added "*Water availability is a major factor when it comes to biogas*". Hence, water is a key factor for implementing BES and should be given consideration. However, organisations interested in BES will prefer a separate and adequate water supply system as against hand dug. Hence, this factor will not affect these communities but perhaps will increase capital and operational cost of the facility.

14.5.4.2 Local Know-How Requirement

Local people's participation to operate and maintain the facilities is very necessary for sustainable usage. Interviewee 5 says that "*when you just employ workers from cities and send them to rural areas, they are certain that at the end of the month, whether the plant works or not, they are going to receive their salary*". Furthermore, "*Whenever we have a pilot project, we usually include training of local people for operation, minor repairs and maintenance"*. Interviewee 10 supported this view that "*What Sokoto energy research centre (SERC) did when they wanted to popularise improved fuel wood stove, was to teach the communities how to produce it"; adding that "If they have to depend on far away person from the city to come down and develop it for them, instead of domesticating it, then, it will not be sustainable*".

14.5.4.3 Appropriate Technology

Interviewee 6 agreed that "*The technology cannot be everywhere, but should be used where it has economic advantage and with little or no hindrances in terms of implementation". "If you site it where they have to source for resources, then in the end, you will be left with no result*" interviewee 5 added. Interviewee 5 further suggested that "*When setting this technology, the policy should be based on adequate raw materials availability in a particular location*".

14.5.5 Theme 3: Constraints of BES in Rural Areas

14.5.5.1 Supply Chain Issue

Available literature such as IRENA (2012) reported that supply chain difficulty is among the major problems of BES. This is particularly the same in this study because all the interviewees agreed with the above instance. Typically, interviewee 5 opined that "*Our people are used to easy technologies, the protocol of collecting these resources and mixing them to utilise the gas may prove difficult*". Similarly, interviewee 6 recounted that "*Supply chain difficulty has to be put into consideration, because it's a fundamental problem". "I know we have a lot of biogas digesters in the energy centre, though not all of them are working because of fuel issue*" interviewee 8 opined.

14.5.5.2 Massive Land Requirements

Interviewees were of the view that an enormous land requirement is a key constraint to BES use. According to interviewee 8 "*When BES becomes operational, waste procurement may prove difficult and there may be need for energy plantation, which utilise large amount of land"*. While interviewee 3 opined that "BES *plantation requires massive land requirement and high water needs during energy plantation and generation*". This finding agrees with Manish et al. (2006) "*Land availability may constrain sustainability of biomass based systems*".

14.5.5.3 Lack of Local Content and Engagement

Typical problem of RETs in Nigeria is the absence of local community content. Interviewee 10 says that "*The way government is operating rural electrification is unsustainable; this is because they just dump the RETs facilities and it costs the community nothing*". Interviewee 5 suggested that "*People should pay for what they consume through community development organisation. There should be business case behind it*". It is clear that if there is nothing behind it that makes it sustainable, it will be abandoned. Therefore, community engagement is necessary, where they will be paying a stipend for services, managing, and operating the system. Interviewee 11 commented on the current strategy in the country that "*Immediately we finish implementing the project, we hand it over to the local government; we involve the traditional rulers, so that they will not damage the project*". He added "*We even arrange for them to contribute money for the operation and maintenance, so they know it is their property not a gift*". This finding is in agreement with Sunderbans India solar PV (2003) "The most effective partnerships have been forged between the state and the community. In these relationships, the village committees have been successful in managing the entire scheme under the technical supervision of the state". However, in Nigerian rural context, caution need to be exercised regarding managing the project by dedicated and trustworthy committee members, and if possible with some form of economic incentives such as free electricity to them.

14.6 Discussion

Given all the analyses above, renewable energy technologies (RETs) can provide sustainable electricity in Nigerian rural areas particularly using biomass and solar energy systems. It is also indicative that RETs practitioners in Nigeria differs from the country's NEP that relegated BES. This is because BES happened to be the most selected RETs in Nigeria. The findings similarly reflect that far reaching consultation with RETs practitioners was not undertaken during the development of NREEEP (2015) and other energy policies in the country; otherwise, the difference noticed would not have been wide. These findings agree with Mahapatra and Dasappa (2012) and Garba and Kishk (2015) that BES is suitable for providing electricity to rural areas. However, disagrees with Evans et al. (2010). Currently, BES is not among Nigerian energy mix; however, it can be adopted, utilised, and even develop its local capacity provided there is political will in developing the system. The use of good incentive strategy can assist in developing RETs system even where natural energy resources are poor, and the use of plant straws for energy generation, respectively, in the case of Germany and Denmark.

14.6.1 Limitation

Proliferation of quack practitioners in the Nigerian RETs industry is the major drawback experienced in this study, as it was difficult to identify real RETs practitioners. In fact, three appointments were cancelled due to information received in respect of this problem particularly at exploratory study stage. Also, space constraint forced the authors to only provide summary of the findings at exploratory interviews section. It is also noteworthy that, some of the interviewees still see BES as biogas system only, which eventually leads to limitation of their responses. They did not look at the bigger picture which includes thermo-chemical conversion systems (gasification and combustion, with over 90% of current biomass electricity globally).

14.7 Conclusion and Way Forward

Following relegation of BES among RETs in the SSA' NEP (Nigeria inclusive), the findings reveal that biomass has been identified as the most suitable RETs for rural areas electrification and wind energy is the least system. Also, 77% of the interviewees agreed that biomass utilisation in the country rural areas are both appropriate and desirable. Even the interviewees that opposed BES utilisation, their objections have been covered by renewable energy policy in the country (NREEEP 2015) and biomass energy conversion systems advancement. Interviewees unanimously agreed that Nigerian rural areas have adequate biomass resources, and the technologies are not so complex to manage even though the country is yet to commence electricity generation from this energy system. Also, they proposed that the policy prerequisite in setting biomass plant should be based on adequate availability of biomass and water resources in these rural communities and should be utilised for villages far from the grid. The communities should be allowed to operate and manage the facilities rather than employing persons from far places. Business case should be introduced by paying a stipend for what they consume (to ensure sustainability). Furthermore, the rising energy demand in the country as a result of population increase, particularly in the rural areas vis-à-vis long gestation period of thermal stations can be mitigated through the use of BES in meeting the immediate and long-term energy needs of rural communities. BES can also serve as an alternative energy source as against fossil fuel system application causing youth restiveness in the Niger delta region of the country given the supply disruption to the country's thermal plants. This is because BES resources can be found everywhere in the country's rural communities in one form or the other. Hence, Nigerian rural communities' electricity needs can be met through application of BES. Furthermore, this study recommends that the country's existing rural areas energy policies be reviewed. Such that BES will be evaluated with inform knowledge and wider consultation among RETs experts for appropriate positioning in the country's energy mix.

References

Dasappa, S. (2011). Potential of biomass energy for electricity generation in sub-Saharan Africa. *Energy for Sustainable Development, 15*(3), 203–213.
Demirbas, A. (2001). Biomass resource facilities and biomass conversion processing for fuels and chemicals. *Energy Conversion and Management, 42*(11), 1357–1378.
Energy Commission of Nigeria and United Nations Development Programme (ECN–UNDP). (2005). *Renewable Energy Master Plan: Final Draft Report*. Retrieved February 22, 2015, from http://www.iceednigeria.org/REMP%20Final%20Report.pdf
Evans, A., Strezov, V., & Evans, T. J. (2009). Assessment of Sustainability Indicators for renewable energy technologies. *Renewable and Sustainable Energy Reviews, 13*(5), 1082–1088.
Evans, A., Strezov, V., & Evans, T. J. (2010). Sustainability considerations for electricity generation from biomass. *Renewable and Sustainable Energy Reviews, 14*(5), 1419–1427.
Fellows, R., & Liu, A. (2008). *Research methods for construction*. Chichester, UK: Wiley-Blackwell.
Garba, A., & Kishk, M. (2014). Renewable energy technology means of providing sustainable electricity in Nigerian rural areas: A review. In *Proceedings of the 30th ARCOM (Association of Researchers in Construction Management) Conference, 2014*. ARCOM.
Garba, A., & Kishk, M. (2015). Economic assessment of biomass gasification technology in providing sustainable electricity in nigerian rural areas. In C. Gorse (Ed.), *01 Annual SEEDS Conference.2015* (SEEDS, Vol. 1, pp. 554–565). Leeds, UK: Leeds Beckett University.
Gray, D. E. (2004). *Doing research in the real world*. London: Sage.
Ikeme, J., & Ebohon, O. J. (2005). Nigeria's electric power sector reform: What should form the key objectives? *Energy Policy, 33*(9), 1213–1221.
International Energy Agency. (2010). *World energy outlook*. Paris: Author.
IRENA. (2012). *Renewable energy technologies: Cost analysis series*. IRENA Working Paper.
Kothari, C. (2009). *Research methodology: Methods and techniques*. New Delhi: New Age International.
Mahapatra, S., & Dasappa, S. (2012). Rural electrification: Optimising the choice between decentralised renewable energy sources and grid extension. *Energy for Sustainable Development, 16*(2), 146–154.
Manish, S., Pillai, I. R., & Banerjee, R. S. (2006). Sustainability analysis of renewables for climate change mitigation. *Energy for Sustainable Development, 10*(4), 25–36.
Mohammed, Y. S., Mustafa, M. W., Bashir, N., & Mokhtar, A. S. (2013). Renewable energy resources for distributed power generation in Nigeria: A review of the potential. *Renewable and Sustainable Energy Reviews, 22*, 257–268.
Naoum, S. (2007). *Dissertation research and writing for construction students*. New York: Routledge.
National Renewable Energy and Energy Efficiency Policy. (2015). *Approved by Federal Executive Council for the electricity sector*. Retrieved from http://www.power.gov.ng/. NREEE Policy.
Nnadim, O. (2016). *3,132 MW of electricity lost in one day*. Retrieved April 26, 2016, from http://www.punchng.com/3132mw-of-electricity-lost-in-one-day-osinbajo/
Ohunakin, O. S., Ojolo, S. J., & Ajayi, O. O. (2011). Small hydropower (SHP) development in Nigeria: An assessment. *Renewable and Sustainable Energy Reviews, 15*(4), 2006–2013.
Oseni, M. O. (2012). Improving households' access to electricity and energy consumption pattern in Nigeria: Renewable energy alternative. *Renewable and Sustainable Energy Reviews, 16*(6), 3967–3974.
Owen, M., van der Plas, R., & Sepp, S. (2013). Can there be energy policy in Sub-Saharan Africa without biomass? *Energy for Sustainable Development, 17*(2), 146–152.
Patton, M. Q. (1990). *Qualitative evaluation and research methods*. London: Sage.
REN21. (2015). *Renewables 2013 global status report*. Worldwatch Institute, Paris.

Sambo, A. S. (2009). Strategic developments in renewable energy in Nigeria. *International Association for Energy Economics, 16*(1), 15–19.

Sunderbans India solar PV. (2003). Retrieved April 23, 2015, from enviroscope.iges.or.jp/contents/APEIS/RISPO/inventory/db/pdf/0022.pdf

UNICEF. (2011). *Nigeria statics at a glance*. Retrieved January 17, 2015, from http://www.unicef.org/infobycountry/nigeria_statistics.html#119

Vaismoradi, M., Turunen, H., & Bondas, T. (2013). Content analysis and thematic analysis: Implications for conducting a qualitative descriptive study. *Nursing & Health Sciences, 15*(3), 398–405.

World Bank. (2005). *World resource Indicators*. Washington DC. Retrieved May 13, 2014, from http://www.worldbank.org/afr/findings

Chapter 15
Local Energy Mapping Using Publicly Available Data for Urban Energy Retrofit

Rajat Gupta and Matt Gregg

Abstract There is an urgent need to improve the energy performance of the built environment, so as to help alleviate fuel poverty, meet national carbon targets, and improve the local economy. This is why local authorities have targets to reduce carbon emissions and fuel poverty and to create long-term, high-quality jobs in their areas. Large-scale energy retrofit schemes can address these objectives but they need to be better targeted, more cost-effective and result in a higher uptake. This chapter investigates how publicly available datasets on housing and energy can be used to plan mass retrofit and provide targeted low carbon measures across a city, in order to address the challenges of having: incomplete data on which homes could benefit from which retrofit measures and the inability to aggregate private sector housing retrofit activities to minimise installation costs. Energy-related assessments are preformed using publicly available national and local data throughout Bicester, Oxfordshire, and presented using a GIS platform. Key datasets include Ordnance Survey (OS) Mastermap, OS Address-point, Energy Performance Certificate data (EPC), and Sub-national energy statistics. The EPC data (6000 properties) and sub-national data for Bicester are used to identify areas with high energy consumption, fuel poverty, and those in need of wall and roof insulation. Interestingly, when the entire EPC dataset for Bicester was compared to the entire town of Bicester's sub-national figure, the values were only off by ~800 kWh. On the other hand at a house level, there appears to be an overestimate of between 3000 and 4000 kWh/yr. in the mean energy figure for the EPCs, as compared to sub-national data.

15.1 Introduction

The 2011 UK Carbon Plan states that "By 2050, all buildings will need to have an emissions footprint close to zero" (HM Government 2011, p. 5). Specifically, the UK is legally committed to an 80% greenhouse gas emissions (GHG) reduction

R. Gupta (✉) • M. Gregg
Low Carbon Building Group, School of Architecture, Oxford Brookes University, Oxford OX3 0BP, UK
e-mail: rgupta@brookes.ac.uk

M. Dastbaz et al. (eds.), *Building Information Modelling, Building Performance, Design and Smart Construction*, DOI 10.1007/978-3-319-50346-2_15

target for 2050 and to 5 year carbon budgets in the interim set by the Committee on Climate Change. To meet this target, retrofit of existing buildings will be required as 28 million homes in the UK are amongst the least energy efficient in Europe (ACE 2015), and are responsible for over a quarter of the UK's annual GHG emissions (Palmer and Cooper 2013). Along with an inefficient housing stock, the UK has one of the highest fuel poverty ratings in Europe, though it is a country with some of the lowest energy prices (ACE 2015). It is the poor state of the housing condition which is the lead cause for inefficiency and fuel poverty (ACE 2015). Following the recent shift of involvement and action from the central government to local government and community-based groups (Wade et al. 2013), local government and community groups now require the tools to assess their local housing stock in order to improve it.

It is within this context that the overall aim of this research is to turn this urban challenge into a marketable product that can facilitate wider roll out of retrofit measures, cheaper delivery, and more effective outreach to vulnerable residents. To meet this challenge, the research will investigate the feasibility of data mapping for identification, analysis, and communication of highly vulnerable areas, and test the service in practice through a pilot in Bicester, Oxfordshire, England. This chapter specifically demonstrates the process of isolating an appropriate area of Bicester for focus based on the greatest need and investigates how publicly available datasets on housing and energy can be used to plan mass retrofit and provide targeted low carbon measures across a city, in order to address the challenges of having: incomplete data on which homes could benefit from which retrofit measures and the inability to aggregate private sector housing retrofit activities to minimise installation costs.

15.1.1 Domestic Energy Mapping

Geographical information systems (GIS) provide a platform for analysing and presenting findings in an aggregated form which can be visually effective in communicating results to householders, community groups, and local government decision-makers. The authors have extensively explored the use of GIS in this way through the use of DECoRuM©. DECoRuM is a GIS-based toolkit for carbon emissions reduction planning with the capability to estimate energy-related CO_2e emissions, and effectiveness of mitigation and adaptation strategies in existing UK dwellings. To communicate findings, DECoRuM can aggregate and map the results on a street, district, and city level. Previously, research has been applied to a number of neighbourhoods to create bottom-up energy assessments (Gupta and Gregg 2012, 2014) and/or assess climate change risk (Gupta and Gregg 2012) for the purpose of mitigation and adaptation recommendations. In contrast, this work assesses in a top-down manner, freely available public datasets with the purpose of identifying an appropriate area of further (bottom-up) investigation based on consumption and fuel poverty.

A number of GIS-based studies focus on energy use estimations using a top-down approach. One such method involves remotely sensed anthropogenic heat to serve as a proxy to derive the spatial pattern of energy use (Zhou et al. 2012). Another method is combining location, demographic and end-use data to enable energy consumption to be calculated and mapped (Pereira and Assis 2013). Finally, another method involves estimating consumption for a city by downscaling via a multiple linear regression model, large datasets including housing characteristics and aggregated energy consumption (Mastrucci et al. 2014). In contrast, this study uses top-down sources to identify an area for further study. The next step will be to define more detailed energy consumption and CO_2e emissions, and identify hot-spots of high energy consumption or specific retrofit need of the study area using DECoRuM (bottom-up carbon mapping) (Gupta and Gregg 2014) as opposed to downscaling through deduction.

15.1.2 Energy Performance Certificate

The research utilises domestic energy consumption datasets, one of which is the Energy Performance Certificate (EPC). The EPC is suited for retrofit advice as it identifies specific measures in place, e.g. wall or roof insulation, and suggests economically justifiable improvements and the potential level of EPC which could be attained. They are designed to help homeowners make informed, incremental improvements. This is however unlikely to influence homeowners unless the improvements are linked to other structural or functional defects that would be highlighted in property inspections or surveys and which are already affecting market behaviour (Davis et al. 2015). A study in Germany found that EPCs are not helpful for understanding the financial implications of energy efficiency and for a number of reasons they have only played a limited role in purchasing decisions. Simply put, translating information on an EPC into energy bills is difficult for a homeowner (Amecke 2012). Likewise, a study in Denmark found the EPC insufficient in encouraging homeowner retrofits; furthermore, though homeowners considered the EPC reliable and accurate, only a few consider it useful as a source of information on how to reduce energy consumption (Christensen et al. 2014). For these reasons, it is theorised that managing the retrofit process using EPC data on a larger implementation scale, community or otherwise, can be helpful and more successful. Community driven and focussed retrofit has been shown to be effective (Gupta et al. 2014).

15.2 Research Review and Methodology

The authors have been working with project partners (Bioregional and Cherwell District Council) to define the appropriate case study area in Bicester. Approximately 500 existing dwellings are identified wherein further research will evaluate

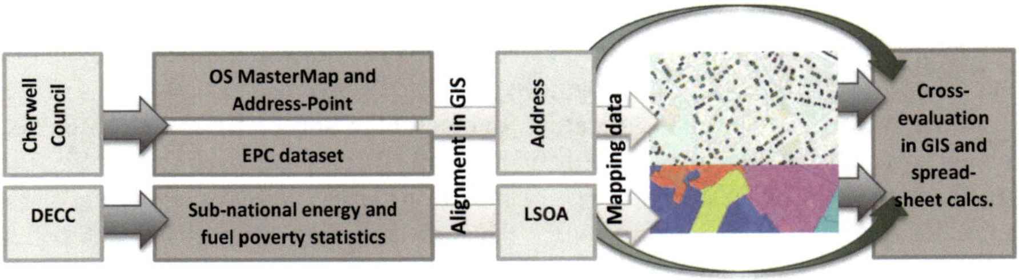

Fig. 15.1 Research method flow diagram

consumption and retrofit suitability with the final intent to create a platform which can facilitate wider roll out of retrofit measures, cheaper delivery, and more effective outreach to vulnerable residents.

The aim was to identify an appropriate neighbourhood case study area for targeting energy retrofitting measures. The rationale for using the following town-wide datasets was to ensure that the method could be performed easily by local authorities or community groups:

- *Ordnance Survey (OS) MasterMap Topography layer and OS Address-Point: OS* MasterMap Topography layer and Address-Point are needed to identify dwelling characteristics (e.g. building form) and to visually identify CO_2e emissions, fuel poverty rating on a house-by-house scale.
- *EPC dataset,* over 6000 dwelling EPCs in a single spreadsheet, obtained from the Department for Communities and Local Government (DCLG n.d.) via Cherwell Council: A free publically available dataset which includes dwelling energy-related information (e.g. wall type, insulation, heating system, annual energy use) compiled through domestic energy assessments at address level by trained individuals. The data collection process began in 2008 and is ongoing. EPC data is currently the most detailed and accurate publically available option for displaying energy-related aspects for the domestic sector at a dwelling level.
- *Sub-national energy consumption statistics* (DECC 2016) and *sub-national fuel poverty statistics* (DECC 2015a) obtained from the Department of Energy and Climate Change: Sub-national datasets are free to use and publically available datasets of metered consumption collected from fuel transporters (DECC 2015b). The data are aligned with Lower layer super output area (LSOA), other defining boundaries include middle layer super output area (MSOA) and postcode level.

15.2.1 Research Method

The research method included the following (Fig. 15.1):

1. Gathering datasets, evaluate usefulness of datasets, data cleansing
2. Aligning datasets with MasterMap and Address-Point data points

3. Creating GIS maps for each individual dataset
4. Cross-evaluating each dataset

Firstly, the previously mentioned datasets were considered useful, specifically based on their provision of energy CO_2e emissions figures and/or dwellings details. The following datasets were assessed for usefulness but considered repetitive or not specifically relevant to the task: the National Energy Efficiency Data-Framework (NEED) and the Energy Saving Trust's (EST) Home Analytics Tool which are based on EPC data already obtained (CSE 2015), and DECC's National Heat Map (DECC n.d.) which is intended to identify areas for further investigation with eventual planning for heat distribution networks. Data cleansing was necessary as the EPC dataset contained invalid data points in a number of categories, e.g. over 700 addresses have the incorrect energy and carbon emissions figures possibly due to data entry failures. For this reason, some dwellings had to be excluded from the set.

Secondly, the EPC dataset is a spreadsheet with addresses aligned with the data collected and calculated for the EPC. Because the OS Topography layer does not provide the postal addresses for the dwelling location points, the Address-Point dataset was required to bridge the two datasets in GIS so that the EPC data could be mapped.

Thirdly and finally, sub-national datasets and EPC data were cross-evaluated as EPCs represent dwelling specific but modelled data and sub-national datasets represent actual but aggregated data. Maps were created to evaluate and demonstrate the appropriate location for further study. Though the datasets could be analysed without mapping, the mapping process assists in communicating the message to stakeholders, e.g. community groups and local councils.

To isolate areas for further detailed study, the analytical sequence progressed from the entire town of Bicester, to four quadrants of the town (Fig. 15.2), down to specific LSOAs. The sequence evaluated the data for high energy consumption, high fuel poverty, and greater need for wall and roof insulation. At the town level, sub-national datasets were prioritised for their greater level of completeness and metered status. As the analysis mined further down, EPCs were used to verify the sub-national findings and to evaluate details not available in sub-national statistics, e.g. wall and roof insulation. Figure 15.2 shows the dwellings for which there are EPC data points. To divide the town into manageable areas, the town was split into four sections called "quads".

15.2.2 Limitations

Most OS products are not free, particularly those used in this study. This is a limitation for community groups. These groups can however theoretically collaborate with local councils to access these data or work with free OS datasets such as OS Open Map—Local which shows figure-ground forms at street level but has the limitation of lacking outline boundary definition for terraced and semi-detached dwellings.

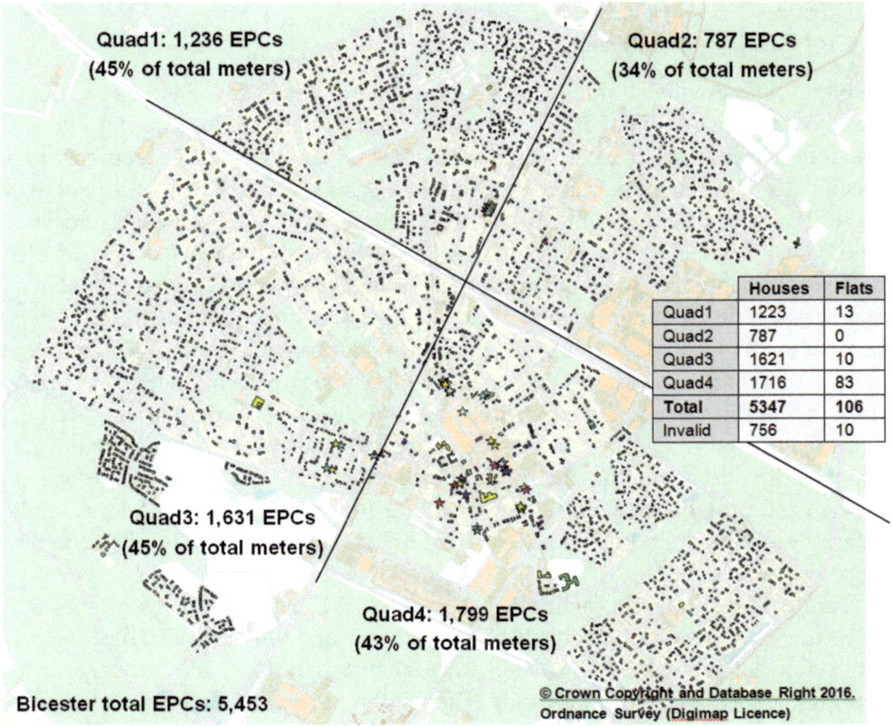

	Houses	Flats
Quad1	1223	13
Quad2	787	0
Quad3	1621	10
Quad4	1716	83
Total	**5347**	**106**
Invalid	756	10

Fig. 15.2 EPC dwellings identified

Though EPC data is considered the best available option, there are limitations:

- As mentioned above, the dataset required cleansing.
- EPCs are valid for up to 10 years (BRE 2012), i.e. some EPCs can be out of date.
- Because a full Standard Assessment Procedure (SAP) for EPC ratings is too complex, the Reduced Data SAP (RdSAP) was developed to rate existing dwellings for this purpose. RdSAP reduces data collection and deduces a large amount of missing data. However, EPCs are required to be within ±5 SAP points (assessed through quality monitoring) (BRE 2012). Examples of calculations for data reduction:
- The model assumes that occupants heat their houses to 21 °C (living rooms) and 18 °C (other rooms). However, many households are likely to heat their homes to different temperatures (CSE 2015).
- Appliance and hot water requirements are made using simplified equations relating to the number of people in a household (CSE 2015).
- EPCs represent only the dwelling; occupant behaviour, living patterns, and economic status can greatly affect real consumption (Sunikka-Blank et al. 2012).

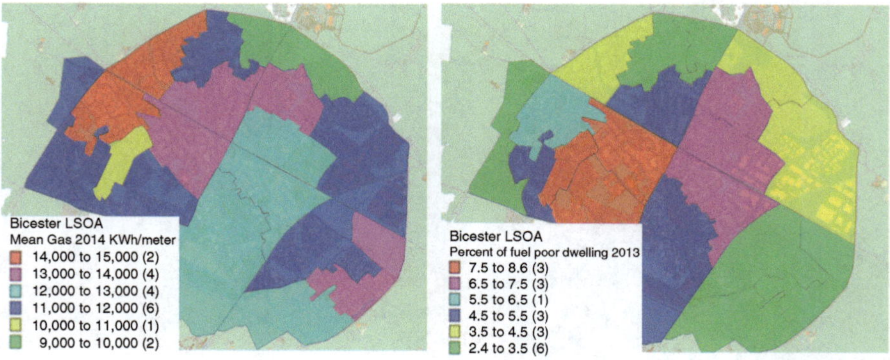

Fig. 15.3 Mean gas consumption by LSOA (*left*); percent of fuel poor dwellings by LSOA (*right*). Crown copyright and Database Right 2016, Ordnance Survey (Digimap Licence)

15.3 Research Results

The results provided a more complete picture of the areas and homes with higher energy consumption, fuel poverty, and of which homes could benefit from specific retrofit measures. The analysis enabled the isolation of the area for focus in most need of energy reduction and fuel poverty relief.

15.3.1 Consumption and Fuel Poverty

Based on the sub-national data at LSOA level, Quad 3 was found to have one of the two highest mean gas consumption zones (coloured red and purple—Fig. 15.3, left) and one of the two highest mean electricity consumption zones (not shown). Quad 3 also had the three highest values of fuel poor dwellings percent (coloured red—Fig. 15.3, right).

15.3.2 Dwelling Characteristics

There are almost 500 uninsulated cavity wall dwellings and almost 250 uninsulated solid wall dwellings in need of insulation in the entire EPC dataset for Bicester.

- Quad 3 has the most uninsulated cavity wall houses (198, 40% of all uninsulated cavity walls in Bicester); 47 solid wall houses with no insulation (21% of all in Bicester).
- Quad 4 has the most uninsulated solid wall houses (130, 58% of all uninsulated solid walls in Bicester); but overall less total uninsulated dwellings than quad 3.

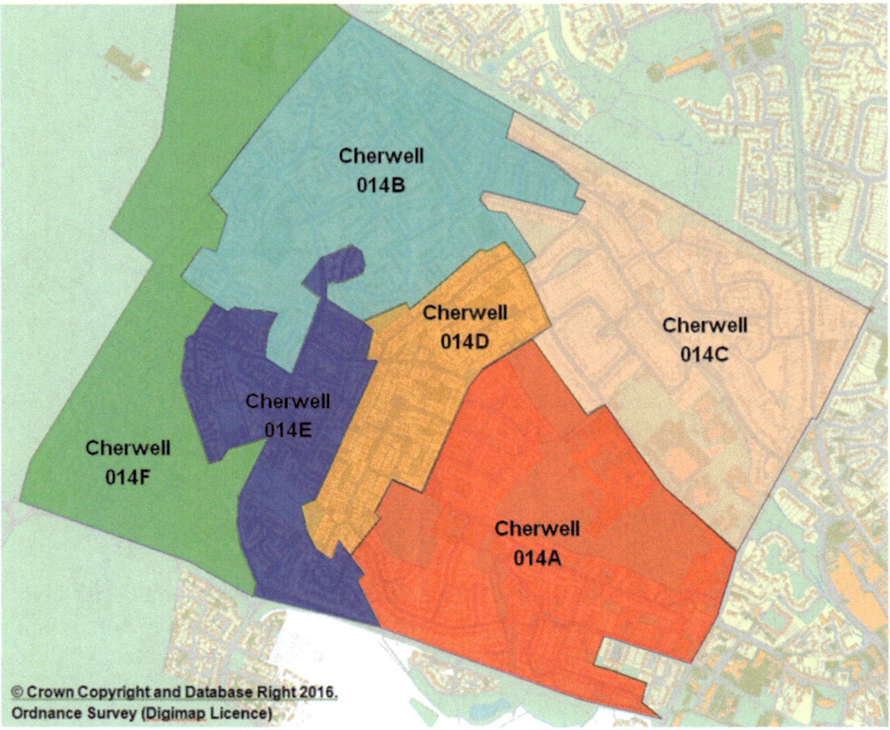

Fig. 15.4 LSOA division of Quad 3

Over 50% of the dwellings with known roof insulation levels in the EPC dataset for Bicester have less than or equal to 150 mm of roof insulation; these dwelling could potentially double their insulation levels.

- Quad 3 represents 34% of uninsulated.
- Quad 4 represents 44% of uninsulated.

Though Quad 4 indicated some dwellings with higher levels of missing insulation, Quad 3 is still important due to the higher levels of energy consumption and fuel poverty. As the analysis focusses in, Fig. 15.4 shows the LSOA divisions of Quad 3.

Figure 15.5 indicates that in Quad 3, LSOAs 014B, 014C, and 014D have the greatest energy consumption per dwelling area. Particularly, LSOA 014C has the greatest percentage of dwellings with energy consumption above 300 kWh/m^2.

When the entire EPC dataset for Bicester was compared to the entire town of Bicester's sub-national figure, the values were only off by ~800 kWh. On the other hand, at a house level, there appears to be an overestimate of between 3000 and 4000 kWh/yr. in the mean energy figure for the EPCs, as compared to sub-national data.

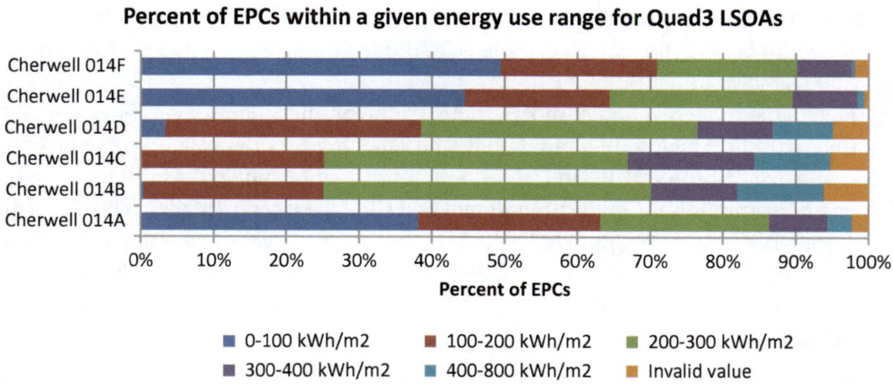

Fig. 15.5 Percent of total energy consumption for each Quad 3 LSOA

Table 15.1 Three LSOAs compared with EPC data

	Cherwell 014B	Cherwell 014C	Cherwell 014D
No. of meters (electricity)	636	691	489
No. of EPCs	212	231	206
% of area covered by EPC	33	33	42
	kWh		
Mean total energy consumption (sub-national)[a]	19,452	17,301	14,761
Mean total energy consumption (EPC)[a]	22,645	21,208	17,556
EPC overestimate (kWh) and (%)	+ ~3100 (+14)	+ ~3900 (+18)	+ ~2800 (+16)
No. of EPCs with mean total energy consumption over mean sub-national energy consumption value	**140** (66% of EPCs in LSOA)	**153** (66% of EPCs in LSOA)	**143** (69% of EPCs in LSOA)
Bicester mean 2014 (sub-national)[a]	16,181		
Bicester mean EPCs[a]	16,929		
Total mean EPC for the three LSOAs[a]	20,470		
Average UK consumption[a] (DECC 2015c)	**2012**	**2013**	**2014**
	19,841	19,581	16,406

[a]Consumption figures include both gas and electricity

Table 15.1 shows statistics for the three LSOAs with the greatest energy consumption. According to the comparison of EPCs with the LSOAs, there appears to be an overestimate of between 3000 and 4000 kWh/yr. (16%) in the mean energy figure for the EPCs.

15.4 Discussion

Based on the results above, Cherwell 014C is selected as a centre of detailed dwelling focus for the following reasons (based on its relationship to the other quad 3 LSOAs): highest percentage of dwellings with energy consumption >300 kWh/m^2/yr.

(EPC), most dwellings in need of solid wall insulation (EPC), second highest mean total energy consumption (sub-national), and third highest percentage of fuel poor dwellings (sub-national).

Due to the modelled nature of EPC data, the incomplete dataset was not expected to align with the (metered) sub-national data. This expectation was found to be correct; however, LSOAs 014B and 014C both represent the highest overall total mean energy consumption for both sub-national and EPC datasets in quad 3. The fact that the three LSOAs, evaluated in Table 15.1, all have EPCs greater than the sub-national figures suggests that the EPCs may overestimate energy consumption. Also, however, a few stipulations should be considered with regard to the mismatch in the datasets:

- The EPC dataset can have EPCs up to 8 years old at this point. As compared to the sub-national energy data which is for a single year, 2014. It is possible that dwellings with old EPCs could have made upgrades and reduced consumption. In addition, dwellings could have changed occupants (including change in tenure, family size, behaviour) over time since the EPC was registered which can change the consumption results.
- The EPC data are calculated using a reduced model whereas the sub-national energy data are annualised estimates of consumption for all Meter Point Reference Numbers (MPRNs) in the specified sub-national boundary (DECC 2015b)
- The sub-national domestic energy datasets are only available for gas and electricity; whereas EPCs can include oil, coal, and biomass heated dwellings.
- Due to the nature of the EPC process (only required when a dwelling is built, sold, or rented (GOV.UK 2015)), EPCs only represent approximately one-third of the addresses (i.e. electricity meters) in each LSOA.

Nonetheless, the EPC dataset is the most detailed publically available option for displaying energy-related aspects for the domestic sector at a dwelling level and is deemed useful for decision-making in the retrofit process.

As indicated by the findings of the Association for the Conservation of Energy (2015), fuel poverty and high energy consumption were expected to align when considering consumption from a dwelling efficiency standpoint. Though the third quadrant could be isolated as having both the highest energy consumption and fuel poverty in Bicester, Fig. 15.6 indicates that very little (albeit positive) correlation ($r = 0.11$ and 0.12, respectively) can be seen between these two indicators when all LSOAs in Bicester are evaluated (Fig. 15.6). In this case, LSOA may be too large of sub-region to be comparing fuel poverty and consumption, as LSOA areas can cover at least 2000 dwellings (DECC 2015b) and represent a large variation in dwelling types and housing tenure. It may be that a smaller area such as postcode would reveal a greater correlation. This would likely be explored as the research focusses in on more detailed, bottom-up analysis.

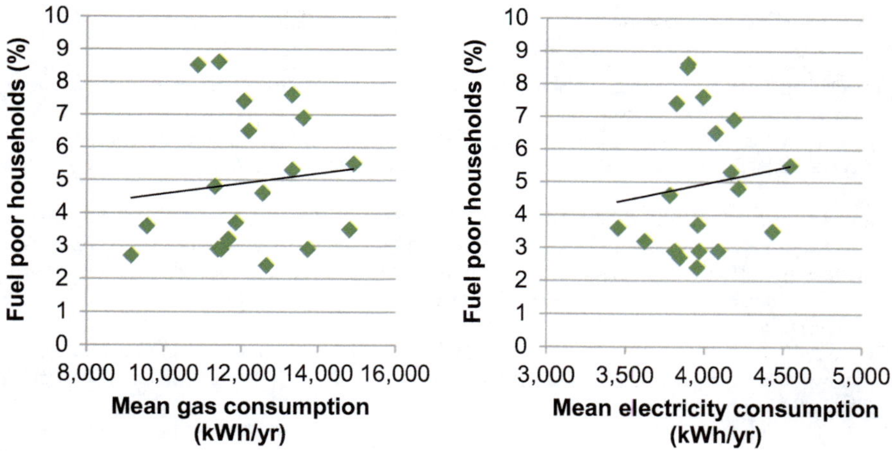

Fig. 15.6 Percent of fuel poor dwellings and energy consumption for each LSOA in Bicester

15.5 Conclusion

The chapter demonstrated the process of isolating an appropriate area of Bicester for detailed mapping and potential retrofit focus based on the high energy consumption, high rate of fuel poverty, and insulation need. The challenge of incomplete data on which homes could benefit from which retrofit measures is met with the EPC dataset through the provision of data on wall types (with or without insulation), roof insulation thickness, glazing type, heating system type, quantity of low energy lights, etc. Each of these dwelling characteristics can be mapped for a specific area including the entire town allowing an individual measure based or package retrofit focus on a mass scale. The above information also assists in the aggregation of private sector housing retrofit activities to minimise installation costs.

Though the chapter has shown how publicly available datasets on housing and energy can be used to plan mass retrofit and provide targeted low carbon measures across a city, there is the current challenge of lack of policy support for retrofit (Sellwood 2015). Nonetheless, the access to data and the process provided would allow retrofit/Energy Company Obligation (ECO) providers to target specific areas for mass retrofit based on these datasets. The process, as demonstrated is relatively simple as long as an organisation interested in performing the task is working with, or for a local authority to gain licence and data access. As EPCs have been registered across the country and likewise sub-national data are available throughout, this research could easily be performed for any city or county throughout the UK.

Following this stage in the project, the identified area of study will undergo bottom-up carbon mapping assessment which rapidly evaluates a selected area on a house-by-house level in order to improve the large-scale access to affordable retrofit with the end goals to improve energy efficiency of the housing stock, reduce CO_2e emissions and fuel poverty, and create long-term, high-quality jobs.

References

Amecke, H. (2012). The impact of energy performance certificates: A survey of German home owners. *Energy Policy, 46*, 4–14.

Association for the Conservation of Energy [ACE]. (2015). *The cold man of Europe—2015*. Association for the Conservation of Energy. Retrieved May 17, 2016, from http://www.ukace.org.

Building Research Establishment [BRE] (2012) *Energy Performance Certificates for existing dwellings: RdSAP manual, Version 8.0*. Retrieved May 17, 2016, from http://www.bre.co.uk.

Christensen, T. H., Gram-Hanssen, K., de Best-Waldhober, M., & Adjei, A. (2014). Energy retrofits of Danish homes: Is the Energy Performance Certificate useful? *Building Research & Information, 42*(4), 489–500.

Centre for Sustainable Energy [CSE] (2015) *Mapping energy performance certificate data by parliamentary constituency: Feasibility report to Citizens Advice*. Retrieved March 7, 2016, from https://www.cse.org.uk.

Davis, P. T., McCord, J. A., McCord, M., & Haran, M. (2015). Modelling the effect of energy performance certificate rating on property value in the Belfast housing market. *International Journal of Housing Markets and Analysis, 8*(3), 292–317.

DCLG. (n.d.). *Domestic Energy Performance Certificate Register*. Department for Communities and Local Government. Retrieved March 7, 2016, from https://www.epcregister.com.

DECC. (2015a). *Fuel poverty sub-regional statistics*. Department of Energy & Climate Change. Retrieved March 7, 2016, from https://www.gov.uk.

DECC. (2015b). *Sub-national consumption statistics: Methodology and guidance booklet*. Department of Energy & Climate Change. Retrieved March 7, 2016, from https://www.gov.uk.

DECC. (2015c). *Average domestic gas and electricity consumption, UK, 2008 to 2014, table 3.07*. Department of Energy & Climate Change. Retrieved March 7, 2016, from https://www.gov.uk.

DECC. (2016). *Sub-national consumption statistics*. Department of Energy & Climate Change. Retrieved March 7, 2016, from https://www.gov.uk.

DECC. (n.d.). *About the National Heat Map*. Department of Energy & Climate Change. Retrieved March 7, 2016, from http://tools.decc.gov.uk.

GOV.UK. (2015). *Buying or selling your home*. GOVE.UK. Retrieved June 17, 2016, from https://www.gov.uk/buy-sell-your-home/energy-performance-certificates.

Gupta, R., & Gregg, M. (2012). Using a mapping-based simulation approach to rapidly investigate the potential for adapting English homes for a warming climate. In Z. Zhai, X. Li, & H. Wang (Eds.), *Proceedings of the Second International Conference on Building Energy and Environment (COBEE 2012). 1–4 August 2012, Boulder, Colorado, USA*. pp. 744–751.

Gupta, R., & Gregg, M. (2014). A quiet revolution: Mapping energy use in low carbon communities. In R. Rawal, S. Manu, N. Khadpekar (Eds.), *Proceedings of the 30th International PLEA Conference: Sustainable habitat for developing societies—Choosing The Way Forward; December 16–18, 2014; Ahmedabad, India.*

Gupta, R., Barnfield, L., & Hipwood, T. (2014). Impacts of community-led energy retrofitting of owner-occupied dwellings. *Building Research & Information, 42*(4), 446–461.

HM Government. (2011). *The carbon plan: Delivering our low carbon future*. London: DECC.

Mastrucci, A., Baume, O., Stazi, F., & Leopold, U. (2014). Estimating energy savings for the residential building stock of an entire city: A GIS-based statistical downscaling approach applied to Rotterdam. *Energy and Buildings, 75*, 358–367.

Palmer, J., & Cooper, I. (2013). *United Kingdom housing energy fact file*. London: DECC.

Pereira, I., & Assis, E. (2013). Urban energy consumption mapping for energy management. *Energy Policy, 59*(2013), 257–269.

Sellwood, P. (2015) *Energy Saving Trust CEO responds to Green Deal cuts*. Energy Saving Trust. Retrieved March 7, 2016, from http://www.energysavingtrust.org.uk.

Sunikka-Blank, M., Chen, J., Britnell, J., & Dantsiou, D. (2012). Improving energy efficiency of social housing areas: A case study of a retrofit achieving an "A" energy performance rating in the UK. *European Planning Studies, 20*(1), 131–145.

Wade, J., Eyre, N., Hamilton, J., & Parag, Y. (2013). Local energy governance: communities and energy efficiency policy. In T. Lindstrom (Ed.), *eceee Summer Study proceedings, 2–7 June 2013, Belambra Les Criques, Presqu'île de Giens, Toulon/Hyères, France*.

Zhou, Y., Weng, Q., Gurney, K., Shuai, Y., & Hu, X. (2012). Estimation of the relationship between remotely sensed anthropogenic heat discharge and building energy use. *ISPRS Journal of Photogrammetry and Remote Sensing, 67*(2012), 65–72.

Part IV
Smart Construction

Chapter 16
Things Change: Exploring Transformational Experiences Within the UK Construction Industry

Alison Pooley

Abstract Our built environment has a significant impact on energy consumption, resource depletion, and ecological degradation—reducing this impact is imperative. Existing policies and research are dominated by the assumption that increased regulation, and an improvement in professional skills and knowledge, will address these issues. Conventional attempts at improving dissemination of good practice have been found wanting as the construction industry makes slow progress towards environmental responsibility. Environmental responsibility is defined by the author as being accountable for one's actions that in turn affect the conditions under which life is developed.

This chapter argues for looking beyond a technical or regulatory fix, by exploring the potential opportunities for change that lie within the relationships between experience, learning, and the transformation of individual and professional perspectives. The emphasis of the research is placed on examining current practice in order to respond to profligate energy and material use, whilst addressing the wider ethical and environmental responsibilities of equity and fairness. The chapter outlines research based on twenty-two intensive interviews with individuals who worked on building projects with a clear commitment to an environmental agenda, beyond that required by building regulation. Individual experiences are explored through nine key emergent themes, which in turn inform opportunities for future transformations in perspective and practice.

Understanding individual learning is critical in creating future environments, where organisations and individuals can experience a transformation beyond acquiring skills and knowledge. Building on existing theories of transformative learning, social learning, and pro-environmental behaviour, this research informs the role non-formal learning can play in developing strategies orientated towards

A. Pooley (✉)
Department of Engineering and the Built Environment, Anglia Ruskin University, Bishop Hall Lane, Chelmsford CM1 1SQ, UK

Welsh School of Architecture, Cardiff University, King Edward VII Ave, Cardiff CF10 3NB, Wales, UK
e-mail: alison.pooley@anglia.ac.uk

M. Dastbaz et al. (eds.), *Building Information Modelling, Building Performance, Design and Smart Construction*, DOI 10.1007/978-3-319-50346-2_16

sustainable development and aims to engender a deeper understanding of the potential for learning and transformation within the construction industry, the emphasis being on greater environmental responsibility.

16.1 Introduction

> The best way to understand is to do. That which we learn most thoroughly, and remember the best, is what we have in a way taught ourselves (Kant 1900:80).

The debate regarding resource depletion, environmental degradation, greenhouse gas emissions, and climate change continues to gather pace, and many now see the threat of climate change and associated feedbacks as *the* imperative facing humanity, and one that we remain reluctant to address (Hillman 2014; Marshall 2014). The transformation of the construction industry is of utmost importance in these times of ecological fragility and increasing certainty of the anthropogenic impact on climatic behaviours (IPCC 2013). Towards the end of the last century, much of the debate around environmental performance of buildings related to energy efficiency and conservation, with legislation mainly addressing new build projects. Whilst retaining a strong focus on energy conservation, the emphasis has shifted from new buildings to refurbishment of existing with projects such as Retrofit for the Future (TSB 2014) and industry awards specifically targeting retrofitting projects (AJ 2016). Focusing on energy conservation and related reductions in emissions remains an imperative, however leading the drive for greater environmental responsibility solely through regulation is neglecting the motivations that lie behind behavioural change and the desire to construct in an environmentally responsive and responsible way. As Moore (2010:9) argues "unsustainability is not a scientific or technological problem it is a social one".

Building on recent quantitative studies, this research emphasises the need for the construction industry to respond to the environmental agenda in ways that allow for learning and transformation beyond skill acquisition (Hartenberger et al. 2012:61), exploring how environmental practice within the construction industry can be influenced through changes in individual practice by understanding "more fully the essences and meanings of human experience" (Moustakas 1994:105). The research moves away from a focus on energy consumption, legislation, and regulation, towards examining individual values and commitment, acknowledging that where individual values are influenced by experience and learning this creates an opportunity for transformation. The concept of transformation adopted in this research is largely informed by the early work of Mezirow (1991) and the subsequent dialogues within transformation theory that reach beyond cognitive processes into deeper learning experiences (Cranton 1996; Dirkx 2006). Transformation theory and transformative learning are not adopted wholesale but are used to inform a position and are examined alongside other adult learning theories relating to workplace learning, and learning as a social experience (Illeris 2011; Jarvis 2012).

This research places the construction industry in a specific framework hinged around three main realms of investigation: environmental responsibility, learning,

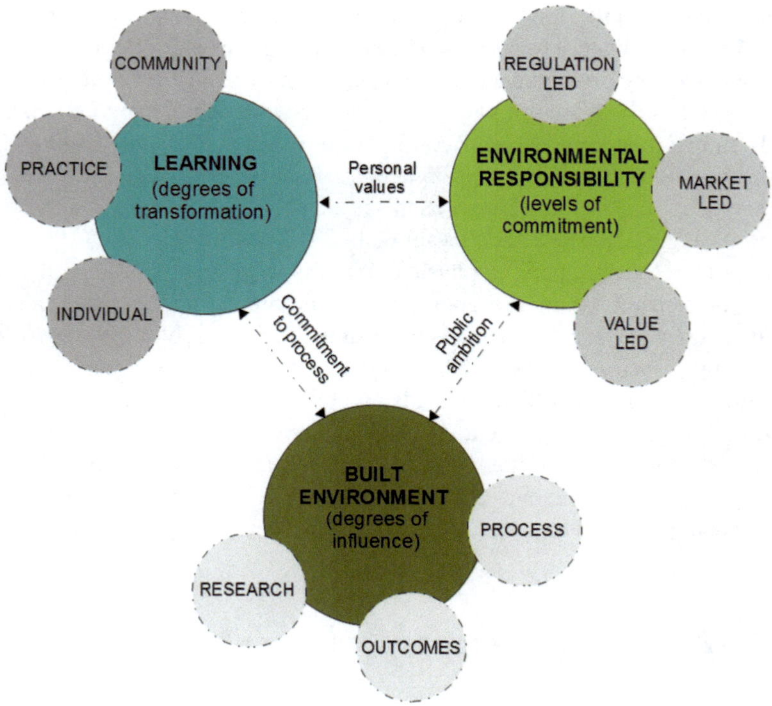

Fig. 16.1 Realms of investigation, spheres of influence, and connecting motivations

and the built environment. The interconnected nature of these three realms is described in Fig. 16.1 below. This model starts to suggest potential dynamic links between personal values, commitment to processes, and perceived ambitions and motivations. Each realm has associated spheres of influence exerting a force on practice. As an example, a level of commitment to environmental responsibility can be exerted through regulation—in the form of building control, by economics—in the form of market forces, or by values—the ecological values held by a client. The model is interpretive through which connections and relationships between the defined realms can be explored.

16.2 Defining the Realms

16.2.1 Environmental Responsibility and Levels of Commitment

Communicating our environmental predicament has preoccupied activists, researchers, and campaigners for many decades. The question of how to engage in dialogue about our environmental impact remains a current one (Marshall 2014). Hillman and Fawcett (2004) has long argued that our "predicament" is so catastrophic that we need to re-evaluate what scares us more, losing our way of life or losing the planet. He has long promoted a top-down approach to change by calling for carbon emissions to be controlled through international governmental intervention, as individuals en masse are incapable of coming to the "right" decision (Hillman 2014). Communicating the need for environmental responsibility is complex, we cannot scare people into behaviour change, as this is unsuccessful and can, in some circumstances, have the opposite effect (Evans et al. 2013; Maio et al. 2001).

Fox (2009:16) argues the biophysical realm is more important than the human realm, which in turn is more important than the human constructed realm. This leads us to question whether humans have an inalienable right use the biosphere to feed/fuel the constructed realm. The human right to life compromises nearly all other life on the planet, which demands an ethical, value led consideration of the rest of the "natural" environment. It is argued here that the current unsustainable situation in terms of the built environment is a moral and social one, rather than a technological or scientific one. Our approach towards environmental responsibility is influenced by our values, as well as market and regulatory pressures.

16.2.2 Learning and Degrees of Transformation

Learning occurs throughout and across our lives and is not confined to formal learning contexts such as school, college, and university. Learning as an adult is likely to occur in more informal or non-formal situations, taking place as part of everyday life and outside of any intentional process (Alheit 2009:117). This research adopts an approach to adult learning, and in particular learning within the workplace, where experience and context is critical (Illeris 2011). Learning in this sense is influenced by our community, our practice, and our individual actions and is not confined to formal learning situations or established routes to professional qualification (Eraut 2000).

Understanding and promoting transformative learning outside of formal education is under-researched. There is a perceived lack of connection between what is discussed as theory and what is applied in practice (Fisher-Yoshida et al. 2009:1). All learning has the potential to be transformative, and whilst there are shared concepts it is important to draw a distinction between transformation as change, and transformational learning as a theoretical basis for perspective transformation (Mezirow 1991). Kegan suggests "as the language of transformation is more widely assimilated, it risks losing its genuinely transformative potential" (Kegan 2009:41). Learning here is concerned with the developing theories of transformation and social learning and the ability for learning to engender change (Dirkx et al. 2006; Wals and van der Leij 2007), as well as aiming to develop the debates centred on built environment education and professional practice in the construction industry. In this sense, learning is influenced by our community, professional practice, and individual action.

16.2.3 Built Environment

This research questions the current problematic issues within the built environment professions where profligate energy and material use is often rewarded and awarded (Hatherley 2014), and where wider ethical responsibilities appear to be neglected for short-term gain. Issues of inconsistent performance and lack of dissemination and limited adoption of good practice, form a vision of an industry that is lacking meaningful acknowledgement of its wider ethical impacts. Given an increasingly urbanised population, it is vital to address the environmental impacts of this realm (Girardet 2008) through processes, outcomes, and future research. An example is the increasingly problematic issues around building performance where divergent assessment outcomes at design and occupation stage reveal a huge discrepancy in predicted and actual energy performance (Bordass and Leaman 2012). Predicting a building's impact on the environment and using that to inform and transform industry practice becomes all the more elusive. The few existing exemplar projects appear to have had little influence on the construction industry as a whole (Tofield 2012).

Regulation has gone some way in changing the industry to rely on it alone will never be enough to implement the deep transformation required. Additionally, external drivers for industry change are unstable and subject to short-term political and economic forces that are beyond the control of the individual. As Pultar (2000:157) suggests, the industry is under pressure to respond to the forces exerted upon it at all stages, by both environmental factors and cultural factors. The established strategies to develop environmental responsibility within the industry remain the acquiring of skills and knowledge, rather than time spent exploring ethical frameworks or how future practice might be informed by past experiences. As Wals and van der Leij argue:

> What is clear by now is that to break deeply entrenched, unsustainable patterns (assumptions, behaviours and values) requires a new kind of thinking inspired and informed by powerful learning processes (Wals and van der Leij 2007:17)

The powerful processes talked about above could encapsulate reflecting on and learning from experience, leading to a potential perspective transformation, the realm in which this research sits.

16.3 Research Method

> To be human is to engage in relationships with others and with the world. It is to experience that world as an objective reality, independent of oneself, capable of being known (Freire 2008:3).

The research is based on the experiences of twenty-seven construction industry professionals, explored through twenty-two intensive semi-structured interviews. The interviews were conducted in two phases: phase 1 consisting of eighteen interviews with the architects, engineers, clients, and contractors who worked on one of the four designated building projects, plus a further five interviews with architects working on projects with a clear environmental agenda. The five interviews with architects formed the scoping projects for the research. Phase 2 consisted of interviews which

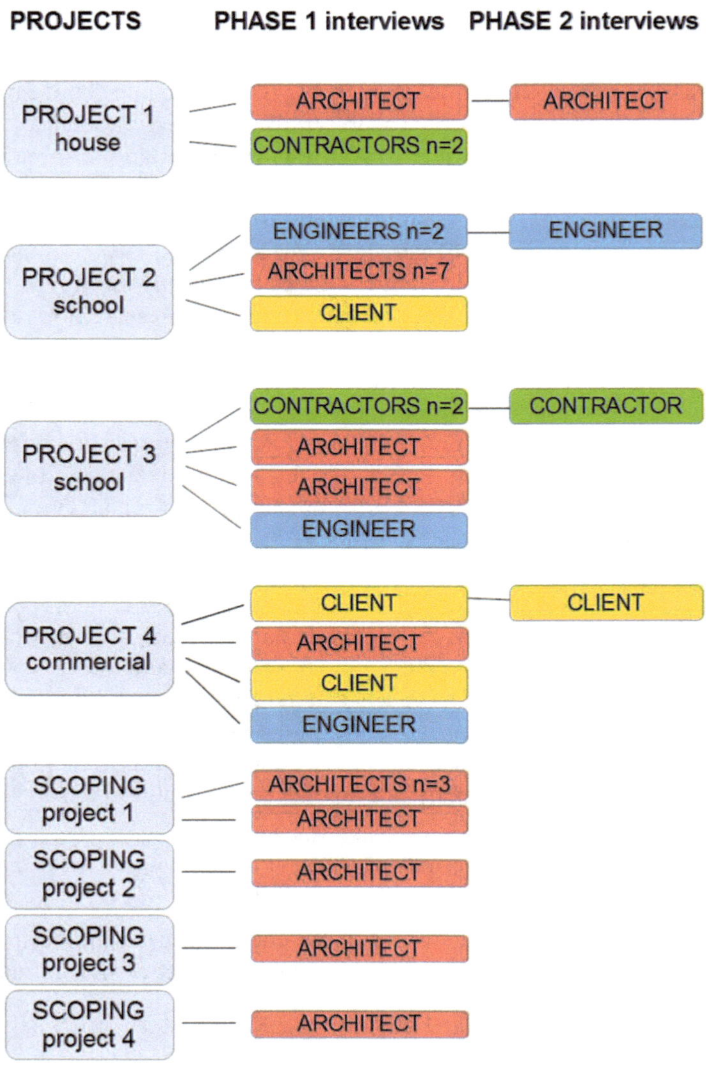

Fig. 16.2 Research interviews—main projects, scoping projects, and professions

took place three years after phase 1, where four of the original participants further reflected on the project they had worked on and the industry as a whole. Each phase 2 participant was selected to represent one of the four building projects and one of the four industry professions (see Fig. 16.2).

The four building projects used in the main body of the research—one commercial, one residential, and two educational buildings—were all completed post 2008 and selected for their building performance assessment at design stage as well as wider environmental ambitions. The architect(s) for each of the four projects were identified as the primary lead, and they then provided contact information for subsequent interviews with

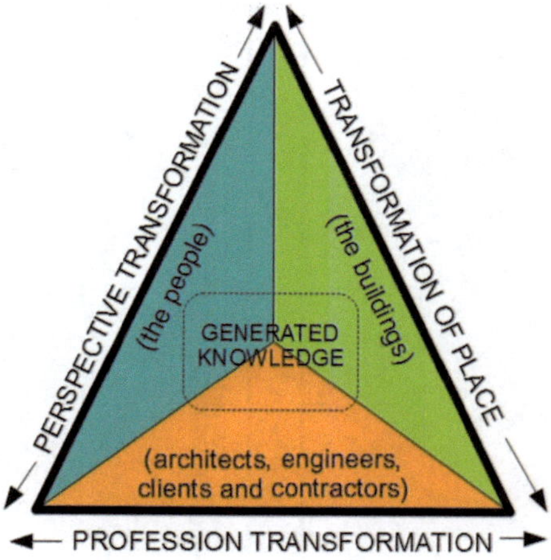

Fig. 16.3 Analysis of the generated knowledge from three perspectives

the client, contractor, and engineer(s).[1] Thus, making use of what Creswell (2013) refers to as chain sampling, with each interview leading to the next link.

The interview transcripts and subsequent analysis form the generated knowledge which is rooted in reflection on experience, with the participants bearing witness to change within the industry over an extended period of time. Seamon (2000) emphasises the nature of environmental behaviour and experience, arguing that a phenomenological approach is an innovative way for looking at the person–environment relationship. The professionals interviewed had thought about their work, the processes, and the project, but not necessarily why they were doing it and what engaged them in that project. Engaging in reflective opportunities or events tends not to be focused on in the construction industry as "people are immersed in a world that normally unfolds automatically" (Seamon 2000:5). Learning does take place but without being given importance outside of skills acquisition.

The individual experiences—captured through the interviews and transcribed—were then analysed from three perspectives: the individual, the profession, and the building project (see Fig. 16.3). Annells (2006) suggests *adopting three theoretical approaches as a way of redefining the tension* that can exist between the role of the researcher and the generation of text. This approach was reinterpreted as an exploration of the complexity of the experience of built form and arises from the equally complex experiences explored within the interviews. Each rereading of a transcript or relistening of an interview was through a different lens, using each lens as an interpretive tool. In this sense, there is a three-way approach to analysis and interpretation of the generated text.

The emergent themes capture the participants' experiences as they relate to the building project and wider professional experiences of the industry as a whole; these nine themes are gathered under three key descriptors where each theme is mapped against each lens used in the analysis (see Fig. 16.4).

[1] Engineers include services engineers (also referred to as mechanical and electrical engineers) and structural engineers.

OPPORTUNITIES FOR TRANSFORMATION: EMERGENT THEMES MAPPED AGAINST EACH ANALYTICAL LENS – PERSON, PROFESSION, PROJECT

	COMMITMENT			COLLABORATION			COMPLIANCE		
	THE DEEP END	JUST WHO I AM	DOING THE RIGHT THING	SHARING AMBITIONS	FEELING AND BEING SEPARATE	VALUING REFLECTION	PUSHING BOUNDARIES	TICKING BOXES	ANTICIPATING CHANGE
PERSON (the people):	personal experience and those of others count	personally held values and attitudes impact environmental behaviours	deontological ethics – society and duty	working with like-minded people	not fitting in with the perceived norm	opportunity to reflect on work	challenging a business as usual approach	compliance is not enough	future directions and diversification, learning and new challenges
PROFESSION (architects, engineers, clients, contractors):	learning through doing – formal and non-formal routes	professional body regulation and alignment with personal values	adhering to a code of conduct and working beyond it	cross and interdisciplinary boundaries challenged – a no blame culture	frustration with professional practice and teaching	embedding reflective practice in the everyday	new ways of working across disciplines	compliance and creativity – pushing for change	professional roles changing and being challenged
PROJECT (the buildings):	invites questioning and challenges learning	project as learning rather than profit – challenges planning and preconceptions	capacity for responsive cohesion, performs according to design	project embodies physical evidence of aims	aesthetics and sense of place	importance of post occupancy evaluation	challenging the norm – site specific	compliance with regulation and assessment	future proofing fabric and services

Fig. 16.4 The nine emergent themes mapped through each lens against research outcomes

16.4 Research Outcomes

Examining commonalities through colligation, the nine themes represent interwoven axiological tendencies threading through the research and are described briefly and illustrated with extracts from the research interviews below.

16.4.1 The Deep End

The inference being that you learn rapidly when confronted with a new situation, such as being thrown into the deep end of a swimming pool. Illeris (2011) argues that being engaged in challenging work creates engaged learners who learn more and better; however, challenging work can also produce stress or a feeling of lack of control. In some interviews, these experiences were remembered as difficult to endure or cope with but when reflected upon a deeper and more valued experience became evident. As one architect recalled when the roof leaked on their first project:

> suddenly then I thought this isn't, this isn't fun, this isn't something which you put a picture on the wall and you take it down after six months when you're bored with it, it's there for good [...] these buildings that you're creating, you're putting people in to work or to live (Architect).

Here, the external factors of a leaking roof work with internal emotions and values to transform thinking and heighten a sense of responsibility (Hostetler 2011). The deep end creates an opportunity for learning and transformation.

16.4.2 Just Who I Am

Participants were invited to consider how their motivations for environmental actions were influenced, the answer was often that it was just part who they were, that environmental issues had always been important to them:

> it's just an inherent characteristic that I've got I think, it just so happens that it aligns with what [...] the world is facing at the moment (Engineer/services)

This notion of it "just" being part of who they were was a strong thread through the conversations. Acting on intuition is valuable, and having value led action, what Freire (2008) refers to as praxis, appears significant in learning and transformation (Schapiro 2009:103). However, if the profession becomes overly reliant on lone environmental champions there is a danger those individuals may become sidelined, appear as outsiders or environmental pariahs.

16.4.3 Doing the Right Thing

In many ways, this underpins the previous two themes as it embraces ethical motivations. The "right thing" is open to question and is not confined to environmentally responsible behaviours in an ecological sense, but expands through the participants comments to include social responsibility:

> I think sometimes that [pause] perhaps we don't think enough about the, the people who are in that environment, you know the people (Contractor/site)

When environmental performance or the "right thing" is demanded by building users, this can force development in a particular direction. However depending on building occupants, users, or tenants to change the direction of the industry cannot be relied on as a single strategy, as motivations can be fickle, and dependent on economic fluctuations and cultural aspirations. One contractor comments on the importance of individual action:

> it takes people who want to change things, who want to challenge it to make it happen […] there needs to be a bit of a change in mentality I suppose but you know if you're going to do it you could do it for the right reasons (Contractor/site)

16.4.4 Sharing Ambitions

A key element in construction is clear communication: the clients, architects, contractors, and engineers interviewed for this research all talked about communication being at the root of a successful project. This was not confined to communication between team members but included communication and cooperation on an industry-wide basis. Working with a sense of shared environmental ambition was acknowledged as increasingly realisable as awareness shifts and the industry slowly transforms. Specifically for one architect in a phase 2 interview, there was no longer a need to "stand over" contractors as there had been in the phase 1 interview. There is recognition of a transformation in the three intervening year—in ambition and action, and a reversal of prior roles as learning flows both ways:

> there are a number [of smaller contractors] we've worked with now who really are teaching themselves stuff [...] they're actually quite keen to join in and they want to be involved and want to try things and want to learn about things […] there are even a couple who [are] possibly teaching us stuff (Architect)

16.4.5 Feeling and Being Separate

Participants frequently identified with being outside of the "norm" of the profession, actively separating from a business as usual approach. Contractors on one project expressed a desire to separate themselves from "the rest" of the industry; this need was rooted in a dissonance with professional practice:

[as contractors] we aren't the typical, normal of the building or construction industry, so I mean we are quite, we are willing to learn and quite interested in changing (Contractor/ engineer)

These contractors consciously operate on a level they view as being beyond the rest of their profession, at times this was expressed as frustration with the slow rate of change. Other participants referred to this separation as being at the lead, the forefront or outside of the mainstream.

16.4.6 Valuing Reflection

This refers to the value placed by participants on the opportunity to reflect on their work and experience. This theme focuses less on the theory of reflection within professional practice and more on the importance of looking back, thinking, and talking about a completed project. For the participants, the experience of reflecting was a valid process in and of itself, albeit one that involved a reassessing of action and perspective. There are explicit statements within the interviews commenting on the value of the process of looking back over a project, sharing and discussing experiences more widely:

we have talked about doing this in a more formal way and I think it would be good to do that actually, you know lessons learnt […] you do take lessons and you do incorporate it but you don't necessarily incorporate, I don't necessarily incorporate what [a colleague] learnt I incorporate what I learnt (Architect/senior)

16.4.7 Pushing Boundaries

It was important for all participants to push the agenda forward, an example of this professional positioning comes from one of the clients interviewed:

I sort of describe the green agenda as being a bit like a comet […] a comet is effectively a body of energy that's pushing along, and at the front of that comet is somebody or something that's constantly getting bashed up and beaten up […] then there's a ball of energy behind it, and then behind that there's a whole load of hot air and dust and rubbish, and we always wanted to be not in the front, and not in the rubbish at the back, but in that body of energy pushing it forward (Client/developer).

Regardless of the advances in the industry, these were not enough for the client, this desire to push forward with techniques and approaches ran through the interviews, either as a personal motivation or a professional commitment, this pushing is vital for change as for this client the industry largely remains:

a bunch of hairy arsed blokes putting things together […] and construction has always been like that (Client/developer)

16.4.8 Ticking Boxes

As building regulation becomes more demanding in terms of environmental performance building assessment tools and benchmarking may become less relevant (Bordass and Leaman 2012; Tofield 2012). One interview with a contractor indicated a change in attitude towards a particular assessment method (BREEAM[2]). On a recent project, they had chosen to work "in the spirit of BREEAM" rather than pay to go through the assessment process, thus saving their clients the additional cost:

> we are designing in the spirit of BREEAM very good, we're not going to pay to have it assessed [...] because we're not going to submit it to the BRE because there was a saving which the client took (Contractor/builder)

In many ways, ticking boxes is what BREEAM and other assessment methods demand. What is needed is a new way of thinking about sustainability, but in particular resources, designing and building differently, rather than the same thinking with just less "stuff". Here, the contractors move away from a tick box mentality to working with values and ambitions for a project rather than specific performance criteria.

16.4.9 Anticipating Change

During each interview, there were comments and criticisms of current practice and how it might improve as a response to environmental responsibility; sometimes, these comments were focused very specifically on materials, or systems, whilst other comments focused more generally on the industry's perception of change. Those interviewed had obviously given prior thought to the way the industry was moving, what works and what might work better to change behaviours within the industry as a whole. At the time of the phase 1 interviews, the building regulations had started to influence building services, at the time of the phase 2 interviews further revisions to the building regulations had only recently come into effect. Despite the revisions to building regulation, one concern was the retention of knowledge that as the money got tighter some of the environmental measures would be lost, as they were still perceived to be an added cost:

> the danger is you slice off that, you know, all that good work, and you end up back with [...] just purely building regs approved boxes, I think that's the danger (Client)

---

[2] Building Research Establishment Environmental Assessment Method is a tool for calculating building performance. Credits are given for various aspects rated against performance targets and benchmarks, buildings are rated in six categories, from unclassified to outstanding.

16.5 Discussion

> If tinkering reforms are not an adequate response to our plight - and they are not - we must rethink our initial assumptions about learning and the goals of education (Orr 2004:41)

The outcomes discussed here are a summary of a much larger piece of work. The outcomes are both reflective and forward casting, looking to future solutions and professional landscapes by drawing on experiences rooted in the past. The interviews encapsulate not just one day, but the past and future as understood through the present. Reflecting on a project in this way reveals how environmental values and professional actions are dependent on the shifting sands of economic fluctuations, societal expectations, and personal situations. The relationship between values, environmental behaviours, and individual change has been much researched, as Schultz et al. discuss:

> self-transcendence is activated by a general awareness of harmful consequences resulting from environmental problems and a feeling of responsibility for these problems (Schultz et al. 2005:460)

An answer to this is proposed by Hostetler (2011), who recommends living a more philosophical life, focusing on education and well-being, he advocates change within a life through learning:

> Transformation is transcendence, lifting the level of one's self. What greater opportunity is there for doing that than education, people being together in joyfully experiencing universal questions of the human condition? (Hostetler 2011:200)

Murray (2013) reports changed perspectives following workshops based on techniques used in neurolinguistic programming. The accounts of those workshop experiences share parallels with Mezirow's theory and in particular the experiencing of a disorienting dilemma (Mezirow 1991:168). These types of reflective encounters can equally produce a deeper transformation of the soul, which we may not be consciously aware off (Dirkx 2006), and encourage us to confront wider global and ecological issues. Lange (2000) highlights the importance of creating both physical and psychological spaces for transformation, and herein lies the challenge for the construction industry—where is the time in the site hut or design team meeting for reflection and challenging perspectives? Working on an environmental project enabled connections between environmental value and personal identity. As Fox (2009:23) argues, these issues need to be equally challenged through our professional roles:

> we need to be able to say: 'No, you should not want me to design this building (or this urban plan) in this way, and these are the reasons why. We need to work together to achieve a satisfactory outcome in the light of these reasons'— that sort of thing. This represents a transition in the way in which many of us—and I include myself—have seen and might like to see our professional roles

Recognition that change within the industry is necessary in order to address environmental issues is a commonality among the participants. They differ on how much of an imperative it is or needs to become and how it might be achieved. This has been widely discussed, with some calling for global changes in humanities' approach (Marshall 2014), whilst others focus on changes within education generally (Hostetler 2011), professional education more specifically (Hartenberger et al. 2012), or a change in the industry as a whole through a new professionalism (Bordass and Leaman 2012; Farrell 2014). Over the past decade, a number of reports have documented the need for the construction industry to transform; strategies indicating a way forward are often focused on training and skills and there is still a recognised gap in the skills and knowledge required by built environment professionals. Increasingly, the education of future professionals is being considered in terms of higher education and academic programmes rather than existing practising professionals. Research into ESD and embedding sustainability into the curriculum has become more established within formal education. The pan-European EDUCATE research programme (Altomonte 2010) placed an emphasis on professional knowledge of sustainability rather than values or motivations for individual and profession-wide environmental behaviours.

> [...] the industry faces a pressing need for a capable workforce that can deliver transformational change in the next decade. As the wider economy emerges from the recession, construction firms must be able to recruit and retain skilled, hard-working people in sufficient numbers to meet the increasing demand for construction. We must also be able to recruit and develop people with new types of skills (HM Government 2013:44).

There are several pressures exerting on the construction industry: more stringent legislation, increased economic uncertainty, climate change, flooding, fuel pricing, energy efficiency, and finite resources to name a few. All these have implications for a rapid paradigm shift in order to respond to the environmental imperative, an imperative for the industry to reform (Betts et al. 2013) and for the built environment curriculum to transform (Hartenberger et al. 2012).

> Systemic transformation and a change in course will occur, if at all, through changes in the attitudes of many individuals volitionally placing greater emphasis on the study, practice and living of reflective, examined lives; and in such endeavour, we are all beginners (Walker 2014:105).

There is an interdependent relationship between critical reflection and affective learning, where recognising feelings and emotions are integral aspects of learning from experience. An unwillingness to reflect and address emotions or past experience can establish a barrier to transformation.

Figure 16.5 above starts to explore how these changes might manifest through transformative opportunity by taking the emergent themes and reinterpreting them as a transformation. We need to develop ways to embrace a commitment to environmental responsibility and transformation by supporting the professions and developing what Sterling (2009:82) refers to as a "collective connective consciousness and competence". It is imperative not to confine this debate to a curriculum, but to feed into professional practice and learning in non-formal situations such as the workplace or life generally. Jarvis (2012) and Illeris (2011) highlight learning is a social action with many levels of change and transformation which neither ends nor

Fig. 16.5 Generalisations mapped with transformations

begins with formal teaching and learning. The challenge for the built environment disciplines is to move away from prescriptive silo delivery in higher education and develop environmental and ecological learning through reflective, collaborative, and participatory practices within professional learning (Jones et al. 2012).

16.6 Conclusion

> Most of us spend most of our time in, near or influenced by built surroundings. We spend our lives in what were once the thoughts of architects. Today's thoughts make the world of tomorrow—an awesome responsibility (Day 2004:283).

We need to shape the construction industry so that values can become embedded rather than relying on legislation and regulation to drive actions, which appears to be leading to a skills shortage where knowledge is acquired rather than change manifesting from within. One of the constant themes of the environmental movement has been how to get thinking and practical strategies into the mainstream. This research suggests that whilst legislation may require levels of environmental responsiveness, motivation also comes from individual environmental responsibility.

Interrogating the role learning plays in inspiring and transforming engages us in a debate rooted in environmental and social movements from the mid 1960s. Movements concerned with equity, preservation of the world's diverse ecology and resources. Reflecting on professional practice has been found to be transformative and restorative, reinforcing previously held beliefs and validating commitments to both environment and community. One of the challenges faced is using this to inform how and what we teach, how we work, and how we create environments for transcendence and transformation within the professions.

Environmental responsibility is not an adequate definition for the ambition or action required. It carries more precision than sustainable development but the argument remains as something beyond ourselves, our values and our experiences, and begins with learning in its broadest sense. We need to be accountable for our actions that in turn affect the conditions under which life is developed and to build resilience through responsibility within our professions, our communities, and our wider world.

References

AJ. (2016). *AJ Retrofit Awards*. Retrieved May 10, 2016, from https://retrofit.architectsjournal.co.uk/

Alheit, P. (2009). Biographical learning—Within the new lifelong learning discourse. In K. Illeris (Ed.), *Contemporary theories of learning: Learning theorists—in their own words* (1 ed., pp. 116–128). London: Routledge.

Altomonte, S. (2010). State of the art of environmental sustainability in professional practice. *EDUCATE* [online]. Retrieved from http://www.educate-sustainability.eu/state-of-the-art.php

Annells, M. (2006). Triangulation of qualitative approaches: hermeneutical phenomenology and grounded theory. *Journal of Advanced Nursing [online], 56*(1). Retrieved February 15, 2011, from http://dx.doi.org/10.1111/j.1365-2648.2006.03979.x

Betts, M., Robinson, G., Burton, C., Leonard, J., & Shards, A. (2013). *Global Construction 2025 Executive Summary* [online]. London: Oxford Economics & Global Construction Perspectives. Retrieved February 11, 2014, from http://www.globalconstruction2025.com/

Bordass, B., & Leaman, A. (2012). A new professionalism: Remedy or fantasy? *Building Research & Information [online], 41*(1). Retrieved July 3, 2014, from http://dx.doi.org/10.1080/09613218.2012.750572

Cranton, P. (1996). *Professional development as transformative learning: New perspectives for teachers of adults* (1 ed.). San Francisco: Jossey-Bass.

Creswell, J. W. (2013). *Qualitative inquiry and research design: Choosing among five approaches* (3rd ed.). London: Sage.

Day, C. (2004). *Places of the Soul: Architecture and environmental design as a healing art* (2nd ed.). Oxford: Architectural Press.

Dirkx, J. M. (2006). Engaging emotions in adult learning: A Jungian perspective on emotion and transformative learning. *New Directions for Adult and Continuing Education [online], 2006*(109). Retrieved March 29, 2014, from http://dx.doi.org/10.1002/ace.204

Eraut, M. (2000). Non-formal learning and tacit knowledge in professional work. *British Journal of Educational Psychology [online], 70*(1). Retrieved 5, 2015, from http://dx.doi.org/10.1348/000709900158001

Evans, L., Maio, G. R., Corner, A., Hodgetts, C. J., Ahmed, S., & Hahn, U. (2013). Self-interest and pro-environmental behaviour. *Nature Climate Change [online], 3*(2). Retrieved May 20, 2014, from http://dx.doi.org/10.1038/nclimate1662

Farrell, T. (2014). The Farrell Review of Architecture and the Built Environment [online], Farrells/Department for Culture, Media & Sport. Retrieved May 28, 2014, from http://www.farrellreview.co.uk/download

Fisher-Yoshida, B., Geller, K. D., & Schapiro, S. A. (2009). Introduction: New dimensions in transformative education In: B. Fisher-Yoshida, & K. D. Geller, S. A. Schapiro (Eds.), *Innovations in transformative learning: Space, culture, & the arts* (pp. 1–19). New York: Peter Lang.

Fox, W. (2009). Ethics, architecture, responsive cohesion, and the transition to a more habitable future. *Ethics and the Built Environment [online]*. Retrieved from http://www.warwickfox.com/files/2009_eb_nott_conf.pdf

Freire, P. (2008). *Education for critical consciousness*. London: Continuum.

Girardet, H. (2008). *Cities people planet: Urban development and climate change* (2nd ed.). Chichester: Wiley.

Hartenberger, U., Lorenz, D., & Lützkendorf, T. (2012). A shared built environment professional identity through education and training. *Building Research & Information [online], 41*(1). Retrieved July 4, 2014, from http://dx.doi.org/10.1080/09613218.2013.736202

Hatherley, O. (2014). What is it exactly we're protesting about?. de zeen. *7 July 2014*. Retrieved February 22, 2015, from http://www.dezeen.com/2014/07/07/owen-hatherley-opinion-zaha-hadid-heydar-aliyev-designs-of-the-year/

Hillman, M. (2014) *Global warming: Where do we go from here?* [video online]. Retrieved June 13, 2014, from http://www.youtube.com/watch?v=r7uuKrUdLH8&feature=youtube_gdata

Hillman, M., & Fawcett, T. (2004). *How we can save the planet*. London: Penguin.

HMGovernment. (2013). *Construction 2025 [online]*. Retrieved May 20, 2014, from https://www.gov.uk/government/uploads/system/uploads/attachment_data/file/210099/bis-13-955-construction-2025-industrial-strategy.pdf

Hostetler, K. D. (2011). *Seducing souls: Education and the experience of human well-being*. London: Continuum.

Illeris, K. (2011). *The fundamentals of workplace learning: Understanding how people learn in working life* (1 ed.). London: Routledge.

IPCC. (2013). *Climate Change 2013: The physical science basis*. Contribution of working Group 1 to the Fifth Assessment Report of the Intergovernmental Panel on Climate Change [online]. Cambridge University Press: Cambridge. Retrieved February 9, 2014, from http://www.climatechange2013.org/images/report/WG1AR5_Frontmatter_FINAL.pdf

Jarvis, P. (2012). *Adult learning in the social context* (2nd ed.). Abingdon: Routledge.

Jones, P., Pooley, A., & Tinning, B. (2012). Ecology, critical reflection and the praxis of change. In: M. Schwartz, C. Lubell, R. Gall, & A. Omer, (Eds.), *A future for earth: Re-imagining learning for a transforming world: Proceedings of the 10th International Transformative Learning Conference* (pp. 391–395). San Francisco, 1–4 November, 2012. San Francisco: Meridian University.

Kant, I. (1900). *Kant on education: (ueber Padagogik)*. Boston: D. C. Heath & Co. Publishers. Retrieved March 24, 2014, from http://lf-oll.s3.amazonaws.com/titles/356/0235_Bk.pdf

Kegan, R. (2009). What "form" transforms? A constructive-developmental approach to transformative learning. In K. Illeris (Ed.), *Contemporary theories of learning: Learning theorists—In their own words* (1 ed., pp. 35–52). London: Routledge.

Lange, E. (2000). Beyond transformative learning: Work, ethical space and adult education. In: T. J. Sork, V. –L. Chapman, & R. S. Clair, (Eds.), *Annual Adult Education Research Conference* (pp. 230–234). Vancouver, 2–4 June, 2000. Vancouver: University of British Columbia.

Leopold, A. S. (1968). A sand county almanac, and sketches here and there; illustrated by Charles W. Schwartz. London: Oxford University Press. Original work published 1949.

Maio, G. R., Olson, J. M., Allen, L., & Bernard, M. M. (2001). Addressing discrepancies between values and behavior: The motivating effect of reasons. *Journal of Experimental Social Psychology [online], 37*(2). Retrieved May 20, 2014, from http://www.sciencedirect.com/science/article/pii/S0022103100914367

Marshall, G. (2014). *Don't even think about it: Why our brains are wired to ignore climate change*. London: Bloomsbury.

Mezirow, J. (1991). *Transformative dimensions of adult learning*. San Francisco: Jossey-Bass.

Moore, S. A. (2010). Introduction. In S. A. Moore (Ed.), *Pragmatic sustainability: Theoretical and practical tools* (1 ed., pp. 1–12). London: Routledge.
Moustakas, C. (1994). *Phenomenological research methods*. Thousand Oaks, CA: Sage.
Murray, P. (2013). Leading by design: Cultivating self-leadership for sustainability. In: R. Crocker, & S. Lehmann, (Eds.), *Motivating change: Sustainable design and behaviour in the built environment*. Oxon: Routledge. Retrieved May 12, 2015, from http://online.vitalsource.com/books/9781135043841/epubcfi/6/2
Orr, D. W. (2004). *Earth in mind: On education, environment, and the human prospect*. 10th Anniversary ed. Washington, DC: Island.
Pultar, M. (2000). The conceptual basis of building ethics. In: W. Fox, (Ed.), *Ethics and the built environment* (pp. 155–169). London: Routledge.
Schapiro, S. A. (2009). A crucible for transformation: The alchemy of student-centred education for adults at midlife. In: B. Fisher-Yoshida, K. D. Geller, & S. A. Schapiro, (Eds.), *Innovations in transformative learning: Space, culture, & the arts* (pp. 87–109). New York: Peter Lang.
Schön, D. A. (1991). *The reflective practitioner: How professionals think in action*. Aldershot: Avebury.
Schultz, P. W., Gouveia, V. V., Cameron, L. D., Tankha, G., Schmuck, P., & Franěk, M. (2005). Values and their relationship to environmental concern and conservation behavior. *Journal of Cross-Cultural Psychology [online], 36*(4). Retrieved July 1, 2012, from http://jcc.sagepub.com/content/36/4/457.abstract
Seamon, D. (2000). Phenomenology, place, environment, and architecture: A review. *Environmental & Architectural Phenomenology [online], Environmental & Architectural Phenomenology Newsletter*. Retrieved December 12, 2014, from http://www.arch.ksu.edu/seamon/Seamon_reviewEAP.htm
Sterling, S. (2009). Ecological intelligence: Viewing the world relationally. In A. Stibbe (Ed.), *The handbook of sustainability literacy: Skills for a changing world* (220 p). Green: Totnes.
Tofield, B. (2012). *Delivering a low-energy building: Making quality commonplace [online]*. Build with CaRe. University of East Anglia. Retrieved July 3, 2014, from http://www.northsearegion.eu/files/repository/20140331180312_BuildwithCaReResearchReport-DeliveringaLow-EnergyBuildingOct2012.pdf
TSB. (2014). *Retrofit analysis*. Retrieved June 17, 2014, from https://retrofit.innovateuk.org/
UNESCO. (2005). *Education for sustainable development (ESD)*. Retrieved November 1, 2010, from http://www.unesco.org/new/en/education/themes/leading-the-international-agenda/education-for-sustainable-development/
Walker, S. (2014). *Designing sustainability: Making radical changes in a material world*. Oxford: Routledge.
Wals, A. E. J., & van der Leij, T. (2007). Introduction. In A. E. J. Wals (Ed.), *Social learning towards a sustainable world: Principles, perspectives and praxis* (pp. 17–32). Wageningen: Academic.

Chapter 17
Comparison of Transient Hygrothermal Modelling Against In Situ Measurement for Thermal Transmittance

Cormac Flood, Lloyd Scott, and William Gleeson

Abstract Thermal transmittance (*U*-values) of exterior walls represents a source of uncertainty when estimating the energy performance of dwellings. It has been noted in research that the standard calculation methodology for thermal transmittance should be improved. Subsequently, hygrothermal analysis has been used as an accurate building design tool due to its incorporation of climate-specific effects on construction assemblies such as moisture retention and release. In situ measurement of thermal transmittance could also be an effective tool for evaluating the material performance of assemblies of a building. This paper provides the context, research process and analysis of four case studies situated in Dublin, Ireland. The case studies offer an account of the in situ thermal transmittance of exterior walls and link these to hygrothermally simulated comparisons along with more traditional design *U*-values. The findings of this paper identify discrepancies between in situ and design *U*-values, using measurement, hygrothermal simulation and standard method *U*-value calculations. This study can form the basis for further research on retrofit of the Irish housing stock. Furthermore, the paper offers a source of information for researchers and designers exploring the performance of external walls to anticipate best practice detailing and in situ thermal performance values.

17.1 Introduction

Building envelopes are continually subject to fluctuating internal and external environmental conditions such as temperature, moisture, solar radiation and wind. These variations represent key factors that affect and define the actual physical thermal

C. Flood (✉) • L. Scott
Dublin School of Surveying and Construction Management,
Dublin Institute of Technology, Bolton Street, Dublin, Ireland
e-mail: cormac.flood1@gmail.com

W. Gleeson
Coady Architects, Mount Pleasant Avenue, Ranelagh, Dublin 6, Ireland

M. Dastbaz et al. (eds.), *Building Information Modelling, Building Performance, Design and Smart Construction*, DOI 10.1007/978-3-319-50346-2_17

performance and sustainability of the building envelope. As such, all techniques for the prediction of in situ hygrothermal behaviour of building components are issues of great interest in building design where the aim of accurate design is vital. The result should be an envelope that anticipates all internal and external environmental conditions allowing the building to perform to its optimum.

As building designs have developed, energy loss analysis has become more important to accurately predict, a key reason being the implementation of these figures to derive CO_2 reduction targets (Kema 2008). With the understanding that moisture affects the material performance of building assemblies throughout the lifespan of a building, it is vital to implement reliable prediction tools to assess potential thermal performance values.

At present the uniform standard for heat loss definition throughout Europe is the *U*-value. This is a calculation which disregards many environmental factors with the exception of wind speed, although as a non-variable. The single method to assess moisture levels of building assemblies within construction assemblies, referred to in Irish building guidance documents, is the Glaser method. The Glaser method is a one-dimensional, steady-state calculation with many limitations including the inability to handle heat and moisture capacity, air transfer through structures and capillary liquid flow. As a result, structures and assemblies may in reality perform entirely different than expected thermally and hygrothermally. Currently, there are two common measurement techniques to evaluate the thermal resistance in existing buildings: direct measurement of the heat flux (non-destructive method) or direct survey of the fabric layers with direct measure of their thickness (destructive method). The non-destructive method requires the use of a heat flux metre that has to be operated according to ISO 9869.

This paper presents the results of hygrothermal simulations with comparable non-destructive in situ *U*-value measurements and standard calculated *U*-values applied to four case study buildings situated in Dublin, Ireland. The buildings were selected for analysis based on thermal upgrade methods implemented, uninsulated, full fill cavity and external insulation. For all four case studies, a process of data collection was adhered to as follows:

(a) Interpretation of qualitative information from infrared thermography in accordance with ISO 6781 and collection of various data about the properties
(b) Calculation of *U*-values (thermal transmittance values) using the methods in ISO 6946
(c) Measurement of *U*-values (thermal transmittance values) using the methods in ISO 9869 and comparisons between measured and expected *U*-values
(d) Simulation of hygrothermally derived *U*-values (thermal transmittance values) using WUFI software in accordance with EN 15026 and ASHRAE 160P

The calculation method defined in ISO 6946 is the standard for calculating *U*-values of exterior walls, principally based on "ideal" conditions. ISO 6946 accounts for thermal conductivities of materials, geometric effects and some types of air voids; however, it excludes moisture, variable wind speed or solar-related occurrences.

The objective of thermographic imaging was to indicate thermal bridges, cracks or similar sources of irregularities in surface temperatures contra venous to the typical thermal performance of the wall. The result of this was the identification of suitable locations on the wall for installation of heat flux metre (HFM) and thermocouples for in situ *U*-value measurements.

In situ *U*-values have been measured by using the heat flux metre (HFM) method performed in agreement with ISO 9869. Accordingly, measurements have been carried on for at least 72 h (typically 1 week), with an acquisition time lapse of 1 min. The measurements have been conducted during spring. The 80 mm diameter and approximately 5 mm thick HFM was temporarily adhered (using masking tape to edges) throughout the period of measurement away from direct influence of either a heating or a cooling device. No protection was required to the HFM to shield from rain, snow or direct solar radiation as it was placed internally. A trial study revealed that direct solar exposure on thermocouples may significantly distort readings. As a result, the externally positioned thermocouples were fixed within a radiation shield to avoid the effect of direct solar radiation. The measured *U*-values are presented alongside the calculated and simulated *U*-values of matching environmental conditions and construction type to facilitate comparison.

17.2 Thermal Transmittance Through In Situ Measurement

In situ is a Latin phrase that translates literally to "on site" or "in position" denoting the way a measurement is taken in the same place the phenomenon is occurring without isolating it from other systems or altering the original conditions of the test. The measurement of actual thermal transmittance in building assemblies is known as in situ *U*-value measurement. It uses a HFM in combination with internal and external temperature measurements taken over time; in this way an in situ *U*-value is able to take into account thermal inertia (mass) and the effect of temperature change and other climatic conditions (Rye 2010; Rye and Scott 2012). This method proves to be reliable and can also be used for non-destructive tests of the thermal characteristics of buildings. The thermal transmittance of a building element (*U*-value) is defined in ISO 7345 as the "Heat flow rate in the steady state divided by area and by the temperature difference between the surroundings on each side of a system". However, since steady-state conditions are never encountered on a site in practice, such a simple measurement is not possible. But there are several ways of overcoming this difficulty:

(a) Imposing steady-state conditions by the use of a hot and a cold box. This method is commonly used in the laboratory (ISO 8990) but is cumbersome in the field.
(b) Assuming that the mean values of the heat flow rate and temperatures over a sufficiently long period of time give a good estimate of the steady state. This method is valid if:

(c) The thermal properties of the materials and the heat transfer coefficients are constant over the range of temperature fluctuations occurring during the test.
(d) The change of amount of heat stored in the element is negligible when compared to the amount of heat going through the element.
(e) Using a dynamic theory to take into account the fluctuations of the heat flow rate and temperatures in the analysis of the recorded data.

17.2.1 Previous Research Involving In Situ U-Value Measurement

Early research published from 2000 has investigated the requisite for in situ measurement to verify calculated *U*-values used commonly throughout the construction industry. Doran (2000) suggests an international need for a better understanding of air and moisture movement within opaque building elements, while Baker (2008) and Currie et al. (2013) outlined the basic technique required to implement in situ analysis. Since then, various publications have analysed numerous wall assemblies arriving at the conclusion that measurements generally highlight a vast performance gap between design values and in situ results (Doran and Carr 2008; Peng and Wu 2008; Rye 2010; Byrne et al. 2013; Asdrubali et al. 2014; Evangelisti et al. 2015). Baker (2011) and Rye and Scott (2012) reported that within the scope of traditional buildings, *U*-value calculations generally overestimate in situ thermal performance. In other words, uninsulated traditional buildings actually perform better than expected from design values. In contrast to this, Hulme and Doran (2015) argued that depending on the wall structure and insulation levels, the reliance on in situ values varied considerably from overestimation to underestimation of design *U*-value. Rhee-Duverne and Baker (2013) then went on to claim that if the thermal conductivity values are known, calculations made using software programmes can be in reasonable agreement with the actual measured *U*-values, suggesting that much of the unreliability of calculating *U*-values lies with the low quality of input data.

With all of the above taken into consideration, in situ analysis of the *U*-value is certainly a practical option to establish the actual performance of external walls. However, the idea within the scope of this research is to establish a method, whereby hygrothermal simulation can be verified as a method to predict thermal performance as an accurate reflection of in situ performance, thus replacing ISO 6946 standard method *U*-value calculations. To do this, a link between in situ measurements and hygrothermal simulations was made.

17.2.2 Review of Methods and Tools

Two methods may be used for analysis of the data in accordance with ISO 9869: the so-called average method, or the dynamic method. Ahmad et al. (2014), Li et al. (2015) and Rasooli et al. (2016) have reflected on the average method with proposals to modify this for more precise outputs. For the purposes of this research however, these modified techniques are too undefined and experimental for use at this stage. The measurements in this research are presented as direct comparisons between the simulated *U*-values and the *U*-values using ISO 6946 standard calculation methodology. This averaging approach is valid if the following conditions apply:

(a) The thermal properties of the materials in the element are constant over the range of temperature fluctuations.
(b) The change in the internal energy of the element is negligible if compared to the amount of heat going through the element.

Following analysis of existing literature, the average method is identified as applicable for similar styles of wall construction as those in this research: solid and cavity masonry. It is assumed that the assemblies here are sufficiently homogeneous or made of sufficiently homogeneous layers to use a HFM.

17.3 Methodology

The methodology used in this phase of the research is modelled around multi-methodological design, incorporating some qualitative research for a fuller and more in-depth piece of research (Creswell 2009). Data collection and analysis through past and present research by others (along with policy design standards, recorded climate data, housing figures, common external wall constructions, standard design calculation methodologies and non-standard design calculation methodologies) correspond well with and suit the theory of a quantitative methodological approach (Corbetta 2003; Maxwell 1998, 2012); the research is structured, performing a series of calculations and recording performance data to produce results which clarify the question. A qualitative approach was used to develop an understanding of the problem and improve methods for the quantitative element of research.

Searches were undertaken of recognised relevant academic and specialist building conservation literature databases through a number of journals and websites of the statutory bodies responsible for the protection of the Irish, UK and European environment. Using the technical indices and Technical Guidance Document Part L, ISO 6946 is referenced to specify the method of calculating *U*-values. The *U*-value calculation was then evaluated, and the exclusion of environmental conditions was identified as the main fault. This error was identified to be addressed using hygrothermal simulation through WUFI 5.3. Verifying this research, the wall assemblies within the case studies were assessed using in situ thermal transmittance measurements in accordance with ISO 9869. The existing wall structures were verified through documentation provided and inspection through a bore scope with measurements using the metric system (mm) as an internationally agreed decimal system of measurement. Thus, the following external wall assemblies were assessed for this study (see Figs. 17.1, 17.2, 17.3, 17.4, 17.5, 17.6 and 17.7).

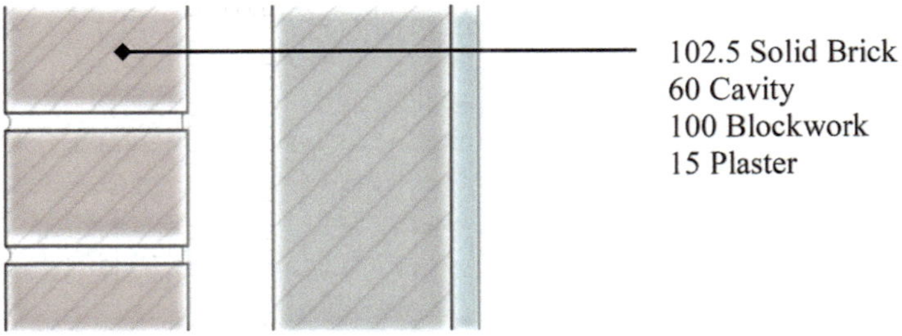

Fig. 17.1 Uninsulated cavity wall

Fig. 17.2 Case Study 1

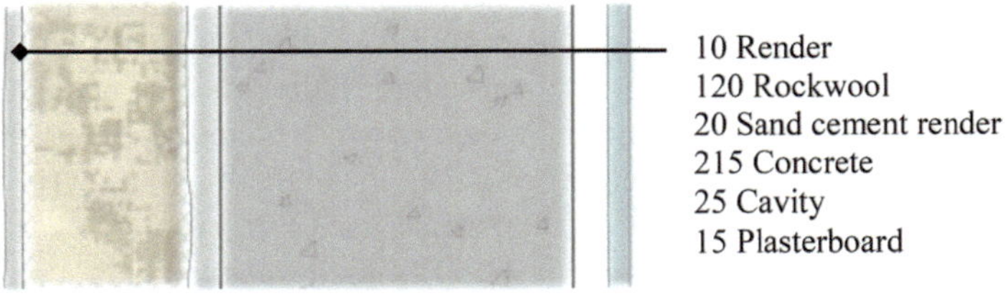

Fig. 17.3 Externally insulated solid wall

Fig. 17.4 Case Study 2

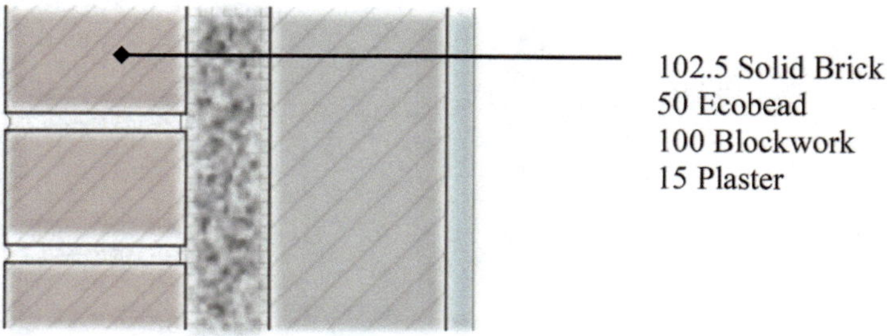

Fig. 17.5 Cavity fill cavity wall

Fig. 17.6 Case Study 3

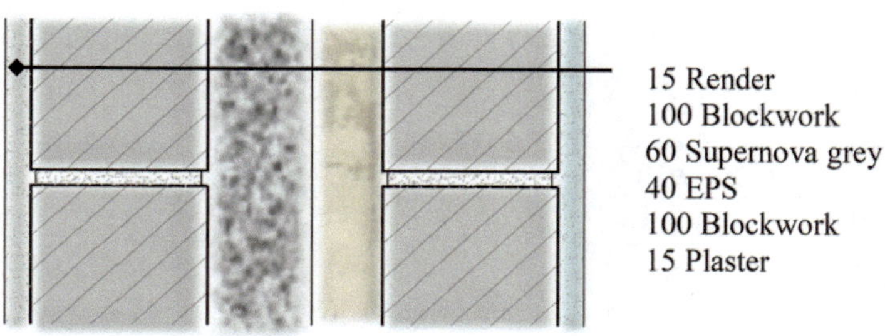

Fig. 17.7 Cavity fill cavity wall

With the aim of measuring the in situ *U*-value of an assembly, it is essential to record the heat flow, internal temperature and external temperature continuously over a sufficiently long period of time. In this project, a Hukseflux HFP01 HFM sensor was employed to measure heat flow, and RS Pro T Type Thermocouples with a 2 m probe were used to record a temperature-dependent voltage to measure internal and external temperatures (see Figs. 17.9, 17.10 and 17.11). A Campbell CR1000 datalogger (see Fig. 17.11) was used to record the measurements of the HFM and thermocouples allowing for cold-junction compensation of the latter.

Fig. 17.8 Case Study 4

Fig. 17.9 HFP01 HFM

Fig. 17.10 T Type Thermocouples

U-values were determined by comparing the heat flow through the element with the temperature difference across it over a minimum of 7-day period. In an ideal situation, the internal and external temperatures would be constant, giving a stable and accurately determined U-value. In practice steady-state conditions do not arise, however, and attention must be given to the variations in temperatures and heat

Fig. 17.11 Datalogger rested on windowsill for duration of study

Fig. 17.12 Image of internal wall surface

flows before the U-value can be determined reliably. Since most building structures have a significant thermal mass, variations in internal or external temperatures lead to large fluctuations in the heat flow either into or out of the element, and it was necessary to measure the heat flows and temperatures over several days in order to arrive at a reliable result.

ISO 9869 recommends thermographic analysis prior to the installation of any HFM. The purpose of the thermography is to establish potential thermal bridges, cracks or similar sources of error in the internal surface temperature near to the potential HFM location. Large variations in surface temperature would indicate that the selected measurement point was uncharacteristic of the typical function of the wall and therefore should not be selected. Multiple thermographic images were taken to ensure accuracy of results and verify that glazing did not distort larger image results. Figures 17.12 and 17.13 are results from thermographic surveying Case Study 2.

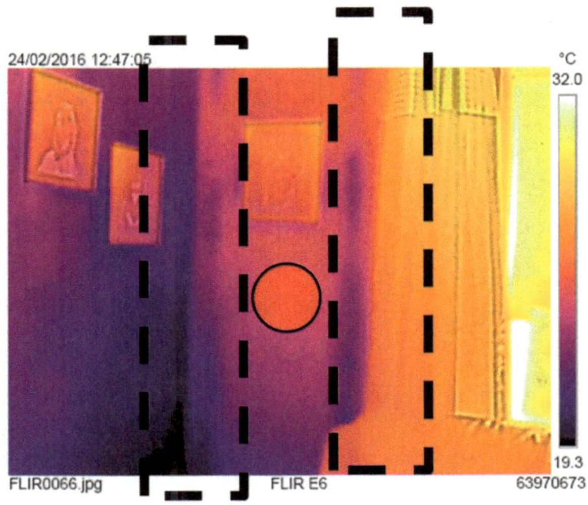

Fig. 17.13 Thermographic image of internal wall surface

Figure 17.12 shows the basic image of an internal wall surface, while Fig. 17.13 is the corresponding thermographic image. While there would not be a significant variation across the wall surface, boxed are what appear to be studs behind the finish plasterboard. The result of this finding was that the sensor was placed between the studs (marked with red circle) to record the typical wall assembly. This typical wall assembly would then relate directly to the calculated and simulated values. The entire schedule of data acquisition composed prior to analysis was invaluable to ensure participants were fully aware of the dates and times associated with each element of research.

In all cases, thermal paste/grease was applied on the wall side of the HFM to ensure full connection to the wall surface. The HFM was then fixed to the wall surface using a masking tape to the edges away from the metre within the plate, to minimise any effect to the heat flux readings.

The probes used for monitoring internal temperatures were usually located approximately 50 mm from the internal wall surface and were located at the same height as the adjacent HFM and positioned to face the room (i.e. to receive a similar radiant temperature to that of the room interior). For the external air temperature, the probes were positioned (housed within a hanging tube shielding to reduce the effect of direct solar radiation) about 50 mm from the external wall surface, fixed to the wall surface using 9 mm round cable clips to provide anchoring. For each dwelling the elemental *U*-values were determined by recording the heat flow through the element together with internal surface and external air or surface temperature. This was done by logging differential voltage from the heat flux transducers and temperature from calibrated T Type Thermocouples (resistance) continuously over 1 week. The signals were measured every 60 sec (Figs. 17.14, 17.15 and 17.16).

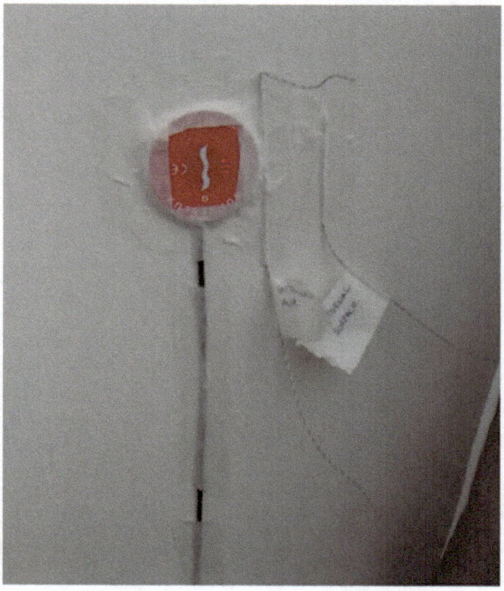

Fig. 17.14 HFM and internal thermocouple fixed

Fig. 17.15 External thermocouples fixed

17.4 Results and Discussion

In situ data was administered by means of the progressive average procedure that is based on the idea that the average of instantaneous ratios between heat flux and temperature differs on a gradually increasing timescale levelling out the oscillations leading to the steady-state value of the thermal transmittance (see Eq. (17.1)).

Fig. 17.16 External thermocouples fixed

Table 17.1 Case study wall type abbreviations

Wall type	Description	Year of construction	Year of thermal upgrade
WT 1	Case Study 1	1970	N/A
WT 2	Case Study 2	1975–1978	2010
WT 3	Case Study 3	1960s/early 1970s	2012
WT 4	Case Study 4	1983–1984	2011

ISO 6891-1 formula

$$U = \frac{\sum_{j=1}^{n} qj}{\sum_{j=1}^{n} (Tij - Tej)} \tag{17.1}$$

17.4.1 Thermal Performance of the Analysed Walls

For the purpose of this research, the wall types for each case study investigated have been assigned with abbreviations for table listings as per Table 17.1:

Standard guidance calculations and simulations were carried out using assembly descriptions and data outlined in Table 17.2:

In accordance with ISO 9869, the analysis was carried out over a period of 7 days at least. Longer recording times would be ideal, but unachievable in this research project. Figure 17.17 is the progressive average *U*-value procedure for WT 1 West façade:

Table 17.2 Material data for calculations and simulations

Wall types	Material (mm)	Conductivity (W/mK)	Specific heat capacity (J/kgK)	Bulk density (kg/m^3)	Porosity (m^3/ m^3)	Water vapour diffusion resistance factor	Insulation location
WT1	15 plaster	0.2	850	850	0.65	8.3	Uninsulated
	100 blockwork	1.33	1000	1900	0.2	15	
	60 cavity	0.071	1000	1.3	0.999	0.73	
	102.5 solid brick	0.77	850	1700	0.24	10	
WT2	15 plasterboard	0.2	850	850	0.65	8.3	External insulation
	25 cavity	0.071	1000	1.3	0.999	0.73	
	215 concrete	1.6	850	2200	0.18	0.92	
	20 sand-cement render	1.2	850	2000	0.3	25	
	120 rock wool	0.038	1030	135	0.953	1.1	
	10 render	0.8	850	1900	0.24	19	
WT3	15 plaster	0.2	850	850	0.65	8.3	Cavity fill insulation
	100 blockwork	1.33	1000	1900	0.2	15	
	50 ecobead	0.031	1200	11.5	0.95	60	
	102.5 solid brick	0.77	850	1700	0.24	10	
WT4	15 plaster	0.2	850	850	0.65	8.3	Cavity fill insulation
	100 blockwork	1.33	1000	1900	0.2	15	
	40 EPS	0.038	1130	14.8	0.95	60	
	60 supernova grey	0.038	1200	11.5	0.95	60	
	100 blockwork	1.33	1000	1900	0.2	15	
	15 mm render	0.8	850	1900	0.24	19	

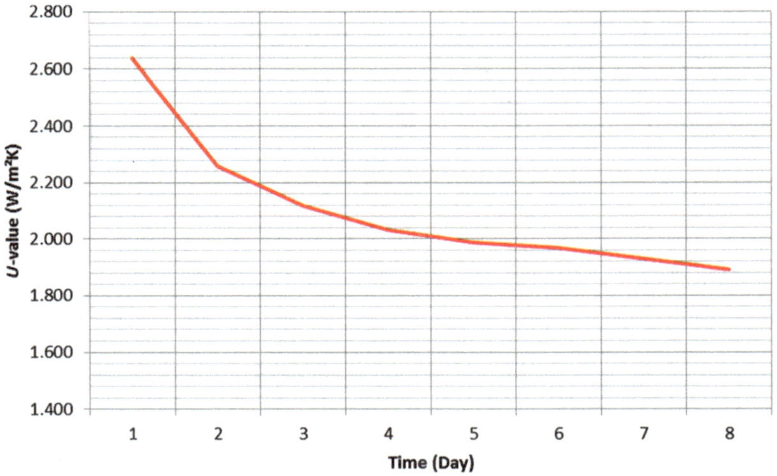

Fig. 17.17 Progressive *U*-value measurement of WT 1 West façade

Table 17.3 Calculated, simulated and measured thermal transmittance values

Wall types	Orientation	Calculated ISO 6846 (W/m^2K)	Simulated WUFI (W/m^2K)	Measured in situ (W/m^2K)
WT1	North	1.688	–	–
	South		–	–
	East		–	–
	West		1.913	1.891
WT2	North	0.267	–	–
	South		0.292	0.315
	East		0.282	0.277
	West		0.297	0.430
WT3	North	0.508	–	–
	South		–	–
	East		0.609	0.603
	West		0.637	0.841
WT4	North	0.312	0.352	0.556
	South		0.354	0.651

WT1–WT3 were all analysed with the same protocol as Fig. 17.17. These results were then compared with hygrothermal simulations implementing corresponding environmental conditions and ISO 6946 standard method *U*-value calculations. The results of these are assembled in Table 17.3:

From analysis of the data, all wall assemblies perform entirely differently depending on orientation, as suggested in previous research by the authors (Flood et al. 2016). It should also be noted that standard ISO 6946 calculations do not align

Table 17.4 Calculated, simulated and measured U-values in chart form

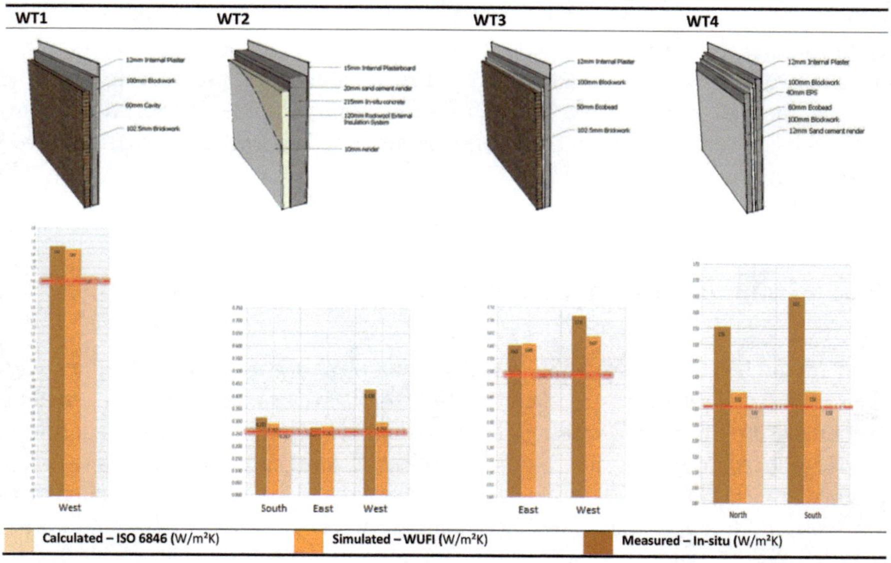

with the in situ measurements in any case, regardless of orientation. All simulated values align much closer with the in situ recorded data. For visual contrast, figures within Table 17.3 have been charted below in Table 17.4 marking constant ISO 6946 calculations for each wall with red lines.

Table 17.4 confirms the inconsistency between the standard ISO 6946 calculations and in situ measurements. This discrepancy appears to have been reduced through the use of simulated values, something linked to orientation – incorporating wind speed, relative humidity, rain and solar transmittance. In situ results of WT 4 highlight a large inconsistency compared to calculated and simulated values. This has been identified as a problem curtailing from the use of incorrect wall assembly data for WT4. These simulations will be corrected following an investigation to verify the wall assembly.

17.5 Discussion

The findings of this stage of the research confirm that orientation has a significant impact on the thermal performance of an external wall, regardless of the overall assembly as previously suggested through hygrothermal simulation (Flood et al. 2016). Orientation dictates the level of exposure the wall is open to, specifically wind speed, rain count, relative humidity and solar transmittance. This means that when designing an external wall, designers should focus the design parameters around each façade considering the variation in associated external conditions. Hygrothermal performance appears to be a step in the right direction towards a progressive thermal transmittance prediction technique in Ireland. It is clear that the existing thermal transmittance calculation methodology is imbalanced with a number of flaws in its composition. This could be addressed using the knowledge derived from this research.

17.6 Conclusions/Further Research

This research has reviewed ISO 9869 in situ *U*-value measurement along with hygrothermal simulations and standard ISO 6946 *U*-value calculations as a method to increase credibility and validity of conclusions resulting from further experimental research. A key supposition is that hygrothermal simulation can be used to derive thermal transmittance comparable to in situ results. This research is intended to serve as an introduction to issues emanating from a larger research project in order to encourage researchers to understand and further explore the topic.

Understanding the fundamental criteria necessary for carrying out a simulation or calculation is critical to achieving accurate results. The reliability and validity of results are largely dependent on the quality of data used. To that end, a recommendation for further research includes the necessity to carry out on-site investigation of wall assemblies prior to simulation/calculations.

The realm of heat transfer and building physics is a question throughout the Architectural, Engineering and Construction (AEC) sector, particularly within retrofit and refurbishment. This has been confirmed through an examination of previous research in the field, accompanied by personal experience. The understanding gained regarding the influence of external and internal environmental conditions has already and continues to enhance the product of this research. Adopting hygrothermal simulations along with accurate material data analysis has allowed a more concise and defined format of information to be assessed. By searching through previous literature available on AEC research, comparable precedent has been established to set a benchmark for results generated from this research.

The findings of this paper identify discrepancies between in situ and standard method *U*-value calculations, proposing to bridge this gap with more representative hygrothermally simulated values. The effect of rain, relative humidity, wind speed and solar radiation may cause the thermal performance gap illustrated in the assemblies. Orientation is a key factor influencing thermal performance which is directly connected to the rain count, prevailing wind and solar exposure. Accordingly, facades with a westerly orientation have been verified as the least thermally efficient in Ireland. Thus, this research offers a source of information for researchers and designers exploring the performance of external walls to anticipate best practice detailing and in situ thermal performance values.

Modelled wall assemblies with different porosities, moisture storage capacities and liquid water transport coefficients along with accurate climate data result in different moisture contents and, correspondingly, a corrected *U*-value. It is clear that if advanced hygrothermal models such as WUFI are to be used to carry out routine assessments of moisture conditions and *U*-values in building structures, considerably more construction material data must be made available by manufacturers to achieve realistic simulation results.

References

Ahmad, A., Maslehuddin, M., & Al-Hadhrami, L. M. (2014). In situ measurement of thermal transmittance and thermal resistance of hollow reinforced precast concrete walls. *Energy and Buildings, 84*, 132–141.

Asdrubali, F., D'Alessandro, F., Baldinelli, G., & Bianchi, F. (2014). Evaluating in situ thermal transmittance of green buildings masonries—A case study. *Case Studies in Construction Materials, 1*, 53–59.

Baker, P. (2008). *Preliminary results*. Scotland: Historic Scotland.

Baker, P. H. (2011). *U-values and Traditional Buildings: Historic Scotland Technical Paper 10*. In situ measurements and their comparisons to calculated values. Edinburgh: Historic Scotland.

Byrne, A., Byrne, G., Davies, A., & Robinson, A. J. (2013). Transient and quasi-steady thermal behaviour of a building envelope due to retrofitted cavity wall and ceiling insulation. *Energy and Buildings, 61*, 356–365.

Corbetta, P. (2003). *Social research: Theory, methods and techniques*. London: Sage.

Creswell, J. W. (2009). *Research design. Qualitative, quantitative and mixed methods approaches* (3rd ed). SAGE Publications, 2455 Teller Road, Thousand Oaks, California.

Currie, J., Williamson, J. B., & Stinson, J. (2013). *Monitoring thermal upgrades to ten traditional properties*. Edinburgh: Historic Scotland.

Doran, S. (2000). *Field investigations of the thermal performance of construction elements as built*. Glasgow, Scotland: BRE.

Doran, S., & Carr, B. (2008). *Thermal transmittance of walls of dwellings before and after application of insulation*. Glasgow, Scotland: Energy Saving Trust.

Evangelisti, L., Guattari, C., Gori, P., & Vollaro, D. R. (2015). In situ thermal transmittance measurements for investigating differences between wall models and actual building performance. *Sustainability, 7*(8), 10388–10398.

Flood, C., Scott, P. L., & Gleeson, W. (2016). *Application of transient hygrothermal modelling to assess thermal transmittance: A case study in Dublin, Ireland*. CIB World Building Congress 2016, Tampere, Finland.

Hulme, J., & Doran, S. (2015). *In-situ measurements of wall U-values in English housing*. Watford, England: BRE.

Kema. (2008). *Demand side management in Ireland: Evaluating the energy efficiency opportunities: Sustainable energy authority Ireland.*

Li, F. G. N., Smith, A. Z. P., Biddulph, P., Hamilton, I. G., Lowe, R., Mavrogianni, V., et al. (2015). Solid-wall *U*-values: Heat flux measurements compared with standard assumptions. *Building Research & Information, 43*(2), 238–252.

Maxwell, J. A. (1998). *Qualitative research design: An interactive approach.*

Maxwell, J. A. (2012). *Qualitative research design: An interactive approach*. Applied social research methods series (3rd ed). Thousand Oaks, CA: Sage.

Peng, C., & Wu, Z. (2008). In situ measuring and evaluating the thermal resistance of building construction. *Energy and Buildings, 40*(11), 2076–2082.

Rasooli, A., Itard, L., & Ferreira, C. I. (2016). A response factor-based method for the rapid in-situ determination of wall's thermal resistance in existing buildings. *Energy and Buildings, 119*, 51–61.

Rhee-Duverne, S., & Baker, P. (2013). *Research into the thermal performance of traditional brick walls*. London, English Heritage.

Rye, D. C. (2010). *U-value report*. London: SPAB.

Rye, D. C., & Scott, C. (2012). *U-value report revised*. London: SPAB.

Chapter 18
It's Housebuilding But Not as We Know It: The Impact of Neighbourhood Planning on Development in England

Quintin Bradley

Abstract Community opposition to housebuilding has been cited as one of the key factors in the decline in new housing supply over the last decade. The policy of neighbourhood planning was introduced to England in 2011 to overcome this opposition by devolving limited powers to communities to influence development. It was anticipated that giving communities the right to draw up neighbourhood development plans would secure their compliance with a pro-growth agenda and increase the number of sites allocated for housing. This chapter explores the impact of neighbourhood planning in England on housing development and analyses its lessons for the state strategy of localism. It argues that neighbourhood planning is emerging as the proponent of sustainability and social purpose in the English housing market, in conflict with the corporate interests of liberalised housing development.

18.1 Introduction

The devolution of governance to communities is an integral component of the state strategy of localism but risks conflict with policies dedicated to the liberalisation of economic growth (Clarke and Cochrane 2013; Davoudi and Madanipour 2015). This has been particularly true in the localisation of planning decisions since new housing developments often generate collective public opposition because of their anticipated environmental impacts (Sturzaker 2011). Community resistance to housebuilding has been cited as one of the key factors in the decline in new housing supply over the last decade and the collective action of citizens' groups in land-use or development planning is routinely dismissed as illegitimate, selfish and ignorant (DeVerteuil 2013).

The emergence of neighbourhood planning in England after 2011 was unusual both in its devolution of statutory planning policy to community organisations and

Q. Bradley (✉)
Senior Lecturer in Planning & Housing, Leeds Beckett University School of the Built Environment and Engineering, Leeds LS2 9EN, UK
e-mail: q.bradley@leedsbeckett.ac.uk

M. Dastbaz et al. (eds.), *Building Information Modelling, Building Performance, Design and Smart Construction*, DOI 10.1007/978-3-319-50346-2_18

in its explicit intention to reshape the protectionist opposition of citizens into enthusiastic support for housing development (Bradley 2015; Parker et al. 2015). It was anticipated that giving communities the right to devise neighbourhood development plans would secure their compliance with a pro-growth agenda and increase the number of sites allocated for housing. Neighbourhood planning was an experimental policy at the seismic juncture between localism and the liberalisation of housing growth (Young and Burcher 2014:1). Making this policy work meant giving local people real influence over the scale and shape of development and at the same time enabling the volume housebuilding corporations that dominate the industry in England to access land and gain planning approval more easily.

The aim of this chapter is to assess the impact of localism in England on community support for housebuilding and to review the spatial practices of neighbourhoods in planning for housing delivery. It is argued that neighbourhood planning created opportunities for communities to advance new socially and environmentally sustainable housing solutions that conflicted with the interests of corporate housebuilders and unsettled the depiction of citizens' groups as protectionist and opposed to all economic growth. The chapter seeks to contribute to housing theory in its analysis of the policy impacts of the engagement of citizens' and residents' groups in planning for housing delivery and, more broadly, it seeks to add to the debate on localism and public participation in housing (Ruming et al. 2012; Cook et al. 2013; Matthews et al. 2015).

It draws on the body of planning policy developed by neighbourhoods across England since 2011 and on fieldwork research with a national sample of 50 neighbourhood plans carried out between 2013 and 2015. It presents specific case studies that evidence the planning pathways followed by neighbourhoods through the drafting and decision-making process and through subsequent legal challenges to illustrate the themes emerging from the sample as a whole. This research was conducted with rural and urban neighbourhoods and involved a preliminary review of draft and final plans and other documentation, including constitutions, applications for designation, council decision papers, minutes of meetings, consultation strategies, and examiners' reports, followed by interviews with the chairs and secretaries of neighbourhood planning committees or forums, observation at meetings and separate interviews with the relevant officers from the planning authority. Participants gave their informed consent on the understanding that the actual place names for neighbourhood plans would be used.

The chapter begins with a review of the liberalisation of governance inherent in the spatial restructuring of localism and introduces neighbourhood planning in the context of housing growth and housing delivery. It explores the conflict that emerged between neighbourhoods and the volume housebuilders and analyses government responses. It then identifies the distinctive spatial practices that are emerging in neighbourhood plans with regard to housing delivery and evaluates the impact of neighbourhood planning on the dominant market model of housebuilding. The chapter concludes with an assessment of the contribution of neighbourhood planning to housing delivery and its significance in understanding the tensions inherent in the state strategy of localism.

18.2 Housing Growth and Localism

The state strategy of localism can be understood as part of a broader liberalisation of governance in which private market interests are promoted in the name of economic growth and community empowerment (Brenner and Theodore 2002; Davoudi and Madanipour 2015). A spatial restructuring of state institutions and services has reduced the regulatory environment for private corporations and increased their opportunities to benefit from the contraction of the public sector. Central to this project has been a rhetorical shrinkage of the space of public responsibility and the corresponding assertion of the 'community' or 'neighbourhood' as a unit of enterprising citizenry (Rose 1999; Hall and Massey 2010). Such a spatial imaginary implies the devolution of limited statutory authority to the neighbourhood as a collective organization of citizens, although any transfer of power is fraught with tension and, as a result, has been confined largely to the realm of the symbolic (Miraftab 2009; Bradley 2014).

Notions of 'citizen control' bring with them a troubling association with oppositional movements and 'counter-publics', and in land-use planning and housing especially, they recall neighbourhood campaigns against development that challenged entrenched networks of power and posited a global failure of trust in the legitimacy of state processes (Arnstein 1969; Fraser 1997; Mihaylov and Perkins 2015). Studies of citizen engagement in planning decisions have undermined the rationale for the use of the acronym NIMBY (not in my back yard) to condemn the planning objections of community groups as selfish and ill-informed (Burningham 2000; Wolsink 2006; Devine-Wright 2012). They have demonstrated that resident opposition to housing may be guided by broader societal concerns such as sustainability and social justice (Matthews et al. 2015). Resistance to, and citizen participation in planning decisions manifests an empowered public claiming to speak for societal concerns and protect the public good (Ruming et al. 2012). These publics appear to assert connections between housing development, the enhancement of democracy and residents' rights to have a say over neighbourhood change (Cook et al. 2013).

The public concerns asserted by neighbourhoods in development planning address what the spatial theorist Henri Lefebvre called representational or lived space, 'space as directly lived through its associated images and symbols' (Lefebvre 1991:39). Lefebvre's classification of space into three elements, as conceived, perceived and lived, enables distinctions to be made between the exchange values or use value of housing development (McCann 1999). The development of market housing is motivated by exchange values, where the price of land determines scheme viability and subsequent affordability—or lack of it (Ball 2003; Bramley and Watkins 2016), and neighbourhood opposition to housing growth has been similarly related to socio-economic status and house values (Taylor 2013; Taylor et al. 2016). But neighbourhoods appear additionally motivated by use values and the potential for land to meet local need and contribute intangibly to place identity and sense of belonging. The claim to local knowledge of resident groups can be also then understood as an articulation of the use values of lived space, as space that serves a range of public functions and civic needs and that has emotional resonance as well as practical purpose. Lefebvre's spatial dialectic emphasises the relationship between social identity and residential space and provides a conceptual framework through which land-use can be understood as central to the political questions of social reproduction and sustainability. Lefebvre contrasted the production of space as a homogenous abstract, commodified as parcelled units of exchange, with its social construction as emplaced labour, mapped into the practices and passions of everyday life. The engagement of neighbourhoods in development planning appears through this lens as a struggle for space, where housing market decisions may transgress local definitions of the public sphere and raise political dilemmas of social purpose (Clark 1994; Mihaylov and Perkins 2015). In this context, the devolution of planning powers to neighbourhoods might provide opportunities to enhance the social benefits of housing development and produce greater specificity and diversity in housing market decisions.

18.3 Neighbourhood Plans and Housebuilding

The introduction of neighbourhood planning to England from 2011 was experimental in its devolution of statutory planning powers to community groups and its expectation that greater collective influence over development would induce citizens to support new housebuilding. One of the principal objectives of neighbourhood planning was to increase the rate of growth of housebuilding by enabling communities 'to exercise real power in respect of the design and precise location of the development that takes place in the neighbourhood' (DCLG 2011:10). Neighbourhood planning was accompanied by financial incentives to sweeten this behavioural change and the policy was expected to boost the number of land sites allocated for housing over and above those already apportioned by higher-level plans. One of the key indicators of the success of this policy would be a reduction in the number of rejected planning applications, planning appeals and legal challenges.

It is important to note that localism takes specific form within the devolved UK, where separate planning systems now exist, and this chapter is concerned with policy enacted only in England by the UK Parliament. Neighbourhood plans, or to give them their full significant title, Neighbourhood Development Plans, were brought in as part of a radical programme of spatial deregulation that aimed to generate economic growth in the context of a chronic mismatch between housing need and housing supply (NHF 2011). Neighbourhood planning powers were set out in a new National Planning Policy Framework that enshrined a presumption in favour of sustainable development binding on local authorities and their neighbourhoods (DCLG 2012). The regional tier of planning that had previously set housing targets was abolished and local planning authorities were instead required to provide more than five years' worth of specific, developable housing land sites and identify broad locations for new housing up to 9 years ahead. Neighbourhood plans had to be in general conformity with these strategic policies as set out in the Local Plan drawn up by unitary and district authorities. They had to contribute to the achievement of sustainable development and 'plan positively to support local development' especially housing development (DCLG 2012 Paragraphs 15–16). They could not promote less development than stipulated in the Local Plan or undermine its strategic policies and they had to have regard to national policies and be compatible with EU obligations. A neighbourhood plan could be made by a Town or Parish Council or, in urban areas, by a community group establishing a Neighbourhood Forum (Davoudi and Cowie 2013; Wills 2016). These 'qualifying bodies' must apply to the local planning authority to be designated as a neighbourhood area. They were responsible for assembling an evidence base from community engagement, and for writing planning policy, and the resulting neighbourhood plan went through a statutory consultation process and was formally examined. To win community support, the neighbourhood plan must be approved in a local referendum and receive more than 50% of the vote of those registered and taking part in the ballot.

While neighbourhood planning was to be evaluated according to its success in increasing housing numbers, the community empowerment imperatives of localism provided space for a more qualitative assessment of housing growth to emerge. Neighbourhood plans were presented in the Localism Act (2011) as a package of 'community rights' and were accompanied by additional powers that enabled neighbourhoods to allocate land and grant planning permission for small-scale building of community-led affordable housing and for co-operative ventures like community land trusts (DCLG 2013). Even though neighbourhood plans were predetermined to a housing growth agenda, they could circumvent the local planning authority in non-strategic matters. They had to pass a 'light touch' examination that contained none of the tests of soundness applied to the Local Plan and once a neighbourhood plan was in force, its policies took precedence over existing non-strategic policies in the Local Plan. Importantly, a neighbourhood plan could be developed in the absence of an up-to-date Local Plan and in these cases it could take precedence over non-strategic land-use policies.

In the rhetoric of localism and devolution, the purpose of neighbourhood planning was to do more than provide land sites for speculative building; it was also intended to help local people 'develop a shared vision for their neighbourhood' (DCLG 2012 Paragraph 183) and 'take control of the look and feel of the places where they live' (Clark 2011). Place attachment and place identity were passions enlisted by government as leverage to achieve the behavioural shift desired in neighbourhood planning and neighbourhood groups were expressly directed to protect green space, prevent sprawl and safeguard heritage (Locality 2014). The government's Community Rights programme promoted community asset ownership and community-run services (DCLG 2015b). Participants in the early neighbourhood plans demonstrated a strong desire for more control over local decisions and many of them were motivated by previous conflicts with the local planning authority and with housing developers (Parker et al. 2014; Bradley 2015).

The next section begins to explore the tension that emerged between the citizen empowerment aims of neighbourhood plans and their induced compliance with deregulated housebuilding.

18.4 Neighbourhoods and the Developers

Early interest in neighbourhood planning came from rural parish councils and market towns under pressure from housing development in the south of England (Parker et al. 2014). More affluent neighbourhoods were better able to mobilise the voluntary resources required by the long and complex process of plan-making. The uneven geography that emerged was only partially offset by government support packages, training programmes and municipal strategies aimed at ensuring the participation of urban and less affluent areas. Deprived areas with well-developed community organisations were prominent early frontrunners, and around 18% of plans made by the end of 2015 were in the least affluent neighbourhoods, although the distribution was still weighted towards the southern counties (Parker 2015; Wills 2016).

The first neighbourhood plans to be successful at referendum accepted the need for more housebuilding and accommodated it within a strong sense of place. In the sparsely populated post-industrial landscape of the North Pennines, the Upper Eden neighbourhood plan demonstrated in 2012 that the devolution of planning could liberalise housing delivery and provide an approach to meeting housing need that was founded on local knowledge and a more intimate feeling for place. It overturned the settlement hierarchy of Eden District Council to allow new affordable self-build housing in small hamlets and remote locations (Upper Eden Community Interest Company 2012). The plan was carefully tailored to provide a sustainable future for a marginal rural economy and it established an enabling framework for housing that ensured its continuing affordability. The next neighbourhood plan to be approved at referendum was for the South Oxfordshire market town of Thame that planned positively for housing growth despite local opposition. Working with a received target of 7750 new homes, the town council of Thame rejected the intention of the planning authority to concentrate this development in one place and instead split the proposed new housing across seven sites in an attempt to retain the compact character of the market town and mitigate adverse impacts from unbalanced expansion (Thame Town Council 2012). Out of the first 75 neighbourhood plans to become part of the development framework over half allocated sites for housing, and 90% had policies on housing with most specifically about affordable housing (DCLG 2015a). Many neighbourhood plans made explicit their opposition to the dominant housing market model and the speculative approach of the volume housebuilders.

Government strategies to rectify 'a broken housing market' (NHF 2011) and increase supply have focused on the role of the volume builders who dominate the development industry. Ten companies produce 44% of all new homes built in England and there has been a long-term market concentration in the housebuilding industry where the largest company has 8% of the market share (Adams et al. 2009). The restricted supply of land in the UK creates incentives for firms to combine both land development and housing construction functions. The larger sized firms that are created can therefore employ strategies to influence local land markets through their housing strategies. Acting strategically, large housebuilders are selective about when they release properties onto the market to control the supply, and therefore prices, so that they can gain a competitive advantage in purchasing or securing options on more land (Ball 2003). The price of land in England has risen by 300% in the last 10 years and there is considerable revenue to be generated by trading on the increased value that accrues once planning permission for housing has been granted (Hetherington 2015). The exponential growth in the revenues of the big five housebuilders (Barratt Developments PLC, Taylor Wimpey PLC, Persimmon PLC, Berkeley Group Holdings PLC, and Bellway PLC) has not been matched by a comparable increase in housebuilding output but is evidenced in the size of their growing land banks (Ruddick 2015).

Since the global financial crisis of 2008, the volume housebuilders have adopted strategies to maximise value over volume, producing detached housing on greenfield sites for an affluent market, and maintaining high prices by retaining substantial banks of land (Archer and Cole 2014). The liberal planning regime, in which neighbourhood plans emerged, dramatically boosted planning approvals for housebuilding in England. The requirement on planning authorities to specify 5 years supply of land resulted in a string of appeal victories for developers who succeeded in winning access to greenfield sites. If the Local Plan did not provide the requisite housing land, it would be found 'out of date' enabling speculative building that did not reflect assessments of housing need (Burroughs 2015). Neighbourhood planning became a significant obstacle to the volume builders when a Local Plan was found to be 'out of date' but the neighbourhood plan passed examination and subsequently took precedence over local housing policies. The potential for communities to devise their own housing plans signalled to developers that neighbourhood planning posed a threat to their interests and they determined to challenge it through the system of planning appeal and through subsequent legal action (Peters 2014).

18.5 Challenges to Neighbourhood Housing Plans

In November 2015, with over 100 neighbourhood plans in place and a further 1700 underway, the Housing and Planning Minister announced the success of the policy in increasing housebuilding. Analysis of selected neighbourhood plans showed they had allocated more sites for housing than required in the strategic Local Plan (Mountain 2015). The findings were presented as confirming the impact of neighbourhood planning in encouraging communities to support housebuilding. The Minister Brandon Lewis said:

> *'We are scrapping the broken old planning system that pitted neighbours and developers against each other, and cornered people into opposing any development in their back yard. Our approach of getting the whole community working together is paying off, and breaking through local opposition'* (DCLG and Lewis 2015).

Far from ending a system that pitted communities against housebuilders, however, the policy of neighbourhood planning had, if anything, exasperated this antagonism. The first legal challenge from housebuilders came when the neighbourhood plan for the parish of Tattenhall in rural Cheshire set a ceiling of no more than 30 homes per site in the built-up part of the village. The Tattenhall neighbourhood plan was produced in the absence of strategic housing policies, since the local planning authority Cheshire West and Chester Council were still preparing their own Local Plan and could not evidence a 5 year land supply. It was approved at examination and Tattenhall neighbourhood plan was successful at referendum in September 2013 on a convincing 52% turnout. Housebuilders Barratt Homes and Wainhomes sought a judicial review of Cheshire West's decision to go to referendum arguing, among other grounds, that the neighbourhood plan sought to restrict the delivery of housing and therefore did not comply with pro-growth national planning policy. Mr. Justice Supperstone dismissed the housebuilder's case ruling that the neighbourhood plan did not seek to limit the overall number of homes and had established its case for housing development at a scale that reflected the existing character of the area (Barratt Homes, and Wainhomes Developments v Cheshire West, and Chester Borough Council, Stephen Robinson, and Tattenhall, and District Parish Council 2014).

The right of neighbourhood plans to identify their own sites for housebuilding was also challenged by the volume housebuilders in legal action. Larkfleet Homes sought a judicial review of the Uppingham neighbourhood plan in Rutland on the grounds that housing site allocation was a strategic matter and the responsibility of the planning authority. Uppingham had allocated three sites for at least 170 homes but Larkfleet had a commercial interest in land not included in the plan and demanded that the referendum result be quashed. Mr. Justice Carter dismissed the claim since neighbourhood planning regulations expressly allowed neighbourhoods to allocate sites for development (Larkfleet Homes v Rutland County Council and Uppingham Town Council 2014). Larkfleet appealed to the High Court and the Court of Appeal and when defeated tried to take their challenge to the Supreme Court only to be refused again. In July 2014, the Minister for Planning issued a Ministerial Statement that brooked no doubt on the government's intentions to intervene in this emerging conflict between the housebuilders and communities (DCLG and Bowles 2014). His edict made decisive revisions to the legislation on planning appeals to enable the Secretary of State to scrutinise any appeal for housebuilding in a neighbourhood planning area.

These powers were tested in the case of Broughton Astley, a large village near Leicester with a population of around 9000 people, whose neighbourhood plan was agreed in January 2014 after a referendum on a turnout of 38%, with an 89% vote in favour. Prior to the referendum, an application by a developer to build 111 homes on a site not included in the neighbourhood plan was rejected by Harborough District Council, a decision overturned on appeal, but then reinstated by the Secretary of State and endorsed by the High Court (Crane v SSCLG 2015). The developer's central challenge was that the Broughton Astley neighbourhood plan was based on a Local Plan that was 'out of date' and could not demonstrate a 5 year supply of housing. The sites allocated for housing in the neighbourhood plan, while meeting some of the local housing need, therefore might not meet all needs. The Secretary of State's judgement cuts to the heart of the dichotomy between devolution and liberal market growth, and it is worth quoting in some length:

'In this appeal case he considers that the key issue in applying the presumption is whether any adverse impacts of the proposal would significantly and demonstrably outweigh the benefits, when assessed against the policies in the Framework taken as a whole including its policies on neighbourhood planning as well as policy on housing supply.

*Paragraph 185 of the Framework states that, outside the strategic elements of the Local Plan, neighbourhood plans will be able to shape and direct sustainable development. The Secretary of State regards this purpose as more than a statement of aspiration. He considers that neighbourhood plans, once made part of the development plan, should be upheld as an effective means to shape and direct development in the neighbourhood planning area in question' (*DCLG 2014:4*).*

The Secretary of State found that the community empowerment aims of neighbourhood planning were more important than ensuring housing growth. He gave more weight to the neighbourhood's right to determine the location of development than to the fact that no 5 year supply of land could be evidenced. The intervention of the Secretary of State in this and other appeals and legal action by housebuilders over neighbourhood plans evidenced a political will to defend the right of neighbourhoods to exercise a level of autonomy over housing development. A complex framework of case law was established in which draft neighbourhood plans might be given the same substantial weight as those that had been approved at referendum; plans based on out-of-date housing numbers could be upheld, and plans that did not demonstrate site viability or conform to European directives might still be allowed. While neighbourhoods were often unsuccessful in these contests with the housebuilders, the spatial practices of neighbourhood plans with regard to housebuilding were shaped by such judgements and appeared as a result to offer considerable latitude for communities to direct and contain housing development within an overall vision of place.

In February 2016, national planning policy was changed to constrain the remit of neighbourhood plans where the allocation of housing sites in the strategic Local Plan was deemed out of date. The amendments opened up the housing policies of neighbourhoods to further challenge if no 5 year supply of land was evidenced. The volume housebuilders had lobbied government to this effect arguing that 'neighbourhood plans are becoming an instrument only for "limited growth" at a level acceptable to local communities' (Young and Burcher 2014:67). The next section explores the neighbourhood spatial practices of housing delivery objected to by developers.

18.6 Housebuilding, But Not as We Know It

The most common policy in neighbourhood plans was the promotion of local distinctiveness and place identity (DCLG 2015a). This predominant concern was set within what was considered an overall balance of policies between 'protectionism' and pro-development (Peters 2014). Its effect, however, was to shape the spatial practice of neighbourhood plans, so that their housing policies enhanced a sense of place and provided for identified local need. Where neighbourhood plans aimed to increase housing numbers, they prioritised small, previously developed or 'brownfield' sites, where development would cause minimum disruption to environmental quality and local character. They were especially concerned to deliver affordable homes to meet local housing need and favoured resident-led approaches in custom-build and community land trusts that might lock-in affordability for the future.

The delivery of community-led housing plans, with the identification of small sites for affordable homes, potentially custom-built or run by a community land trust, presented a very different equation to the speculative approach of the volume builders, and was sometimes explicitly in opposition to the corporate model of housebuilding. One example was the neighbourhood plan of Slaugham, a parish of 2000 people in Sussex, where in January 2014 the first community right to build orders providing sites for at least 130 new custom-build affordable homes delivered through a parish-run community land trust 'built and owned to meet local affordable needs' went to examination (Slaugham Parish Council 2013:18). The neighbourhood plan for Petersfield, a town in Hampshire of 6500 households, earmarked sites for up to 112 self- and custom-built homes restricted to people with a local connection, and gave its strong support for co-operative housing ventures. The plan stated: 'Self-build dwellings are likely to cost less than the market equivalent and the dwellings that are built will tend to be better quality with more innovative architecture than a standard developer's offering' (Petersfield Town Council 2013:17). The neighbourhood plan for Frome, a town in Somerset, lambasted the standard of new estates built by the volume housebuilders and expressed its desire to rigorously control any such future development, stating: 'There is strong support for self-build and community-led development and…Such housing is likely to be more sustainable, affordable and community focused than conventional development' (Frome Town Council 2014:14). The neighbourhood plan for Arundel, a town of 3650 homes on the Sussex coast, was innovative in securing previously developed or 'brownfield' sites for housing use, stating that it intended to send 'a strong signal to landowners and developers that it [Arundel] will not support speculative planning applications for housing development on the lower-cost, green field sites on the edge of the town' (Arundel Town Council 2014:26).

This preference for 'brownfield' and small-scale custom-build and affordable housing in neighbourhood plans can be evidenced in interviews with neighbourhood planning participants. The north Pennine parish of Allendale included policies for housebuilding in their neighbourhood plan to give their post-industrial community a sustainable future. These plans for growth needed to be tailored to suit the place identity of a wild and remote upland landscape. The parish clerk explained:

> *'What we've said in the plan is that we will support small-scale developments. Now if someone comes along and wants to put up 50-plus houses, we would probably oppose it. The parish council was instrumental in setting up a community land trust and they've managed to obtain a bit of land to build another four houses. And they are doing it the way that we would like to see it done, seeing an opportunity to have three here and have two there, and so on'.*

In the inner-city regeneration area of Holbeck, in Leeds, members of the neighbourhood forum identified sites for residential growth and invited a local co-operative custom-build organisation to give a presentation to the planning group. But their plans foundered against the land banking strategies of 'high profile multi-wealthy' property companies, as two members of the forum (male and female) explain:

M: 'To my mind that would be ideal as a residential area, because the council have got to build, or give permission to build houses within the city to whatever level it is, and I think building down there would be ideal. But the Bank of Scotland or RBS have bought the land to sit on it to wait until the land values.

F: The market changes.

M: So whilst we have a plan in our minds, the kind of brick walls we are meeting are the fact that a lot of pieces of land are owned privately and we have got to persuade them.

F: That there is an alternative stance they can take that will offer the necessary amount of affordable housing that will still make money for them, rather than in the meantime, what is that land doing? Nothing; absolutely nothing'.

Neighbourhood planning's advocacy of small to medium builders, community ownership and custom-build 're-appropriated space' in planning policy for the use values of affordability and sustainability and triggered demands for a market shift in housebuilding (Clark 1994:937). The self-build and community-led affordable housing schemes enabled by neighbourhood plans directed government attention towards a European model of housing delivery where the custom-build market accounted for 50% of new homes (Wilson 2015). A Self-Build and Custom House Building Act was passed in April 2015, followed by a Housing and Planning Bill the same year, that introduced a so-called 'Right to Build' promoting self or custom-build and supporting small- and medium-size developers. The rationale for the Bill signalled government acceptance that the business model of the volume builders was constraining housing development and that enabling 'the smaller builders to increase their output could have the most impact on getting more homes built' (DCLG 2015c:34).

The behavioural intent behind neighbourhood planning was to redirect the behaviour of citizens to support housebuilding in a market system dominated by speculative building practices. In achieving its community empowerment outcomes, neighbourhood planning appeared to have authored new spatial practices of housing delivery that challenged the norms of the industry and resonated with the passion for place at the core of the community localism agenda. Opposition from developers to these spatial practices might suggest that the problems of housing supply lie not with the refusal of communities to support growth, but with the dominant model of housebuilding and its dysfunctional market.

18.7 Conclusion

Neighbourhood planning emerged at the juncture of community localism and the liberalisation of economic growth and it aimed to devolve power to citizens in order to win their support for increased housebuilding. Evaluated according to its success in boosting housing numbers, neighbourhood planning appeared to demonstrate citizen acquiescence to the agenda of spatial liberalism. The hostility of the volume housebuilders, however, suggested the emergence of new spatial practices in neighbourhood planning housing allocations. In its devolution of planning powers to communities, neighbourhood planning gave licence to a model of housebuilding that promoted small and medium sized companies, affordable community-led and custom-build housing on previously developed sites, rather than the greenfield speculative strategies of the volume builders. It aimed to balance the imperatives of housebuilding with the priorities of place identity, heritage and environmental protection and it authored a spatial practice of housing delivery that resonated with the passion for place at the core of the localism agenda. Neighbourhood planning can be seen as a reappropriation of space from the dominant market model with the neighbourhood emerging as the proponent of sustainability and social purpose in housebuilding. The need to evidence community empowerment to ensure the success of this policy meant that government was required to support neighbourhood plans even when that meant acting against the corporate interests of the volume housebuilders. State housing strategies were changed accordingly with new initiatives in support of custom-building introduced to restructure the share of the market enjoyed by the volume builders and support the grassroots initiatives of neighbourhoods. The devolution of planning power to the community had been successful in orienting citizens towards housebuilding but it had achieved a corresponding impact on the model of liberal development promoted through localism. Neighbourhood planning endorsed the spatial practices of lived or representational space in contrast to the exchange values promoted in the current market model. It directed citizens to an awareness of housing needs not the market needs of the volume housebuilders. The balance between community engagement and spatial liberalism appeared to have undergone a similar qualitative shift with empowerment and sense of place emerging as new arbiters of development planning.

References

Adams, D., Leishman, C., & Moore, C. (2009). Why not build faster? Explaining the speed at which British house-builders develop new homes for owner-occupation. *Town Planning Review, 80*(3), 291–314.

Archer, T., & Cole, I. (2014). Still not plannable? Housing supply and the changing structure of the housebuilding industry in the UK in 'austere' times. *People, Place and Policy, 8*(2), 97–112.

Arnstein, S. (1969). A ladder of citizen participation. *American Institute of Planners Journal, 35*(4), 216–224.

Arundel Town Council. (2014). *Arundel Neighbourhood Plan 2014–2029*. Sussex: Arundel Town Council.

Ball, M. (2003). Markets and the structure of the housebuilding industry: An international perspective. *Urban Studies, 40*(5–6), 897–916.

Barratt Homes & Wainhomes Developments v Cheshire West & Chester Borough Council, Stephen Robinson & Tattenhall & District Parish Council. (2014). EWHC 1470 Q.B.

Bradley, Q. (2014). Bringing democracy back home: Community localism and the domestication of political space. *Environment & Planning D: Society & Space, 32*(4), 642–657.

Bradley, Q. (2015). The political identities of neighbourhood planning in England. *Space and Polity*, 97–109. doi:10.1080/13562576.2015.1046279.
Bramley, G., & Watkins, D. (2016). Housebuilding, demographic change and affordability as outcomes of local planning decisions: Exploring interactions using a sub-regional model of housing markets in England. *Progress in Planning, 104*, 1–35.
Brenner, N., & Theodore, N. (2002). *Spaces of neoliberalism; Urban restructuring in North America and Western Europe*. Oxford: Blackwell.
Burningham, K. (2000). Using the language of NIMBY: A topic of research not an activity for researchers. *Local Environment, 5*(1), 55–67.
Burroughs, L. (2015). *Getting Homes Built*. Housing Foresight Paper No. 4. London: Campaign for the Protection of Rural England.
Clark, H. (1994). Taking up space: Redefining political legitimacy in NYC. *Environment and Planning A, 26*, 937–955.
Clark, the Rt., Hon, G., MP. (2011). *Local planning for sustainable development*. Speech to Campaign to Protect Rural England London. Department of Communities & Local Government.
Clarke, N., & Cochrane, A. (2013). Geographies and politics of localism: The localism of the UK's coalition government. *Political Geography, 34*, 10–23.
Crane v SSCLG. (2015). *EWHC 425 (Admin)*.
Cook, N., Taylor, E. J., & Hurley, J. (2013). At home with strategic planning: Reconciling resident attachments to home with policies of residential densification. *Australian Planner, 50*(2), 130–137.
DCLG. (2011). *Impact assessment to the localism bill: Neighbourhood plans and community right to build*. London: DCLG.
DCLG. (2012). *National planning policy framework*. London: DCLG.
DCLG. (2013). *You've got the power. A quick and simple guide to community rights*. DCLG: London.
DCLG (2014). Town and Country Planning Act 1990—Section 78 Appeal by Mr. I. P. Crane—Site at Land South of Hallbrook Primary School, Crowfoot Way, Broughton Astley, Leicestershire [Internet]. Retrieved August 10, 2015, from http://www.planningportal.gov.uk/theplanning inspectorate
DCLG. (2015a). *Notes on neighbourhood planning 16*. London: DCLG.
DCLG. (2015b). *My community rights*. London. Locality [Internet]. Retrieved Novomber 2, 2015, from http://mycommunity.org.uk.
DCLG. (2015c). *Housing Bill 2015/16 impact assessment*. London: Department of Communities and Local Government.
DCLG, & Bowles, N. (2014). *Written statement to parliament: Neighbourhood planning*. London: DCLG.
DCLG & Lewis. (2015). Neighbourhood planning powers boosting plans for housebuilding by more than 10%. Press release 2 November. Internet. Retrieved Novomber 2, 2015, from https://www.gov.uk.
Davoudi, S., & Cowie, P. (2013). Are English neighbourhood forums democratically legitimate? *Planning Theory and Practice, 14*(4), 562–566.
Davoudi, S., & Madanipour, A. (2015). *Reconsidering localism*. London: Routledge.
DeVerteuil, G. (2013). Where has NIMBY gone in urban social geography? *Social and Cultural Geography, 14*(6), 599–603.
Devine-Wright, P. (2012). Rethinking NIMBYism: The role of place attachment and place identity in explaining place-protective action. *Journal of Community and Applied Social Psychology, 19*, 426–441.
Fraser, N. (1997). *Justice interruptus: Critical reflections on the 'post socialist' condition*. London: Routledge.
Frome Town Council. (2014). *A neighbourhood plan for Frome 2008–2028*. Frome Town Council.
HM Government. (2011). *Laying the foundations: A housing strategy for England*. London: Department of Communities & Local Government.
Hall, S., & Massey, D. (2010). Interpreting the crisis. *Soundings, 44*(Spring), 57–71.
Hetherington, P. (2015). *Whose land is our land?* Bristol: Policy Press.

Larkfleet Homes v Rutland County Council & Uppingham Town Council. (2014). *EWHC 4095 Q.B.*
Lefebvre, H. (1991). *The production of space*. Trans. D. Nicholson-Smith. Oxford: Blackwell. Original work published 1974.
Localism Act 2011 (c.20) London: The Stationery Office.
Locality. (2014). *The power of neighbourhood planning*. London: Locality.
Lowndes, V., & Sullivan, H. (2008). How low can you go? Rationales and challenges for neighbourhood governance. *Public Administration, 86*(1), 53–74.
McCann, E. J. (1999). Race, protest & public space: Contextualising Lefebvre in the US City. *Antipode, 31*(2), 163–184.
Matthews, P., Bramley, G., & Hastings, A. (2015). Homo economicus in a big society: Understanding middle-class activism and NIMBYism towards new housing developments. *Housing Theory & Society, 32*(1), 54–72.
Mihaylov, N., & Perkins, D. (2015). Local grassroots environmental activism. *Behavioural Sciences, 5*, 121–153.
Miraftab, F. (2009). Insurgent planning: Situating radical planning in the global south. *Planning Theory, 8*(1), 32–50.
Mountain, C. (2015). *Neighbourhood planning: Progress on housing delivery*. London: DCLG Neighbourhood Planning Team.
NHF. (2011). *Home truths. Fixing our broken housing market*. National Housing Federation: London.
Parker, G. (2015). *The take-up of Neighbourhood Planning in England 2011–2015*. Working Paper in Real Estate and Planning. September. Reading. University of Reading. Retrieved January 2016, from http://www.centaur.reading.ac.uk.
Parker, G., Lynne, T., Wargent, M., & Locality. (2014). *User experience of neighbourhood planning in England research*. Reading: University of Reading.
Parker, G., Lynn, T., & Wargent, M. (2015). Sticking to the script? The co-production of Neighbourhood Plans. *Town Planning Review, 86*(5), 519–536.
Peters, R. (2014). *Neighbourhood plans: Plan and deliver?* London: Turley.
Petersfield Town Council. (2013). *Petersfield's neighbourhood plan*. Petersfield Town Council.
Rose, N. (1999). *Powers of freedom: Reframing political thought*. Cambridge: Cambridge University Press.
Ruddick, G. (2015). Revealed: Housebuilders sitting on 600,000 plots of land. *The Guardian*. Wednesday 30 December. London: Guardian Newspapers.
Ruming, K., Houston, D., & Amati, M. (2012). Multiple suburban publics: Rethinking community opposition to consolidation in Sydney. *Geographical Research, 50*(4), 421–435.
Slaugham Parish Council. (2013). *Slaugham neighbourhood plan*. Slaugham Parish Council.
Sturzaker, J. (2011). Can community empowerment reduce opposition to housing? Evidence from Rural England. *Planning Practice and Research, 26*(5), 555–570.
Taylor, E. (2013). Do house values influence resistance to development?—A spatial analysis of planning objection and appeals in Melbourne. *Urban Policy and Research, 31*(1), 5–26.
Taylor, E., Cook, N., & Hurley, J. (2016). Do objections count? Estimating the influence of residents on housing development assessment in Melbourne. *Urban Policy and Research*. doi:10.1080/08111146.2015.1081845.
Upper Eden Community Interest Company. (2012). *Upper Eden Neighbourhood Development Plan 2012–2025*. Cumbria: Upper Eden CIC.
Wills, J. (2016). Emerging geographies of English localism: The case of neighbourhood planning. *Political Geography, 53*, 43–53.
Wilson, W. (2015). *Self-build and custom-build housing (England)* . London: House of Commons Library.Briefing Paper Number 06784
Wolsink, M. (2006). Invalid theory impedes our understanding: A critique on the persistence of the language of NIMBY. *Transactions of the Institute of British Geographers, 31*, 85–91.
Young, C. & Burcher, J. (2014). *Love Thy Neighbour: An update on neighbourhood plans*. London. No 5 Chambers [Internet]. Retrieved from https://www.no5.com/.

Chapter 19
Integrated Façade System for Office Buildings in Hot and Arid Climates: A Comparative Analysis

Yahya Ibraheem, Poorang Piroozfar, and Eric R.P. Farr

Abstract High solar gain in hot and arid climates causes overheating in summer and increases the demand for air conditioning, energy and carbon footprints of buildings. There are different strategies, tools and solutions to address this problem, but Integrated Façade Systems (IFSs)—façade systems where different technological solutions are integrated to improve performance and to lower the impact of the building—are still an underdeveloped yet a fast-growing field of research. Such systems can reduce solar heat gain, lower air conditioning costs and lessen glare, while maximising the use of natural light and help produce energy if combined with PV technology. Previous research has addressed one or a combination of some of the influential factors on performance of such systems, yet there still is a gap in the state-of-the-art research in comprehensive systematic approach not only to help gauge the impact of alteration of parameters on the IFS performance, but also an approach which can be deployed in other studies where the focus is on façade systems. With a special reference to office buildings in hot and arid climates, this chapter sets out to systematically identify IFS parameters which have potential impacts on energy, lighting, glare and heat gain. Then as a part of a comprehensive ongoing research in this area, this chapter presents a proof of concept to demonstrate the application of such methodology to a parametric study of IFS technology. In doing so, it chooses only one of the parameters indicated in this systematic review and uses building simulation as its core method to investigate the influence and impacts of those variations on performance of IFS. It will indicate how this

Y. Ibraheem (✉)
Department of Architectural Engineering, University of Technology,
P.O. Box: 18310, Baghdad, Iraq

School of Environment and Technology, University of Brighton, Brighton BN2 4GJ, UK
e-mail: Y.Ibraheem@brighton.ac.uk

P. Piroozfar
School of Environment and Technology, University of Brighton, Brighton BN2 4GJ, UK

E.R.P. Farr
NONAMES Design Research Foundation, 1249 F Street, San Diego, CA 92101, USA

M. Dastbaz et al. (eds.), *Building Information Modelling, Building Performance, Design and Smart Construction*, DOI 10.1007/978-3-319-50346-2_19

approach provides high flexibility to adjust or configure any combination of those parameters and to measure, subjectively, how this will result in change in façade performance.

19.1 Introduction

The integrated design has been gaining momentum in built environment and architectural design over the past two decades and façades are no exemption and as such have witnessed major technological improvements in this regard. Integrated Façade Systems (IFSs)—systems where different technological solutions are integrated to improve performance and to lower the impact of the building—are a fast-growing field of research with broad scope for application and significant positive impact on the environment. Such systems can reduce solar heat gain, lower air conditioning rates and lessen glare, while maximising the use of natural light. The technological solutions used in IFS can broadly be classified under three components: (1) High-Performance Glazing (HPG), (2) Shading Devices (SD) and (3) Integrated Photovoltaics (IPV). Moreover, the literature on these systems in fully- or highly-glazed buildings is scarce and even more so for non-residential buildings or buildings in hot and arid climates. When considering IFSs, shading devices can be considered as an effective part of the façade components that plays a significant role in reducing the negative impact of the solar radiation (in the form of heat gain) and help provide an acceptable indoor condition. Besides, solar radiation can be harvested using photovoltaic (PV) cells and fed into the building itself, hence the integration of PV Shading Devices (PVSD). Although previous research has addressed some or a combination of influential factors on the performance of such systems, there still is a gap in the state of the art that is a comprehensive and systematic approach to gauge the impact of alteration of parameters on the IFS performance, with full potential to help best configure such solutions for different design contexts. This chapter aims to establish and demonstrate the application of the systematic approach to the evaluation of the performance of integrated facade systems. To fulfil this aim, this chapter sets out to contribute to the knowledge in the field by answering the research questions as follows:

- Can integrated facade systems contribute to a more environmentally concerned approach to the design of buildings? What are, if there is any, benefits of such integration? And how can they be adjusted, calibrated or configured to maximise those benefits?

To achieve that aim this study attempts to provide data using building simulation as the main method and is aimed at offering a systematic approach to enhance the built environment in developing economies with a research prospect that investigates the influence of those parameters for further in-depth study to achieve the optimum trade-off between the livable indoor environment, energy consumption and carbon footprint.

19.2 Literature Review

As a part of the façade components, shading devices play a significant role in reducing the heat gain and providing acceptable indoor conditions (Alzoubi and Al-Zoubi 2010). Although the application of PV in buildings was introduced in the late 1970s, it was first characterised as a building-integrated component in late 1990s (Patankar 2010) and it was not until 1998 that Yoo and Lee proposed, most probably for the first time, integrated photovoltaics as shading devices. Combining external solar shading devices and photovoltaic panels has many advantages (DGS 2008), such as generating green energy which reduces the use of fossil fuels as well as adding architectural features specific to the design of shading devices when combined with photovoltaic panels. However, the application of PVSD has significant challenges due to the complexity of the system and the adaptability of these systems to different contextual conditions (Lee et al. 2009). Nevertheless, it is important to note that integration of PV panels, what is commonly known as 'Building-Integrated Photovoltaic' or BIPV, is not limited to shading devices only. They can be integrated into any part of the building that can potentially receive a considerable amount of solar radiation like windows, claddings, skylight as well as external shading devices (BCA 2008).

Photovoltaics Shading Devices (PVSD) are usually an external building skin layer that can be applied independently in both new and existing buildings. This technology has the dual advantage of generating electricity directly from the incident sunlight and the normal function of external shading devices in protecting the building from overheating, providing visually comfortable interior space and saving energy (Zhang 2014; Kang et al. 2012). They have proven technical advantages over other types of PV installations like roof stand-alone PV systems (Mandalaki et al. 2014a). Advantages include ease of maintenance, ease of inspection, allowing the roof space for other uses and higher possibilities to integrate kinetic technologies that can track the sun while acting as an interactive solution for optimising solar gain throughout the year. In order to appropriately apply this technology into a building, it is essential to highlight the main influential parameters that affect the performance of buildings with PV shading devices to be able to systematically and comparatively investigate the impact pertaining to adjustments of such parameters. These parameters include but are not limited to building geographical location (Bahr 2013), building orientation (Yoo and Manz 2011), type of shading devices (Mandalaki et al. 2012), inclination angle of louvers (Hwang et al. 2012) and the dimensions of the louvers (Kang et al. 2012), to name a few.

The comprehensive critical review of the literature led to classifying the existing state of the art into three systemic levels divided into three categories i.e. 'design considerations and configurations', 'performance aspects' and 'assessment methods'. This systemic approach was then used as a methodological framework for this study. To be able to stay within the limits, the second two categories i.e. 'performance aspects' and 'assessment methods' will not be reviewed in this chapter and the development of the methodological approach has not been included in this chapter but the approach itself will be explained in the next section (Please see Fig. 19.3).

19.2.1 Design Considerations and Configurations

Studies concerned with design consist of two different subcategories i.e. design considerations and design configurations. Design considerations are the factors over which there is limited to no control but they need to be taken into account when the design process of building or the course of façade is being carried out. Design configurations, by contrast, are those elements which can be adjusted, changed or manipulated by the designer and are accounted for as a part of the project that can be shaped by the design process.

At context level, latitude and geographical location were considered with a direct impact on solar radiation, temperature, sky conditions while other climatic parameters were pointed out as a means to determine type and dimensions of shading (Bahr 2009), and optimal design option (Bahr 2013). Relative layout of the roads to the building shape (Di Vincenzo et al. 2010), diffuse radiation and its correlation with urban design to help decide where to install PVSDs to maximise electricity generation (Tongtuam et al. 2011) were also studied with some others looking into the effect of architectural and technical aspects of the PVSD to predict their performance at urban scale using GIS (Karteris et al. 2014).

At building level, although not commonly agreed down to the detailed levels, orientation is still considered as a key determinant to optimise PV shading devices (Bahr 2009; Bahr 2013; Kang et al. 2012; Yoo 2011). Building functions such as weather proofing, noise reduction, shading, flexibility, transparency, colour and texture can be affected by the application of PVs (Vassiliades et al. 2014). The impacts of building types and proposed PV solutions on potential architecturally-suitable areas of a façade were investigated by Karteris et al. (2014).

At envelope level, variations were studied by Youssef et al. (2015) where a tool was developed, under a specific climate condition for a specific building type. One of the most influential parameters determining the PVSD's performance is the angle of inclination that helps ensure an optimum value for both internal solar gain control and electric generation (Bahr 2013; Hwang et al. 2012; Kang et al. 2012; Kim et al. 2014). Probably one of the most worked over from different perspectives, yet still one of the least agreed upon areas at envelope level, is the tilt angle of the PV panels (Bahr 2009; Bahr 2014; Hwang et al. 2012; Jung 2014; Kim et al. 2010; Yang and Lu 2005). The size and dimension of the PV panels are also effective parameters that have been the focus of several studies (Kang et al. 2012; Mandalaki et al. 2014b; Sun and Yang 2010; Mandalaki et al. 2012) along with distance between louvres and their depth (Bahr 2009; Bahr 2014; Hwang et al. 2012), and the problem of overshading (Yoo and Lee 2002).

In many occasions, a clear link between the parameters at different systemic levels is quite evident.

19.3 Research Design and Methodology

This study is divided into two stages. The first stage outlines, very briefly, the approach developed for this research to facilitate systematic investigation of the influential factors and the second stage provides a proof of concept by simulating a subset of combinations of possible configurations to demonstrate how this study works and to set the scene for rolling out a full-fledged investigation of all possible combinations of such parameters and their impacts on energy and indoor comfort conditions.

This chapter utilises a methodology where the topic is looked into through systems theory. The idea of the building as a system was derived from modern systems theory and the application of building science to building performance (Kesik 2014). Piroozfar (2008) investigates the building envelope as 'the system', the building as 'the supersystem' and the façade components as 'the subsystem' to investigate the trade-offs in mass customisation of envelope systems using off-site production methods. This has then been further developed to investigate the application of BIM for a fully customisable façade system by Farr et al. (2014). A slightly different approach has been used for this study to also include the contextual determinants to facilitate a global systematic approach to the concept of IFSs. This study takes the building as 'the system', the context as 'the supersystem', and the façade as 'the subsystem' (Fig. 19.1). This triad systemic classification can and may be expanded further into next lower level which includes the façade components if and when a closer, more detailed investigation would be needed. This methodological

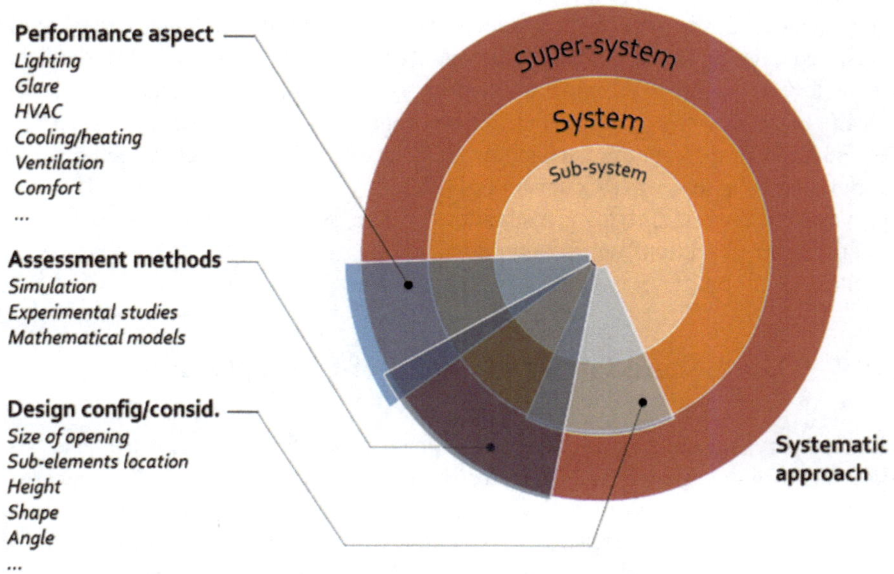

Fig. 19.1 Identified categories in the literature superimposed on the systematic approach

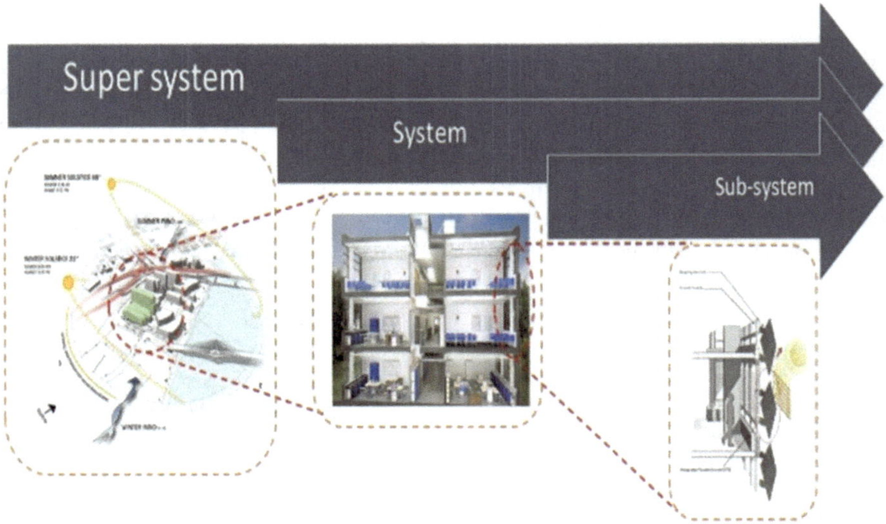

Fig. 19.2 Systematic approach developed for this study

approach has twofold benefits both at theory and practice level. It can facilitate not only the study of the literature on the topics related to that of this research but can also help classify their impacts and further enables the decision support for the course of intervention/action when it comes to the proposition of solutions for practical applications of building façades design.

The literature identified three main categories (Fig. 19.2)—design configurations and considerations, performance aspects and assessment methods—under which it clustered existing literature on PV-integrated shading devices at three different systemic levels, i.e. supersystem: context level, system: building level and subsystem: façade level. Design considerations and configurations are considered (Fig. 19.3).

Findings in the literature, which are related to the chosen factor, feed into the simulation to assess whether sensible results have been achieved or not. If it was proven to be meaningful, the investigations will be carried on. Otherwise, alteration of contextual conditions will be applied.

19.4 Data Generation for the Proof-of-Concept Case

An industry standard application for dynamic building performance analysis, IES-VE, which has been rigorously tested and validated against real-life scenarios was chosen for carrying out simulations for this study. Built-in components of IES-VE—MODELIT, SUNCAST, APACHESIM and FlucsDL—were used for building up the model, shading calculations, dynamic thermal simulation and daylighting analysis, correspondingly.

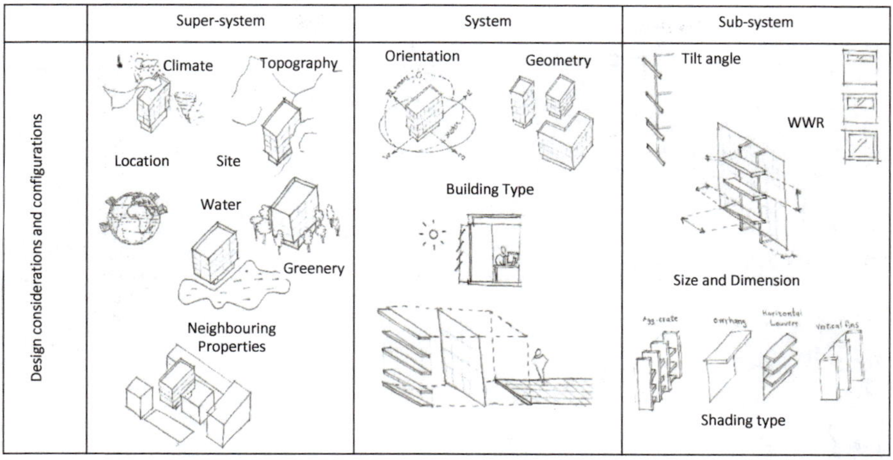

Fig. 19.3 Design considerations and configurations

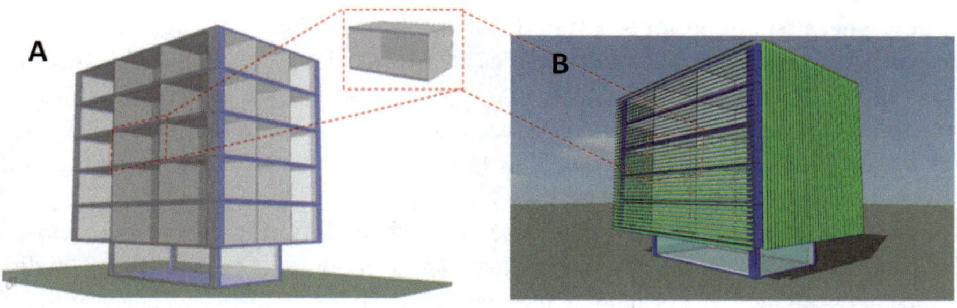

Fig. 19.4 Developed model for this study

19.4.1 Modelling Process in IES-VE and Simulation

Weather data file for Baghdad city (33°13'N, 44°13'E) was utilised for this simulation. Simulations were conducted for the whole year to determine the peak months of total energy consumption and solar gain. To be able to observe the sole impact of changing the angle of inclination, no internal thermal gain or occupants profiles were applied at this stage. Then construction materials were assigned to the building elements. The daylighting analysis was carried out using CIE standard clear sky condition in order to consider the impact of the application of shading devices in all scenarios. To ensure that the simulations are carried out in the most appropriate way, all other variables except for inclination angle of the louvers were kept constant throughout all simulations. Figure 19.4a shows the base case and Fig. 19.4b shows the case study with shading devices which was used to simulate performance under different inclination angle.

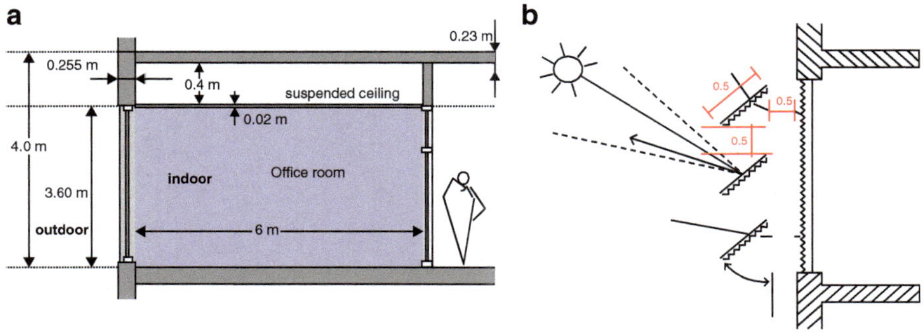

Fig. 19.5 Details of the model

19.4.2 Details of the Model

A simple six-storey office building with an area of 436 m^2 per each floor divided into nine thermal zones [each 9 m × 6 m × 4 m (L×W×H)] was developed for this study. The model represents a significant sample of office buildings in Iraq. The overall opening (i.e. windows) is defined in a form of a percentage of glass to the overall external envelope of the building, otherwise known as Window-to-Wall-Ratio (WWR) as of 80% for highly glazed buildings. In this model, the reflectance of the material used was set to 0.85 for the ceiling, 0.65 for the walls and 0.20 for the floor. Details of the model are shown in Fig. 19.5.

External shading devices were also defined by ModelIT. These devices are defined based on specific characterisation on the module (IESVE 2014). The distance between the louvers and the main façade and between the louvres are both 0.5 m, and the depth of the louvres is also 0.5 m. The louvers cover the full length of the façade. All surfaces of the office unit are considered to be adiabatic. Air conditioning and artificial lighting were not used and the average of daylighting above the minimum illuminance level of 500 LUX was calculated for each case. The glazing type is single glazing with a U-value: 5.1742, Tvis: 0.76 and SHGC: 0.9549.

19.5 Data Analysis

A base case scenario was first simulated to provide a benchmark as the worst possible scenario against which improvements could be measured. In this case, no shading devices were applied on a façade with an 80% of WWR where one thermal zone was simulated. The day selected for the simulation was 15th of June as suggested to be the highest average on record for Iraq based on CLIMATEMPS (2015). The alternative scenarios were simulated where horizontal louvers were applied to south-facing façade and vertical louvers were applied to east- and west-facing facades. The horizontal inclination angle of the louvers on the south-facing façade

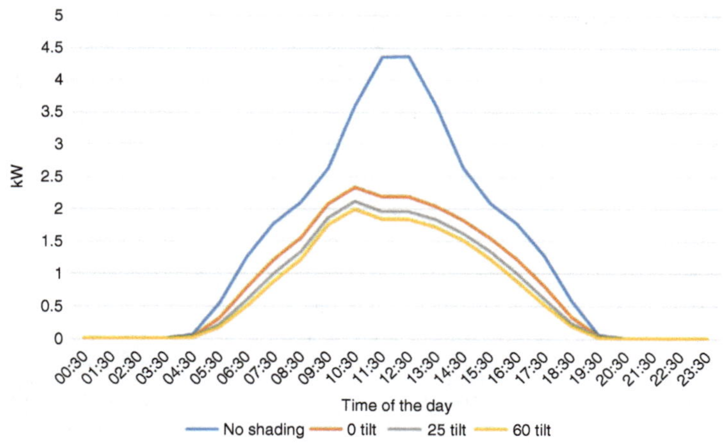

Fig. 19.6 Solar gain in the studied room on 15 June (first run)

was set to 0°, 25° and 60° as suggested by Bahr (2013) while the louvers on both the east- and west-facing facades were kept unchanged. The results were assessed based on the influence of this calibration on solar gain and the monthly cooling plant sensible load. Analysis of the first run of simulations is as follows:

19.5.1 Solar Heat Gain, the First Run

Amongst all 4 simulated cases [the base case and the three selected inclinations as suggested by Bahr (2013)], the highest solar gain was observed in the base case as expected where no shading devices were applied. When applying shading devices, a sharp decrease of solar gain was observed between the base case and the 0° inclination. Then a less significant decrease was observed from 0° inclination onwards. The decrease between 25° and 60° inclination is less than the decrease between 0° and 25° inclination as shown in Fig. 19.6. This observation is not as significant as suggested by Bahr (2013).

19.5.2 Cooling Loads, the First Run

There is a significant difference in the percentage of the monthly cooling loads compared to the base case. However tilting the angle to 25° and 60° was also not having as significant effect as suggested by Bahr (2013). A shift of the peak load was observed when the peak load changes from early August back to early June once shading devices on the south-facing façade were applied. Altering the inclination angle has clearly affected the cooling loads. During moderate seasons, a

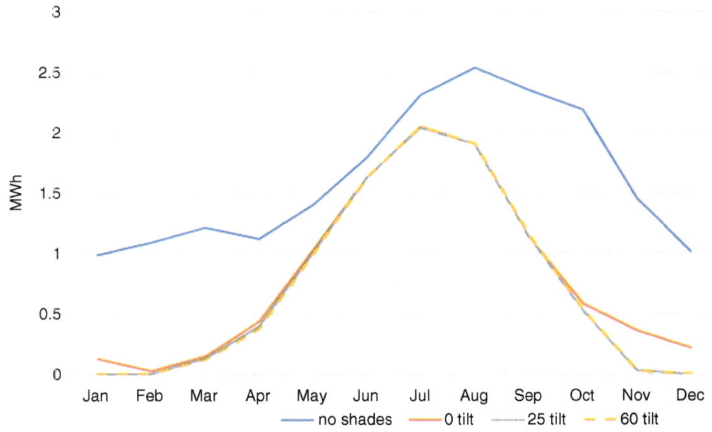

Fig. 19.7 Cooling plant sensible load (first run)

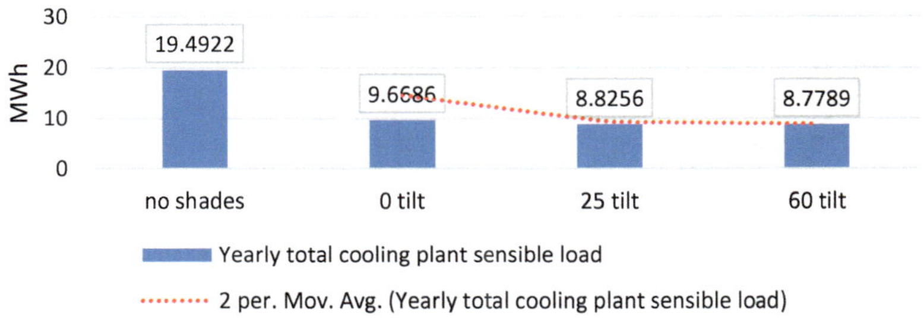

Fig. 19.8 Cooling plant sensible load-yearly totals (first run)

significant reduction was observed in all cases in comparison with the base case. However, 0° inclination has not brought the cooling load down to zero during the cold season compared to other inclination angles, i.e. 25° and 60°. During the hot season (April–October), the load was slightly different even though the inclination angle was kept unchanged or inclined to 25° and 60° (Fig. 19.7).

This effect is due to the low sun angle during moderate and cold seasons on the south façade. Therefore, further investigation to include heating loads as well is needed. On the system level (or the building level), the effect of modifying the inclination angle of horizontal shading devices on the south-facing façade can also be confirmed by looking at the results of the yearly cooling plant sensible load. Significantly high percentage of the yearly cooling plant sensible load was observed in the base case. This was significant as expected.

A significant improvement in the cooling load for the user room was observed when shading devices with 0° inclination were applied compared to the base case where yearly figures were reduced by up to 50%. However, the reduction in the cooling load was not significant when changing the inclination angle to 25° and 60° where the cooling load levelled off (Fig. 19.8). Therefore, the next step will be

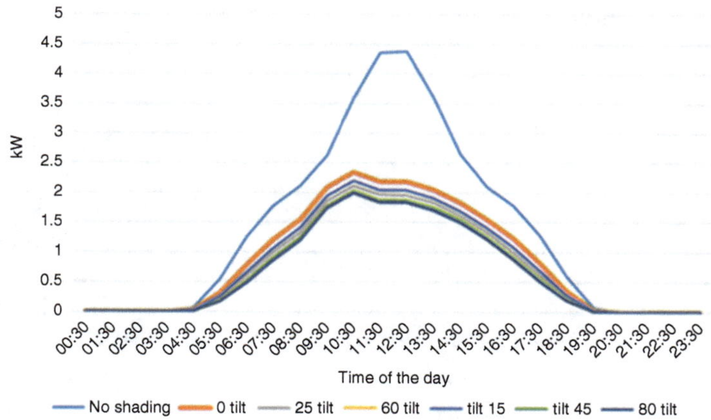

Fig. 19.9 Solar gain in the studied room on 15 June (second run)

adding more variations of inclination angle as suggested by other literature as opposed to Bahr (2013).

The second run of analysis was attempted on by further developing on the findings of literature. Sun et al. (2012) suggest that the optimum inclination angles for different designs vary from 30° to 50°. Since the simulated angles were 0°, 25° and 60°, it is reasonable to provide variations that cover the possible values of the inclination angles of spectrum. Therefore, the angles that will be simulated are 15°, 45° and 80°.

19.5.3 Solar Gain, the Second Run

Amongst the simulated cases in the second run of simulations, a less significant change in the trend of the decrease of solar gain was observed in reference to the first run of simulation results (Fig. 19.9).

19.5.4 Cooling Loads, the Second Run

Figure 19.10 shows that there is hardly any difference in the cooling load between the additional simulated angles. Moreover, when combining all the results, it can be seen that there is no significant change between all alterations (15°, 25°, 45°, 60° and 80°) compared to 0° inclination. In addition, these alterations have considerably reduced the cooling loads during the moderate season and have brought down the load down to zero during the cold season in comparison to 0° inclination.

The results from the first run of simulation can also be substantiated when combining them with the results of the second run of simulations. The same pattern

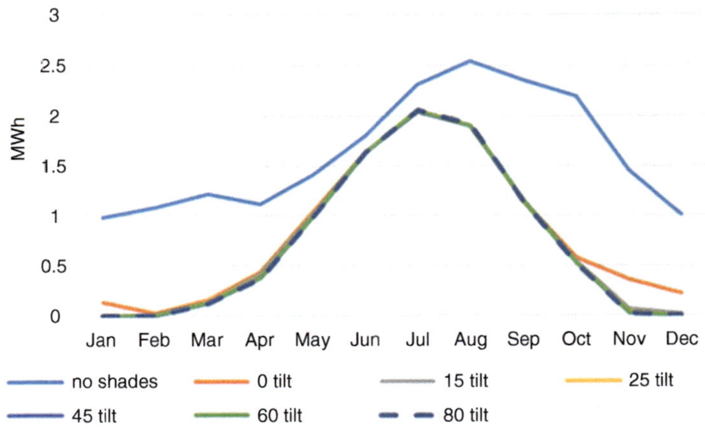

Fig. 19.10 Cooling plant sensible load (second run)

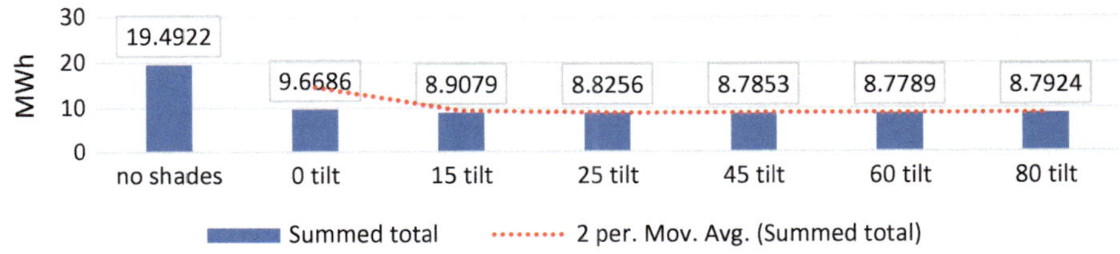

Fig. 19.11 Annual summed cooling plant sensible load for all simulated cases

was observed in the total yearly cooling plant sensible load for all simulated cases (Fig. 19.11). However, a slight increase in the load was observed in the case of 80° inclination which represents a fluctuation in the trend.

19.5.5 Daylighting Analysis

The average illuminance of user room in question was calculated for all cases on the same design day of 15th June at the midday (12:00 pm) using FlucsDL. It can be seen that there is a steady decrease in the average daylight. In the base case, the average daylight was 1350 LUX which exceeds the minimum average illuminance for office spaces that is 500 LUX (ASHRAE reported in O'connor et al. 2013). However, from 15° up to 60° a significant reduction of daylight can be observed due to changes in inclination angle where the space between two louvers is reduced, thereby reducing the daylight passing through to internal space. The case with 80° angle is the worst case where artificial lighting will be inevitable resulting in additional energy consumption not only because of the energy required for the artificial lighting itself but also for the additional heat generated as a result of that.

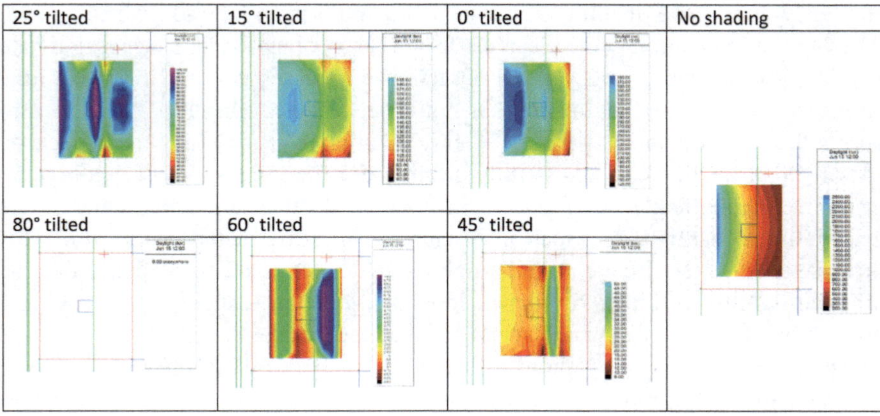

Fig. 19.12 Daylight analysis of user room

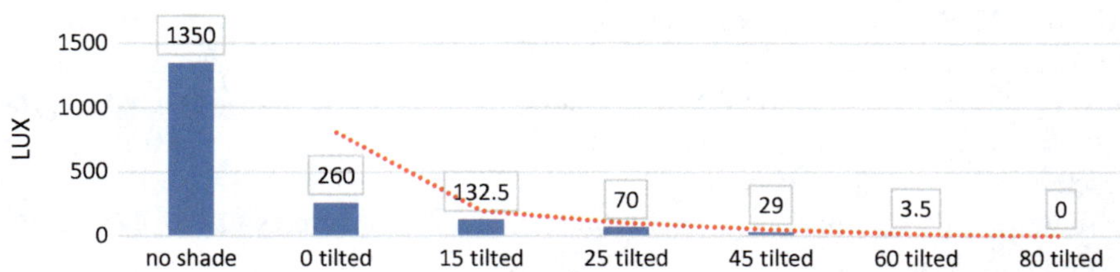

Fig. 19.13 Average illuminance in lux for the simulated cases

Figure 19.12 shows average daylight available in the simulated user room on 15th June at 12:00 pm.

Figure 19.13 shows average illuminance in LUX for all 7 simulated cases.

19.6 Discussion of Findings and Conclusions

Preliminary results of two rounds of simulation were presented where the analysis begins with establishing the scenarios to be simulated. A single variable—the inclination angle of horizontal louvers—was chosen to be adjusted and compared to the base case while the rest of the parameters were kept unchanged. The inputs used from the literature were 0°, 25° and 60° for the first run and no significant changes between angles above 0° were detected. Therefore the second run of simulation with 15°, 45° and 80° was carried out to investigate the effect of a change in angle in more detail. The consequences of each modification were evaluated based on solar gain, cooling load and natural daylighting as assessment criteria as suggested in the literature. Some of the findings were sensible as suggested by the literature whereas others were not. A shift of the peak cooling load was observed between early August and early June once shading devices on the south-facing façade were applied. Altering the inclination angle clearly affected the cooling loads. During moderate seasons, a significant reduction was observed in all cases in comparison with the base case. However, 0° inclination has not brought the cooling load down to zero during the cold season compared to other inclination angles. During the hot season (April–October), the load was slightly different for 0°, 25° and 60°. The alteration of the inclination angle was found to have a considerable impact on reducing the availability of daylighting which may cause an increase in the energy use. This increase is due to additional artificial lighting and additional cooling loads that the artificial lighting will add. All angles show a significant reduction in daylighting especially 60° and above, which nearly blocked the light.

The results of this study showed that it is possible to carry on with full-scale investigations as the approach and the expected outcomes were sensible and meaningful. Detailed additional contextual conditions will be used from the literature to inform the recalibration of the input for simulation. The angles tested at this stage will be used as the basis for the next stage of simulations where other parameters will be added to configurations. The results can help narrow down the number of simulations in the next stages. The major third function of IFSs—electricity generated by the PVSD, which will be calculated in the next stage—is expected to provide a counter-balance parameter with which the full if-then scenarios of this research can be further probed.

Future research will develop a combination matrix to include all possible façade configurations and the simulations for each scenario will be carried out. Results will provide optimised models and guidelines on the practical level with detailed analysis and expected performance/saving.

Acknowledgement This research has been sponsored by Iraqi Ministry of Higher Education and Scientific Research.

References

Alzoubi, H. H., & Al-Zoubi, A. H. (2010). Assessment of building façade performance in terms of daylighting and the associated energy consumption in architectural spaces: Vertical and horizontal shading devices for southern exposure facades. *Energy Conversion and Management, 51*, 1592–1599.

Bahr, W. (2009). Integration of shading devices and photovoltaic panels into existing building facades. In*Green retrofit conference*. Dubai: Dubai Knowledge Village.

Bahr, W.. (2013). Optimal design parameters of the blinds integrated photovoltaic modules based on energy efficiency and visual comfort assessments. *Central Europe towards Sustainable Building 2013*. Prague.

Bahr, W. (2014). A comprehensive assessment methodology of the building integrated photovoltaic blind system. *Energy and Buildings, 82*, 703–708.

BCA. (2008). *GREEN: Handbook-photovoltaic (PV) systems in buildings*. Singapore: Building and Construction Authority.

CLIMATEMPS. (2015). *Climate, average weather of Iraq* [Online]. Retrieved February 5, 2016, from http://www.baghdad.climatemps.com/temperatures.php

DGS. (2008). *Planning and installing photovoltaic systems: A guide for installers, architects and engineers*. London: Earthscan.

Di Vincenzo, M. C., Kesten, D. & Infiel, D. (2010). Assessment of performance of building shading device with integrated photovoltaics in different urban scenarios. *IEEE ICSET 2010*. Kandy, Sri Lanka.

Farr, E. R., Piroozfar, P. A., & Robinson, D. (2014). BIM as a generic configurator for facilitation of customisation in the AEC industry. *Automation in Construction, 45*, 119–125.

Hwang, T., Kang, S., & Kim, J. T. (2012). Optimization of the building integrated photovoltaic system in office buildings—Focus on the orientation, inclined angle and installed area. *Energy and Buildings, 46*, 92–104.

IESVE. (2014). *ShadingDevices*. Scotland: Integrated Environmental Solutions-Virtual Environment.

Jung, S. K. (2014). *Optimal control methods for PV-integrated shading devices*. Ph.D. thesis, University of Michigan.

Kang, S., Hwang, T., & Kim, J. T. (2012). Theoretical analysis of the blinds integrated photovoltaic modules. *Energy and Buildings, 46*, 86–91.

Karteris, M., Theodoridou, I., Mallinis, G., & Papadopoulos, A. M. (2014). Façade photovoltaic systems on multifamily buildings: An urban scale evaluation analysis using geographical information systems. *Renewable and Sustainable Energy Reviews, 39*, 912–933.

Kesik, T. J. (2014). *Building science concepts [Online]*. National Institute of Building Sciences. Retrieved April 24, 2016, from https://www.wbdg.org/resources/buildingscienceconcepts.php.

Kim, J. J., Sung Kwon, J., Yoon Seok, C., & Jeong Tai, K. (2010). Optimization of photovoltaic integrated shading devices. *Indoor and Built Environment, 19*, 114–122.

Kim, S.-H., Kim, I.-T., Choi, A.-S., & Sung, M. (2014). Evaluation of optimized PV power generation and electrical lighting energy savings from the PV blind-integrated daylight responsive dimming system using LED lighting. *Solar Energy, 107*, 746–757.

Lee, E. S., Selkowitz, S. E., Dibartolomeo, D. L., Klems, J. H., Clear, R. D., Konis, K. et al. (2009). *High performance building façade solutions: PIER final project report*. Public Interest Energy Research (PIER) Program. California, USA: Lawrence Berkeley National Laboratory.

Mandalaki, M., Papantoniou, S., & Tsoutsos, T. (2014a). Assessment of energy production from photovoltaic modules integrated in typical shading devices. *Sustainable Cities and Society, 10*, 222–231.

Mandalaki, M., Tsoutsos, T., & Papamanolis, N. (2014b). Integrated PV in shading systems for Mediterranean countries: Balance between energy production and visual comfort. *Energy and Buildings, 77*, 445–456.

Mandalaki, M., Zervas, K., Tsoutsos, T., & Vazakas, A. (2012). Assessment of fixed shading devices with integrated PV for efficient energy use. *Solar Energy, 86*, 2561–2575.

O'connor, J., Lee, E., Rubinstein, F. & Selkowitz, S. 2013. Tips for daylighting with windows: The integrated approach.

Patankar, C. (2010). *BIPV—Building integrated photovoltaic [Online]*. Solar Energy Explorer. com. Retrieved February 4, 2015, from http://www.solarenergyexplorer.com/bipv.html#axzz3QnEPfHCm.

Piroozfar, P. A. E. (2008). Mass-customisation: The application on design, fabrication and implementation (DFI) processes of building envelopes. Ph.D. thesis, University of Sheffield.

Sun, L., Lu, L., & Yang, H. (2012). Optimum design of shading-type building-integrated photovoltaic claddings with different surface azimuth angles. *Applied Energy, 90*, 233–240.

Sun, L. L., & Yang, H. X. (2010). Impacts of the shading-type building-integrated photovoltaic claddings on electricity generation and cooling load component through shaded windows. *Energy and Buildings, 42*, 455–460.

Tongtuam, Y., Ketjoy, N., Vaivudh, S., & Thanarak, P. (2011). Effect of the diffuse radiation reflection from exterior wall on shading device integrated photovoltaic case of Thailand building. *Energy Procedia, 9*, 104–116.

Vassiliades, C., Savvides, A., & Michael, A. (2014). Architectural implications in the building integration of photovoltaic and solar thermal systems—Introduction of a taxonomy and evaluation methodology. *World Sustainable Building*, Barcelona, Spain.

Yang, H., & Lu, L. (2005). The optimum tilt angles and orientations of PV claddings for building-integrated photovoltaic (BIPV) applications. *Journal of Solar Energy Engineering, 129*, 253–255.
Yoo, S.-H. (2011). Simulation for an optimal application of BIPV through parameter variation. *Solar Energy, 85*, 1291–1301.
Yoo, S.-H., & Lee, E.-T. (2002). Efficiency characteristic of building integrated photovoltaics as a shading device. *Building and Environment, 615–623*, 615–623.
Yoo, S.-H., & Manz, H. (2011). Available remodeling simulation for a BIPV as a shading device. *Solar Energy Materials and Solar Cells, 95*, 394–397.
Youssef, A. M. A., Zhai, Z. J., & Reffat, R. M. (2015). Design of optimal building envelopes with integrated photovoltaics. *Building Simulation, 8*(3), 353–366.
Zhang, X. (2014). Building performance evaluation of integrated transparent photovoltaic blind system by a virtual testbed. M.Sc., Eindhoven University of Technology.

Chapter 20
Decarbonising Construction Using Renewable Photosynthetic Materials

Craig White and Oliver Styles

Abstract The need to reduce CO_2 emissions from the operational energy used in buildings is more pressing as we seek to mitigate the effects of climate change. As we do so, the ratio of embodied emissions increases compared to operational energy emissions. The use of bio-based materials in construction might allow us to tackle both operational and embodied CO_2 emissions. The ModCell Straw Technology system achieves this by using the renewable materials timber and straw. The ModCell system has been used commercially on over 40 projects across the UK and secured a whole build system Passivhaus Component Certification in February 2015. ModCell not only uses renewable materials that deliver a 90% reduction in emissions from heating and cooling, it also delivers an effective carbon capture and storage solution (CCS). Bio-based renewable materials are created through photosynthesis. Plants are able to grow through the conversion of atmospheric CO_2 into useable carbon. Powered by photosynthesis, plants absorb atmospheric CO_2, deconstruct the molecule into its component atoms, carbon and oxygen and bank the carbon into complex sugars, the building blocks of cellulose. This natural CCS process allows us to exploit the physical properties of straw and timber to build with carbon. The amount of CO_2 banked into the system is more than is emitted through its making, resulting in a carbon negative footprint. This CCS mechanism means we can consider the built environment as a man-made carbon sink—the planet's sixth carbon sink. This approach is supported at a European level by funding from the EU's Horizon 2020 programme via the www.IsioBioProject.com. The ISOBIO project proposes an innovative strategy to bring bio-based construction materials into the mainstream. The use of bio-based aggregates for construction, which include insulation materials, hygrothermal and moisture buffering materials, binders, sol-gel and bio-based resins, will help create a decarbonised and healthier built environment.

C. White (✉)
Department of Planning and Architecture, University of the West of England,
Frenchay Campus, Coldharbour Lane, Bristol BS16 1QY, UK
e-mail: craig@white-design.co.uk

O. Styles
Oliver Style, Progetic Barcelona Office, C/Ramón Turró 100, 3-3, 08005 Barcelona, Spain

M. Dastbaz et al. (eds.), *Building Information Modelling, Building Performance, Design and Smart Construction*, DOI 10.1007/978-3-319-50346-2_20

20.1 Introduction

Passivhaus is now recognised as the most rigorous energy performance standard in the world. It seeks to deliver lower energy use in buildings by a number of integrated means. They are:

1. Reducing the heat transfer through the fabric of the building envelope by means of U-Value equal to or better than 0.15 W/(m^2 K).
2. A thermal bridge free design for key connection detail equal to or better than 0.01 W/(m^2 K).
3. All internal surface temperatures are maintained at no less than 17 °C when external temperatures are −10 °C and the internal ambient temperature is 20 °C.

Passivhaus certification can be for whole built projects where every part of the design is assessed to ensure that the above criteria can be met. There also exists what is known as Component Certification. If a project is made up of a selection of products and systems that are individually certified, then the journey to project certification can be made easier. Initially this was applied to components such as glazing and ventilation systems, where the manufacturer certifies with the Passivhaus Institute the performance of their product. This is now being more widely extended to the fabric of the whole building to include floors, walls, roof, basement and partition build-ups.

Because the demands of Passivhaus are so stringent, the desire to maximise u-Value performance has led to a preference for the use of foam and expanded polystyrene-based insulation products. While these types of materials are individually very high-performance materials, there are emerging concerns about their dependence on non-renewable and fossil fuel-based materials, their vapour diffusion properties and possible off-gassing of VOCs that may unduly affect indoor air quality.

ModCell is a company that specialises in the use of renewable, bio-based materials such as straw and timber that are vapour diffuse. There are number of benefits in using straw: it is a by-product of growing cereals, predominantly wheat, is a renewable resource, is vapour diffusible, has a very low embodied energy and consequently is low in embodied carbon. Timber and straw also have unique benefits over other materials; they not only have low carbon intensity, they also have the additional benefit of actually banking carbon into their fabric. This banking of carbon is achieved through the photosynthetic action of carbon dioxide absorption at a molecular level of plants. Photosynthesis is the means by which plants absorb atmospheric CO_2 and split it into its component atoms, retaining the carbon atom to make complex organic sugars that are the building blocks of the cellulose, hemicellulose and lignin that make up the cell walls of plants. The oxygen atoms are ejected back to the atmosphere as a by-product of the process. The process of photosynthesis is the building block for all lives on the planet, other than a very small group of species that thrive on chemosynthesis near the thermal stacks recently discovered at the boundaries of ocean tectonic plates.

The subject of this paper is to explain how the ModCell system of construction achieved Passivhaus Component Certification using the bio-based renewable materials timber and straw.

20.2 Background

The construction industry is a key European employer and contributor to the built environment affecting the quality of life and work of all EU citizens.

> 'Buildings use 40% of total EU energy consumption and generate 36% of greenhouse gases in Europe. The construction sector is on its critical path to decarbonise the European Economy by 2050, reducing its CO_2 emissions by at least 80% and its energy consumption by as much as 50%. As the replacement rate of the existing stock is very small (1–2% per year), acceleration is urgently needed. Simultaneously, this offers a unique opportunity for sustainable business growth, provided that products and related services for both new and refurbished buildings are affordable, non intrusive and of durable quality, in line with European Directives'.

Source: European Commission Energy-Efficient Buildings: Multi-annual roadmap for the contractual PPP under Horizon 2020.

The Passivhaus Component Certification came about as a result of a major European Eco-Innovation-funded research project called EuroCell. The Eco-Innovation EuroCell project has sought to provide a solution towards decarbonisation through the development of wider sector uptake of straw bale construction using ModCell: an innovative prefabricated low carbon cellulose-based panel building system designed for use in a wide variety of construction sectors, including housing, schools and retail projects.

Research on materials, product testing and development and monitoring of building performance has proven that buildings built using the ModCell system reduce CO_2 emissions and the cost of heating and cooling by up to 85 %. The inexorable rise in utility costs and the dependency of the EU on imported natural gas set a trend that continues to reinforce the market opportunity for new forms of super low carbon building systems like ModCell.

The EuroCell Project addressed current EU wide market barriers to the mainstream uptake of innovative systems of constructions such as ModCell. Barriers include the lack of product certification, warranty approval, scaling the manufacturing approach and limited market presence. The EuroCell project allowed for two significant certifications to be secured:

1. QMark Certification, that has allowed for what are called 'high street' mortgages to be provided for the purchase of homes built using the ModCell system
2. Passivhaus Component Certification

There are now over 20,000 completed Passivhaus buildings worldwide. As part of the EuroCell project, the ModCell system secured Passivhaus Component Certification involving the design of 25 standardised construction details all of which were modelled to confirm that they met the stringent Passivhaus criteria. Certification provides a guarantee of quality and proof that the system has extremely good thermal performance in the most challenging climate conditions. As such, any building built with ModCell using the standardised details will be highly energy efficient and comfortable, built from natural and renewable building materials. The ModCell system is the first prefabricated straw bale timber construction system to achieve Passivhaus Component Certification in the world, bringing together low embodied energy and carbon-positive materials, with high levels of energy efficiency and thermal comfort.

The principle contributors to the successful Passivhaus Component Certification were:

Oliver Style, Director, Progetic
Barcelona Office
C/Ciutat de Granada, 69b
08005 Barcelona
Spain
www.progetic.com
and
Craig White, Director, ModCell Ltd.
The Proving House
21 Sevier Street
Bristol
BS2 9LB
www.ModCell.com

More can be found out by clicking on http://euro-cell.org.

A similar consortium working on the use of bio-based materials went on to secure one of the first major Horizon 2020 research grants. The project is called IsoBio and more information can be found at http://isobioproject.com.

IsoBio is a radical approach to using natural construction materials at scale. The IsoBio project proposes an innovative strategy to bring bio-based construction materials into the mainstream. A key innovation consists of the use of pretreated bio-based aggregates for construction, which include insulation materials, hygrothermal and moisture buffering materials, binders, sol-gel and resins. IsoBio is supported by the European Union Horizon 2020 Programme, within the 'materials for building envelopes' call for energy-efficient buildings.

IsoBio will combine existing technologies in order to develop bio-based panels and renders with high insulating properties, low embodied energy, low embodied carbon and hygrothermally efficiency. The project objectives are twofold. From a scientific standpoint, IsoBio partners aim to assess and further advance the state of the art in bio-based insulation materials, hygrothermal and moisture buffering materials, binders, sol-gel and resins. From a technological perspective, the focus is on developing new sustainable construction materials that, through a series of prototype-level demonstrations, will show significant social, environmental and financial benefits in the building sphere. Thus, the ultimate goal of the project is to optimise the construction process and create more energy-efficient buildings that will lead to strengthen the competitiveness of the European construction sector in the field of "green" construction technologies.

The key materials used in the EuroCell project to achieve Passivhaus Component Certification are targeted for further enhancement to improve the performance of all of the components of the ModCell system.

20.2.1 ModCell Core Passiv: Components

The typical build-up of the ModCell system is shown in the illustration below. The system is certified using the brand name ModCell Core Passiv (Fig. 20.1).

The principle unique materials used as part of ModCell Core Passiv are:

Timber lightweight construction	Abbreviation	modcell®
Base details external wall on floor slab	Detail_9.0	helping you build a more sustainable future

Construction drawing - vertical section

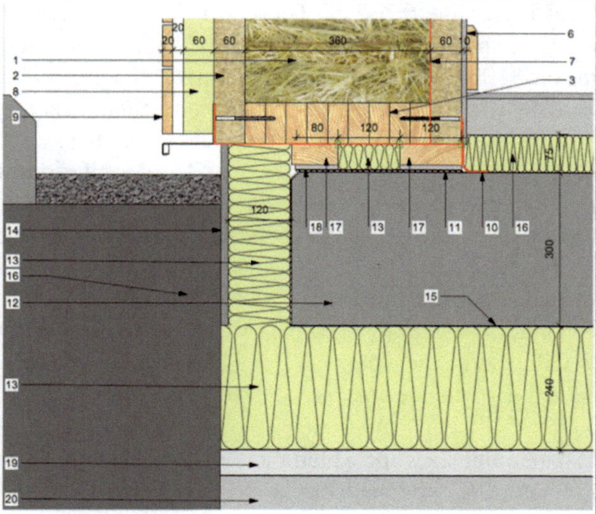

Typical external wall on floor slab connection

Materials used:
1 Straw insulation
2 Compressed straw board
3 Glulam timber
4–
5-
6 Lime plaster
7 Vapour and air-tight membrane
8 Wood fibre insulation
9 Rain screen timber cladding
10 Airtightness tape
11 Damp proof course
12 Reinforced concrete slab
13 Rigid PIR insulation
14 Viroc wood cement board
15 Damp proof membrane
16 Compressed wood fibre insulation
17 Timber sole plate
18 Grout (mortar)
19 Sand blinding
20 Rolled hard-core

Air tightness:
The vapour and air-tight membrane (7) provides the air tightness for each panel. The membrane meets DIN 4108, SIA 180 and Norm B8110-2. Installed in the factory and taped prior to the installation of the straw bales, the membrane is wrapped up the outside of edge each panel to form a compete enclosure to the inside and edges of each panel. The outer edge is then taped at every panel to panel joint. The compressed straw board is sealed with cardboard providing airtightness across its surface.

	From inside to outside	λ [W/(mK)]	Thickness [cm]		From top to bottom	λ [W/(mK)]	Thickness [cm]
Control Component: External Wall (EW_01)				Control Component: Floor Slab (FS_01)			
6	Lime plaster	0.700	1	10	Compressed wood fibre insulation	0.049	7.50
2	Compressed straw board	0.119	6	12	Reinforced concrete slab	2.300	30
1	Straw insulation	0.052	36	13	Rigid PIR insulation	0.024	24
2	Compressed straw board	0.119	6				
Control component:				Other materials (not included in the Control Components)			
				14	Viroc wood cement board	0.220	1.5
				17	Timber sole plate	0.130	-
				18	Grout (mortar)	1.000	-
				8	Wood fibre insulation	0.045	-

Fig. 20.1 An extract from the standardised pro forma used by the Passivhaus Institute for component certification

Straw bale
Compressed strawboard (CSB)

The system also utilises glue-laminated timber as the principle structural material and other sundry materials such as screws, vapour control membranes, etc.

Organic materials such as straw and reed have been used in construction for millennia. The invention of the baling machine in 1872 used twine to secure the straw in rectangular units. This led naturally to people using these straw bricks to form the components of a walling system. The system emerged with most rigour in America on the plains of Nebraska and combined limited timber supply with straw bales to create a systemised method of construction known as the Nebraskan method. As such, straw is a readily available by-product of growing cereals as well as being cost effective. The idea of prefabrication of straw panels has been around almost as long as the Nebraskan method of construction with patents being lodged for such systems as early as 1880. Up until the latter part of the twentieth century, they remained firmly lodged as ideas with no significant commercial application. In 1998, the inventors of the ModCell system, Craig White and Tim Mander developed a system of prefabrication that went on to be the formation of ModCell Limited in 2005. A patent for the system was filed in 2008 and awarded in 2009 as Patent No GB2457891B.

Straw is an excellent insulator through the nature of its dry cellulose stalks, whose form traps air between the stalks within the bale as well as within their internal cellular structure. Straw will conduct heat along its length, but the combination of trapped air and low conductivity provide for an accepted lambda of 0.052 W/(m^2 K) (Fig. 20.2).

Straw bales come in all shapes and sizes for which there is no absolute internationally adhered to standard. In the UK, the more commonly used construction bale is 1 m long, 45 cm deep and 36 cm tall. They can also come in much larger sizes up to 1.2 × 1.2 × 2 m—known in the UK as a Heston Bale. The smaller 1 m bale is becoming less available as farmers move towards more efficient baling methods including the very large round bales. Because conventions vary so much across the world, ModCell adopted the universal 3D coordinate system to describe a bale. This translates length, height and depth into x, y and z coordinates, where a typical bale is described as $x = 1000$ mm, $y = 450$ mm and $z = 360$ mm. It is important to note that the tolerance for the sizing of a bale should assume an accuracy of ± 5 mm and that they will compress, or settle, under load. ModCell has carried out slump and loading testing of bales to account for this in their construction (Fig. 20.3).

There has been some debate about how a straw bale conducts heat, and for the purposes of the Passivhaus Component Certification, the research produced by FASBA (Fachverband Strohballenbau Deutschland http://fasba.de) was relied upon by the Passivhaus Institute. Essentially heat is conducted along the length of stalk of straw more readily than from stalk to stalk that lie adjacent to each other. FASBA contends that a bale produced by a baling machine has strand orientation, which would influence its conductivity. As such the ModCell Core Passiv certification uses straw bales laid on edge in the wall construction so that $y = 450$ is orientated vertically (Fig. 20.4).

Fig. 20.2 A typical small bale

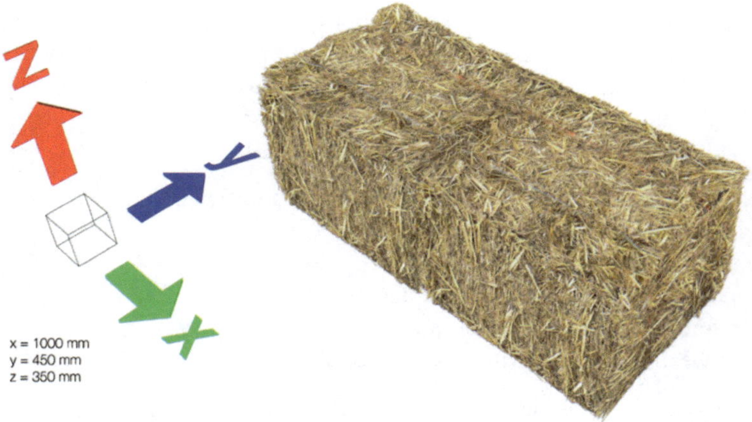

Fig. 20.3 A straw bale showing the convention of describing the bale using *x*, *y* and *z* coordinates

The other unique material used in the ModCell Core Passiv Certification is compressed strawboard (CSB), the product name is LigniCell, and it is supplied by the company Coobio Circular Materials.

CSB has a unique set of features:

Replaces the need for plasterboard
Is load-bearing
Reduces the need for stud work
Has exceptional acoustic performance

Fig. 20.4 The bale on the left is as it emerges from a baler. The bale on the right as installed in the ModCell Core Passiv wall so the *y* coordinate is vertical

Is fast to install
No special fixings required
Reduces waste and cost

The CSB panels can be used as partitions, sheathing boards in external wall structures, roof cassettes, door blanks and furniture. In ModCell Core Passiv, it forms both the internal and external sheathing boards of the wall and roof systems.

The following describes the process of CSB manufacture:

Application of heat and pressure to straw.
Heat causes naturally present moisture to be turned to steam.
Steam liquifies lignin in the cell walls.
Lignin is a long chain molecule.
Long chain molecules act as bonding agents.
Pressure forces materials together into a self-bonded whole.
Straw can become load-bearing board materials.

Lignin is an integral part of the cell walls of plants. Twenty-five percent of a plant's cell wall is made of lignin. After cellulose, lignin is the second most abundant renewable carbon source (Fig. 20.5).

The straw in CSB is strand orientated by the machine carding the straw into one direction prior to being rammed into the board. This strand orientation gives CSB its unique structural properties.

Over and above its unique strength, CSB provides much better thermal breaking performance than conventional sheathing materials. It also reduces the number of vertical joists in a wall construction to a minimum, reducing the frequency of potential thermal bridges to the edges of the ModCell prefabricated panel.

CSB comes in board sizes of 800 and 1200 mm and in two thicknesses of 40 and 60 mm and can be manufactured in lengths up to 3.2 m (Fig. 20.6).

CSB has Lambda value of 0.119 W/(m^2 K).

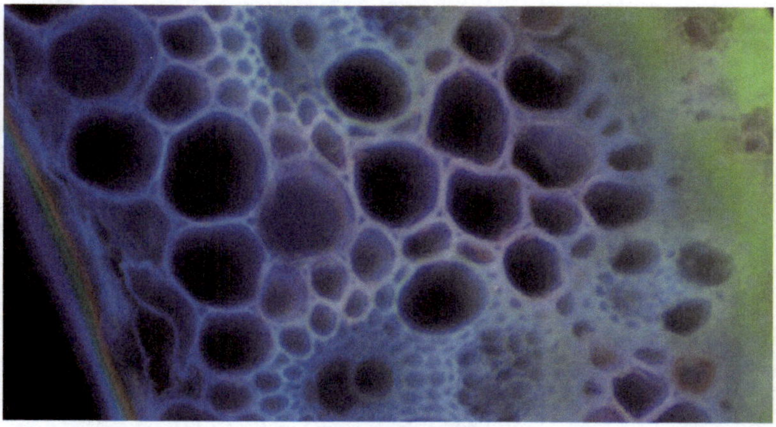

Fig. 20.5 Lignin, stained blue/purple in the cell wall of straw

Fig. 20.6 CSB as finally encapsulated in recycled paper and cross sections cut perpendicular to the straw strand orientation

The net result is a prefabricated building unit made predominantly from bio-based, carbon banking renewable materials, over 95% of the volume of which is made of straw (Figs. 20.7 and 20.8).

20.2.2 Airtightness

Airtightness is provided for by three means:

The CSB is airtight across its face.

The vapour control barrier built into every panel that sits behind the internal CSB sheets, providing a factory fitted continuous and protected barrier.

Localised airtightness tapes at panel to panel joints (Fig. 20.9).

Fig. 20.7 ModCell Core Passiv wall panel as certified as a Passivhaus Component

Fig. 20.8 CSB is used to sheath the construction inside and out, providing a thermal break over the more thermally conductive glulam structure

20.2.3 Thermal Bridging

The thermal bridging strategy for ModCell Core Passiv is to encapsulate all highly thermally conductive structural elements in CSB. This decouples the route for thermal bridging across most of the face of the panel. However, once modelling started, it became clear that panel edges, where the glulam structure is present, were bridging marginally higher than they should be. A number of strategies were looked at including introducing a thermal break in the glulam itself; however, the final strategy was to leave the panel in its simplest form and apply external 'zips' of additional and localised insulation. Initially, the Passivhaus Institute suggested the use of foam or expanded polystyrene-based materials to deal with this localised effect.

Timber lightweight construction External edge of external wall	Abbreviation Detail_5.0	modcell® helping you build a more sustainable future
Construction drawing - horizontal section		

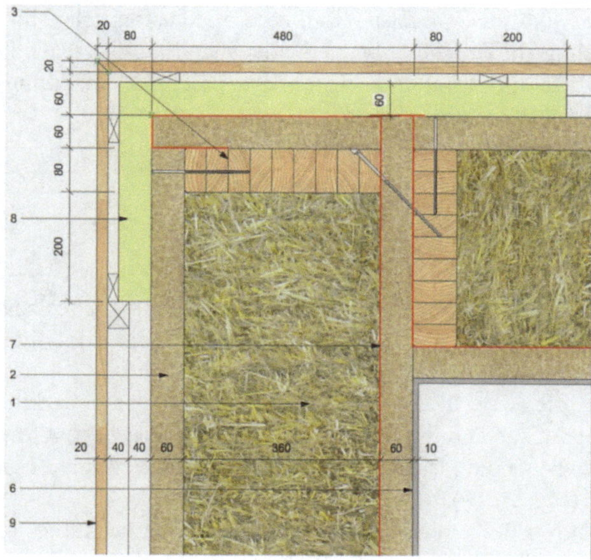

Typical external edge of external wall connection

Materials used:
1 Straw insulation
2 Compressed straw board
3 Glulam timber
4 Ventilated roof
5 Waterproof membrane
6 Lime plaster
7 Vapour and air-tight membrane
8 Wood fibre insulation
9 Rain screen timber cladding

Air tightness:
The vapour and air-tight membrane (7) provides the air tightness for each panel. The membrane meets DIN 4108, SIA 180 and Norm B8110-2. Installed in the factory and taped prior to the installation of the straw bales, the membrane is wrapped up the outside of edge each panel to form a compete enclosure to the inside and edges of each panel. The outer edge is then taped at every panel to panel joint. The compressed straw board is sealed with cardboard providing airtightness across its surface.

	From inside to outside	λ [W/(mK)]	Thickness [cm]
Control Component: External Wall (EW_01)			
6	Lime plaster	0.700	1
2	Compressed straw board	0.119	6
1	Straw insulation	0.052	36
2	Compressed straw board	0.119	6
Control component:			

	From inside to outside	λ [W/(mK)]	Thickness [cm]
Control Component: External Wall (EW_01)			
6	Lime plaster	0.700	1
2	Compressed straw board	0.119	6
1	Straw insulation	0.052	36
2	Compressed straw board	0.119	6
Other materials (not included in the Control Components)			
3	Glulam timber	0.130	-
8	Wood fibre insulation	0.045	6

Fig. 20.9 Airtightness VCL shown in red

This ran counter to the vision of only using bio-based insulation materials. In the end, we were able to use wood fibre insulation with a Lambda value of 0.045.

This moved the modelling forward to the point where the internal surface temperatures were universally at or higher than 17 °C, apart from a very small zone at the internal corners of rooms. It was assumed that the wood fibre board would have to be changed to EPS or foam insulation; however, this did not solve the problem.

Instead, what was needed was a counter-intuitive solution. The internal surface of the finished rooms needed to conduct heat across their face, to even out the temperature to 17 °C. Having spent a long time minimising conduction through using materials that decoupled the movement of heat across the depth of the wall, a highly conductive material was now required on the finished internal face. Initially, the Passivhaus Institute suggested a metal angle at each corner. This did not fit the vision for low carbon materials, and so the ModCell Core Passiv wall is certified using a clay, lime or gypsum plaster as its final finish. These highly conductive materials distribute the heat laterally across the surface of the wall to achieve the desired 17° temperature. Plasters have a lambda value of 0.7 W/(m^2 K).

20.2.4 *Other Components: Roofs and Openings*

By systemising the panels into prefabricated units, the wall and roof cassettes have an identical core thickness of 480 mm. With wall cassettes, the system is modified into a ladder structure. The glulam from the primary span elements and CSB is used as secondary cross spanning elements. The roof cassettes can be manufactured in a variety of widths, but are optimised at either 1.2 m or 2.4 m. This approach allows the means of manufacture to be similar, irrespective of whether the panels are used as wall or roof components (Fig. 20.10).

The illustrations below show a 2.4 m roof cassette. The primary glulams span left to right, and the shorter span is picked up by the CSB cross rails. This allows for the straw bales to be fitted into the chambers created (Fig. 20.11).

This 3D perspective shows the components of the roof cassette and the empty chambers awaiting the insertion of the straw bales. In essence, a wall panel and roof panel are the same, other than for the frequency of the CSB cross rails. Sloped roof profiles simply have modified end sections to accommodate different angles of connection (Fig. 20.12).

The illustration below shows the typical floor connection. This proved the most complex connection to resolve as the floor is simply supported on the wall panel below. This means that the floor joists, shown in the smaller illustration, top right, are in effect acting as conduction elements. This is the only location in the certification where EPS insulation had to be used rather than bio-based insulation materials (Fig. 20.13).

The illustrations below give a flavour of the number of details produced to achieve Component Certification, 25 in all (Fig. 20.14).

20.3 Discussion and Conclusions

ModCell Core Passiv is an optimised whole building system that delivers excellent performance in use, to reduce energy consumptions and CO_2 emissions to a minimum while maintaining a comfortable internal environment year-round. ModCell Core Passiv not only has a low carbon footprint, it also successfully deploys the

Timber lightweight construction Roof connection detail, eaves	Abbreviation Detail_4.0	modcell helping you build a more sustainable future

Construction drawing - vertical section

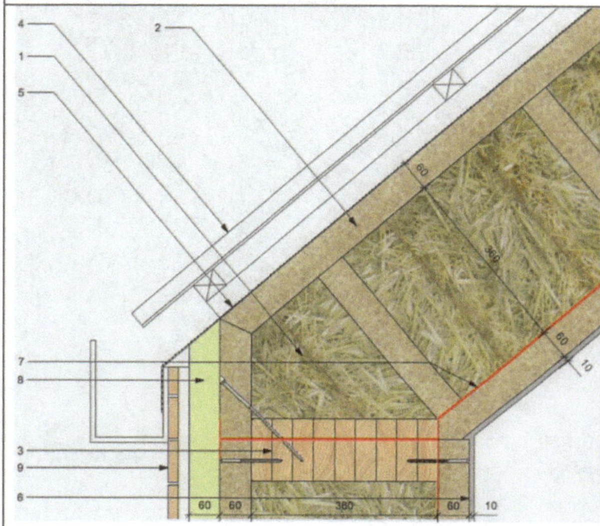

Typical eaves roof connection between roof and external wall

Materials used:
1 Straw insulation
2 Compressed straw board
3 Glulam timber
4 Ventilated roof
5 Waterproof membrane
6 Lime plaster
7 Vapour and air-tight membrane
8 Wood fibre insulation
9 Rain screen timber cladding

Air tightness:
The vapour and air-tight membrane (7) provides the air tightness for each panel. The membrane meets DIN 4108, SIA 180 and Norm B8110-2. Installed in the factory and taped prior to the installation of the straw bales, the membrane is wrapped up the outside of edge each panel to form a compete enclosure to the inside and edges of each panel. The outer edge is then taped at every panel to panel joint. The compressed straw board is sealed with cardboard providing airtightness across its surface.

	From inside to outside	λ [W/(mK)]	Thickness [cm]
Control Component: External Wall (EW_01)			
6	Lime plaster	0.700	1
2	Compressed straw board	0.119	6
1	Straw insulation	0.052	36
2	Compressed straw board	0.119	6
Control component: External Wall (EW_01)			

	From inside to outside	λ [W/(mK)]	Thickness [cm]
Control Component: Roof (R_01)			
6	Lime plaster	0.700	1
2	Compressed straw board	0.119	6
1	Straw insulation	0.052	36
2	Compressed straw board	0.119	6
Other materials (not included in the Control Components)			
3	Glulam timber	0.130	-
8	Wood fibre insulation	0.045	6

Fig. 20.10 Eaves detail of sloped roof

carbon banking capability of materials derived from a photosynthetically powered bio-based materials set to store carbon in its fabric. This carbon banking means that a Passivhaus project built using ModCell Core Passiv would bank 200 kg of CO_2 per m^2 of external wall and roof elevations. This is more than is emitted through the finding, processing making and transporting all of the materials used to make the panels (Fig. 20.15).

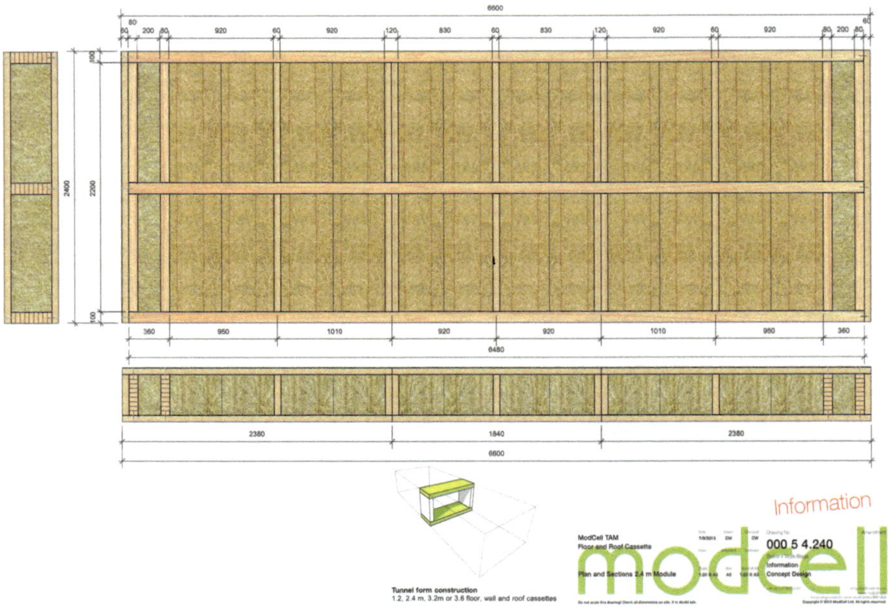

Fig. 20.11 Plan and section of roof cassette

Fig. 20.12 Perspective of roof cassette module

The illustration above shows the amount of atmospheric CO_2 that was banked into the prototype Balehaus @ Bath. It includes timber at 742 kg m^3 and straw at 211 kg/m^3. If compressed straw board is included, the figure rises to almost 50 tonnes of CO_2 per typical 88 m^2 house.

Next steps. The product development will not stop with the currently certified ModCell components. Under the Horizon 2020 research project IsoBio, the ambition is to improve the performance of straw based insulation. This will be achieved by blowing straw. The first tests of this are underway. This will allow for the u-Value

Timber lightweight construction Ceiling supported by external wall	Abbreviation Detail_8.0	modcell® helping you build a more sustainable future
Construction drawing - vertical section		

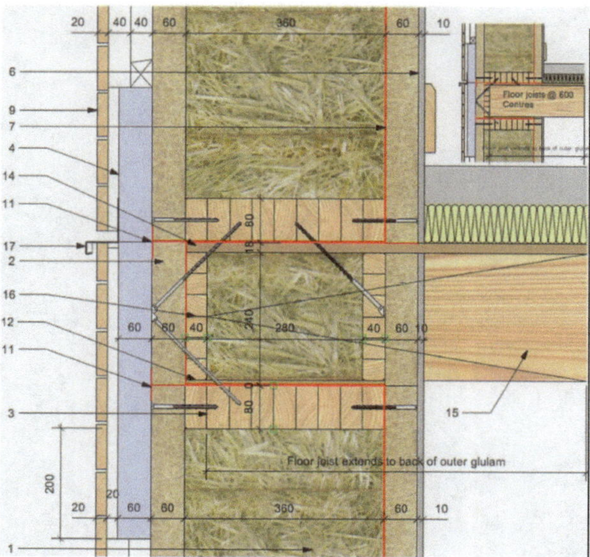

Typical ceiling supported by external wall connection

Materials used:
1 Straw insulation
2 Wood fibre insulation
3 Glulam timber
4 EPS insulation
5 -
6 Lime plaster
7 Vapour and air-tight membrane
8 -
9 Rain screen timber cladding
10 -
11 Airtightness tape
12 OSB panel (10 mm)
13 -
14 OSB panel (18 mm)
15 Solid timber floor joist (240 x 40mm at 600mm centres)
16 Glulam edging to floor cassette
17 Aluminium flashing

Air tightness:
The vapour and air-tight membrane (7) provides the air tightness for each panel. The membrane meets DIN 4108, SIA 180 and Norm B8110-2. Installed in the factory and taped prior to the installation of the straw bales, the membrane is wrapped up the outside of edge each panel to form a compete enclosure to the inside and edges of each panel. The outer edge is then taped at every panel to panel joint. The compressed straw board is sealed with cardboard providing airtightness across its surface.

	From inside to outside	λ [W/(mK)]	Thickness [cm]		From bottom to top	λ [W/(mK)]	Thickness [cm]
Control Component: External Wall (EW_01)				Control component: Internal floor (IF_01)			
6	Lime plaster	0.700	1	15	Timber floor joist	0.130	-
2	Compressed straw board	0.119	6	14	OSB panel 18 mm	0.130	-
1	Straw insulation	0.052	36				
2	Compressed straw board	0.119	6				
Control component:				Other materials (not included in the Control Components)			
				3	Glulam timber	0.130	4
				4	EPS insulation	0.037	-
				12	OSB panel 10 mm	0.130	-

Fig. 20.13 Upper floor connection

to be improved by reducing the density of the straw used from 135 kg m^3 as a bale to an anticipated 100 kg m^3 when blown. This will simplify the process of manufacture and improve the u-Value, requiring less straw so that thinner wall build-ups can be developed.

The opportunity now exists to deliver high-performance buildings that reduce CO_2 emissions in use that have a lower carbon footprint as well as being predominantly

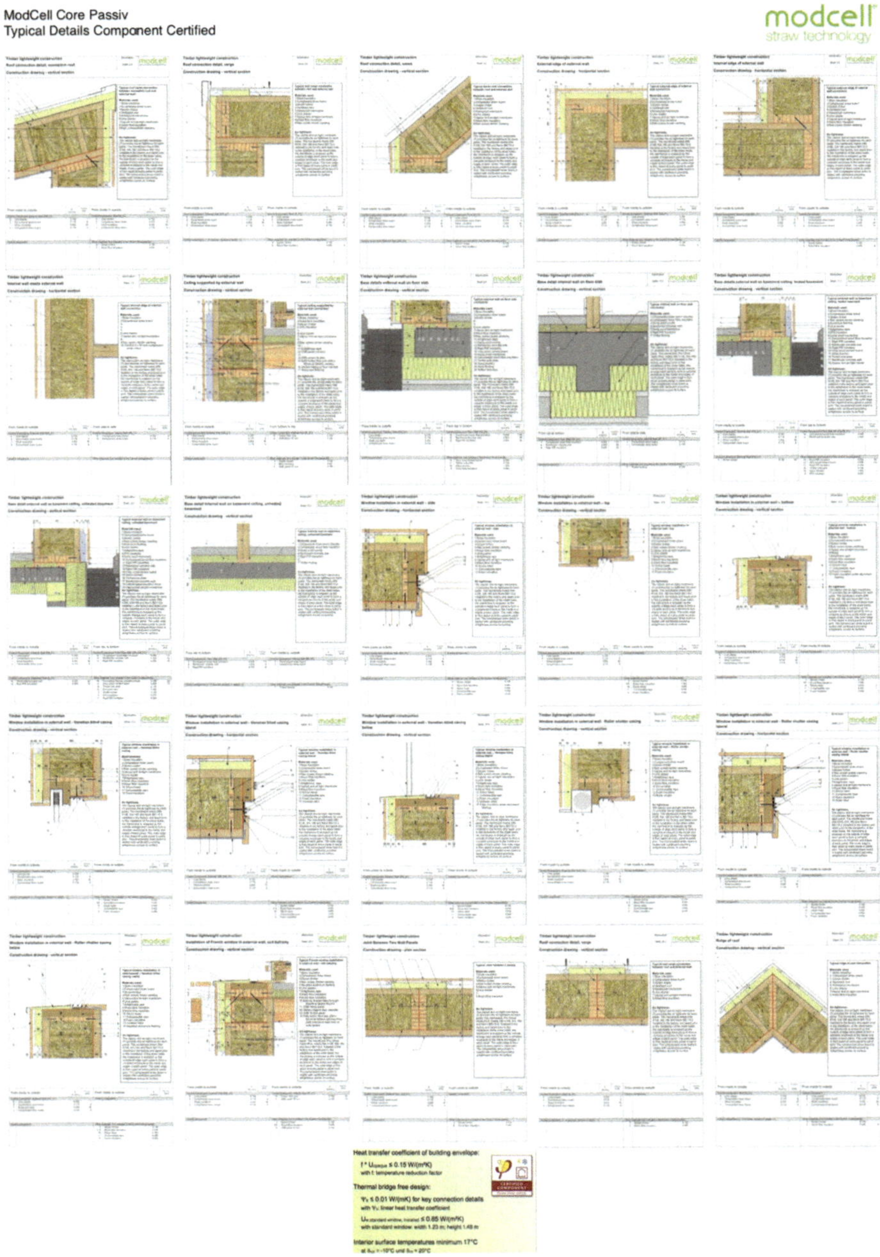

Fig. 20.14 Selection of standardised details

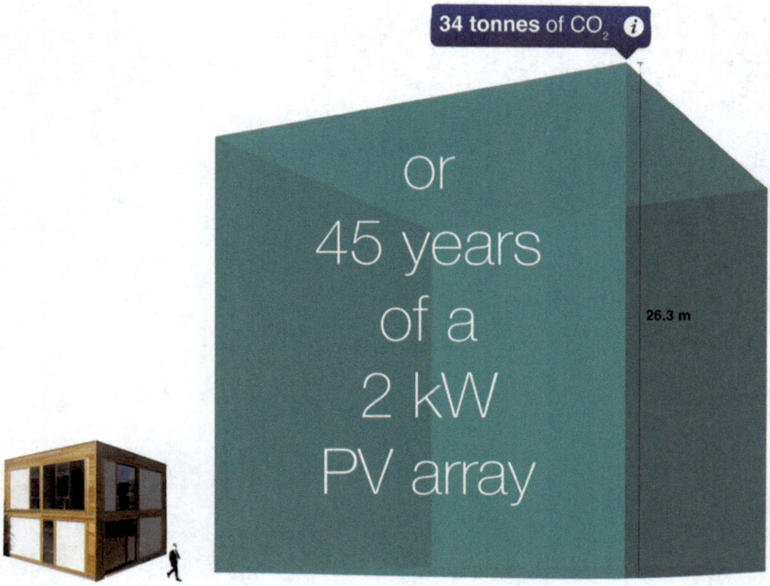

Fig. 20.15 34 tonnes of atmospheric CO_2 as a volume of gas banked in the prototype Balehaus @ Bath

made of banked carbon. The system is scaleable to volume production, and ModCell now has a secured pipeline of work that will rise to 200 homes per year by 2017. We hope that a significant proportion of these will be Passivhaus Certified projects.

20.3.1 A Vision for the Creation of the Sixth Carbon Sink

Currently there are five recognised carbon sinks:

Atmosphere
Oceans
Geology
Biosphere
Soils

Nature has balanced these five sinks using photosynthesis as the means of capturing and releasing carbon. Humankind, through the burning of fossil fuels, is releasing carbon in the form of carbon dioxide into the atmosphere at rate higher than nature has done since long past geological times. Carbon dioxide is what is known as a greenhouse gas, which means it has the ability to retain heat. This increase of CO_2 is changing the performance of our atmosphere leading to it warming over time. It is proving increasingly difficult to regulate these emissions, and so the opportunity to imagine a

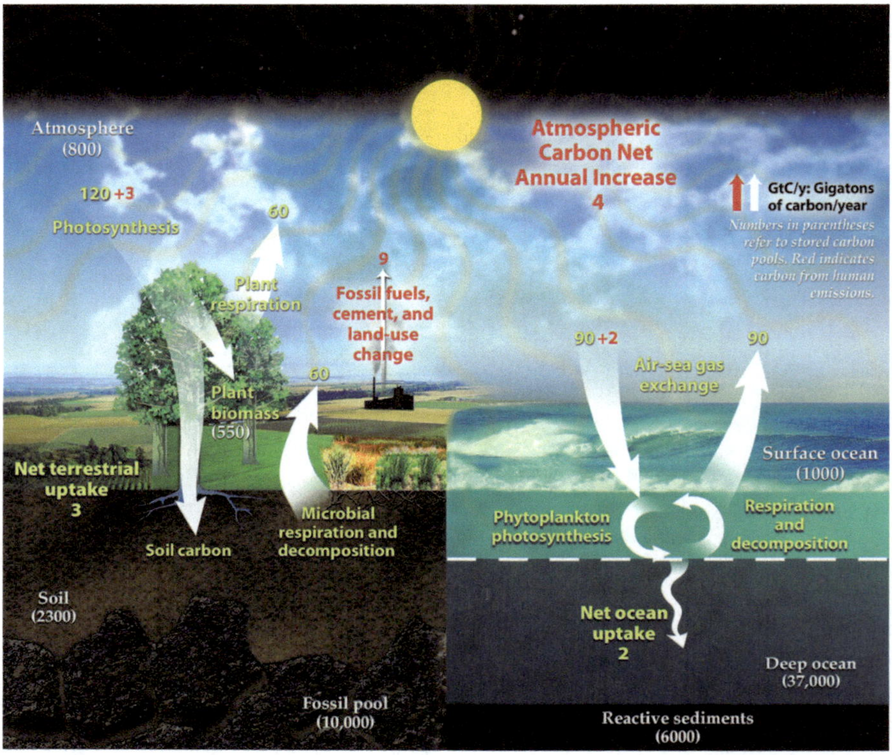

Fig. 20.16 The five natural carbon sinks are atmosphere, ocean, geology, biosphere and soil

man-made sixth carbon sink is intriguing. If volume of bio-based materials were to be increased in construction in all there forms, then it could legitimately be claimed that the built environment could be considered a sixth carbon sink. For example, if only 10% of the houses built each year in the UK were to switch to a ModCell like form of construction, the UK could bank 500 gigatonnes of atmospheric CO_2 per annum.

So why the built environment as a carbon sink?

Buildings in the built environment are assets that sit on our national balance sheet. The built environment is made up of insured and maintained physical assets; consequently, the carbon stored in buildings can be genuinely accounted for rather than simply an intangible offset currently used in carbon trading. Uniquely, carbon stored in our built environment increases in value over time as the value of the property from which it is made increases. If the buildings last more than 100 years, the global warming potential of the original CO_2 emissions from which plants sequestered the carbon is effectively set to zero. The opportunity to build capital assets using the sequestered carbon in renewable materials could offer a systemised way to safely provide an alternative form of carbon capture and storage without the need for economic subsidy (Fig. 20.16).

20.4 Summary

The need to reduce carbon dioxide emissions from the operational energy use in buildings is becoming more and more pressing as we seek to mitigate the effects of climate change. As we do so, however, the ratio of emissions embodied in the materials we use to make our buildings, compared to operational emissions, will become the next target for decarbonising our built environment. The whole building ModCell system of construction secured PassivHaus Component Certification in February 2015 using straw as its primary material. The system uses a unique set of renewable materials that use the power of photosynthesis to bank carbon into their fabric. The amount of carbon dioxide banked into the ModCell system is more than is emitted through its manufacture and transportation, resulting in a carbon negative footprint. This unique approach of working within the natural carbon cycle, rather than upon it, means the journey towards a zero carbon built environment is now achievable. This paper describes how we might create a sixth carbon sink and store carbon in our built environment, rather than emitting it to atmosphere.

References

Wall, K., Walker, P., Gross, C., White, C., & Mander, T. (2012). Development and testing of a prototype straw bale house. *Proceedings of the Institution of Civil Engineers: Construction Materials, 165*(6), 377–384.

Lawrence, M., Heath, A., & Walker, P. (2009a). Determining moisture levels in straw bale construction. *Construction and Building Materials, 23*(8), 2763–2768.

Walker, P., Thomson, A., & Maskell, D., 2016. Straw bale construction. In: Harries, K. A. & Sharma, B., (Eds.), *Nonconventional and vernacular construction materials* (p 127). Woodhead Publishing. (Woodhead Publish Series in Civil and Structural Engineering; 58)

Thomson, A., & Walker, P. J. (2013). Condition monitoring and durability assessment of straw bale construction. In L. Braganca, M. Pinheiro, & R. Mateus (Eds.), *Portugal SB13: contribution of sustainable building to meet EU 20-20-20 Targets* (pp. 791–798).

Lawrence, M., Heath, A., & Walker, P. (2009b). Monitoring of the moisture content of straw bale walls. In R. J. Howlett, C. J. Lakhmi, & H. L. Shaun (Eds.), *Sustainability in Energy and Buildings: Proceedings of the International Conference in Sustainability in Energy and Buildings (SEB'09)* (Vol. Part 3, pp. 155–164). Berlin: Springer.

Thomson, A., & Walker, P. J. (2014). Durability characteristics of straw bales in building envelopes. *Construction and Building Materials, 68*, 135–141.

Maskell, D., Walker, P. J., Thomson, A., Gross, C. D., Beadle, K., & Mander, T. (2014). Structural development and testing of a prototype house using timber and straw bales. *ICE Proceedings Structures & Buildings, 168*(1), 67–75.

Chapter 21
The Replacement of Wood or Concrete in Construction Projects: An Industrial Case Study Demonstrating the Benefits of Intrusion Moulded Waste Plastic

Howard Waghorn and Paul Sapsford

Abstract Intrusion moulding technology can convert waste plastic into a strong, lightweight construction material, for use as an alternative to wood or concrete. Using a combination of published research, independent testing and an industrial case study to demonstrate the benefits, this chapter will propose that the construction industry should consider intrusion moulded waste plastic as a first choice construction material.

Recycled waste plastic is particularly suitable in applications involving contact with earth and/or water as the material does not require chemical preservatives and is resistant to moisture, micro-biological attack and acid soil. From an international perspective, recycled plastic is an option in areas where wood is subject to termite attack and avoids the problems of salt weathering on concrete in the Middle East.

A major benefit of intrusion moulding is the ability to create large cross-sectional and non-uniform shapes, which results in recycled plastic being used in applications as diverse as biofilter floors, underground utility chambers and drop kerbs. The robust nature of the intrusion moulding process also allows different polymers to be blended together in order to obtain the best combination of rigidity and flexibility for specific applications.

The industrial case study refers to the Ecocrib retaining wall, which is manufactured using a blend of post-industrial waste plastics. Ecocrib has been certified by the British Board of Agrément to achieve a Design Service Life in excess of 120 years, which is similar to steel reinforced concrete crib walls and twice that of a wooden alternative; however, the recycled plastic does not require the marine transport or chemical preservatives of timber, whilst avoiding the CO2 emissions of concrete production. Moreover, the Ecocrib headers and stretchers are far lighter than the concrete alternatives, resulting in reduced cost of transport and installation.

Yesterday's waste can become tomorrow's construction material.

H. Waghorn • P. Sapsford (✉)
Hahn Plastics Limited, PO Box 4, Rake Lane, Swinton, Manchester M27 8LJ, UK
e-mail: paul.sapsford@hahnplastics.co.uk

M. Dastbaz et al. (eds.), *Building Information Modelling, Building Performance, Design and Smart Construction*, DOI 10.1007/978-3-319-50346-2_21

21.1 Introduction

As the search for more cost-effective, sustainable construction increases pace, there is a need to raise awareness of the benefits available from the use of alternative materials. Using a combination of published research, independent testing and a specific case study, this chapter will propose that the construction industry should consider intrusion moulded recycled plastic as the first choice material in preference to wood or concrete for an increasing range of applications.

21.2 The History and Use of Plastics

World production of plastics in 2014 amounted to 260 million tonnes with demand in Europe (EU28 plus Norway and Switzerland) being 47.8 m tonnes (PlasticsEurope 2015). Of this latter figure, 20.1% was used in building and construction.

Polyvinyl chloride (PVC) is the main plastic used by the construction industry, with the polymer being used in the production of pipes and window frames. PVC was the third most widely used polymer within Europe, after polyethylene (PE) at 29.3% and polypropylene (PP) 19.3%. These latter two polymers are little used by the construction industry.

The limited use of plastics within construction can be partly explained by the timelines of polymers as compared with other structural materials. The Concrete Network (n.d.) notes that Ancient Romans used a material that is "remarkably close to modern cement" in order to construct the Colosseum and credits Joseph Aspdin as the inventor of modern Portland cement in 1824. In comparison, the British Plastics Federation (2016) notes that polyethylene was discovered by Fawcett and Gibson at ICI more than 100 years later in 1933, with production of high density polyethylene (HDPE) not occurring in the UK until 1955 and polypropylene production by the Italian firm Montecatini following in 1957.

21.3 Plastic Recycling

The major attraction of recycled plastic is that it provides an opportunity to use a material with characteristics that are little changed from those of virgin polymers but at a much reduced cost. Plastic is employed extensively in single-use packaging, with 39.5% of European demand for plastic in 2014 stemming from the packaging industry, according to PlasticsEurope. This latter organisation also reported that 25.8 m tonnes of plastic entered the European waste stream in 2014, of which 29.7% was recycled; 39.5% was incinerated and 30.8% went to landfill. PlasticsEurope noted that the near 8 m tonnes of landfilled plastic would have required around 100 m barrels of oil in its production.

The propensity to record waste by reference to weight, rather than volume hides the extent of the landfill problem as regards plastic. Converting imperial figures produced by the Environment Protection Agency in the USA indicates that compacted HDPE bottles (typically used for milk) weigh 0.093 tonnes per cubic metre. This figure contrasts sharply with the 1.3 tonnes per cubic metre for waste concrete or mortar as reported by EAUC, (The Environmental Association for Universities and Colleges, 2016), with the compacted HDPE bottle figure indicating that waste plastic consumes nearly 14 times the landfill space required by the equivalent weight of waste concrete or mortar.

Recycled single polymers are widely used in injection moulding for the production of thin walled items such as buckets. However, the use of waste plastic to produce large cross-sectional items for the construction industry in place of wood or concrete requires the employment of other plastic moulding processes, such as intrusion moulding.

21.4 Intrusion Moulding

The intrusion moulding technique is similar to the more widely used extrusion moulding, in that molten plastic is forced under pressure to shape the end product. With extrusion, the material is forced through a die to produce items with a continuous profile, such as PVC window frames. In the case of intrusion moulding, the material is forced into and allowed to cool in a mould, enabling the manufacture of a wide range of non-uniform shapes, such as drop kerbs, or more complex shapes, e.g. underground chambers for utilities purposes.

The cooling requirement for large cross sections available from intrusion moulding results in long cycle times. This potential problem is overcome by attaching a number of moulds to one extruder, either by means of a carousel or a manifold, with an empty mould being filled as others cool. One benefit from this situation is that a range of different items or a family of related products can be manufactured concurrently, rather than having to wait for completion of a batch before replacing the die, as would be the case with extrusion.

Unlike injection moulding which mainly uses single polymers, resulting in a choice between flexibility and rigidity, intrusion moulding can combine the benefits of different polymers. For example, polypropylene provides good rigidity and is easy to process, whilst the various polyethylenes offer flexibility and high impact strength. Styrenes can be incorporated to provide greater rigidity but result in increasing brittleness and fillers can be used to reduce cost but with a consequent increase in weight. The key with intrusion moulding is to find the correct combination of polymers appropriate for a specific application.

The graph of figures from WRAP (2008) indicates that domestic waste includes seven types of plastic, creating technical and cost problems in recycling the material. Intrusion moulding allows mixed plastics and some contaminants such as paper and wood to be processed due to the large material inlets into the moulds, potentially leading to the conclusion that intrusion might be the solution to the problem of land-filling post-consumer waste plastic. However, whilst it is possible to manufacture items using post-consumer waste plastic, the different melt temperatures of the various polymers can result in surface finish problems. More importantly from a construction viewpoint, the high melt temperature of polyethylene terephthalate (PET), for example, could result in inclusions of unmelted material within the moulded product, whilst the gases generated by the various polymers, moisture, and contaminants can produce large and uneven voids within the finished product, resulting in hidden weaknesses.

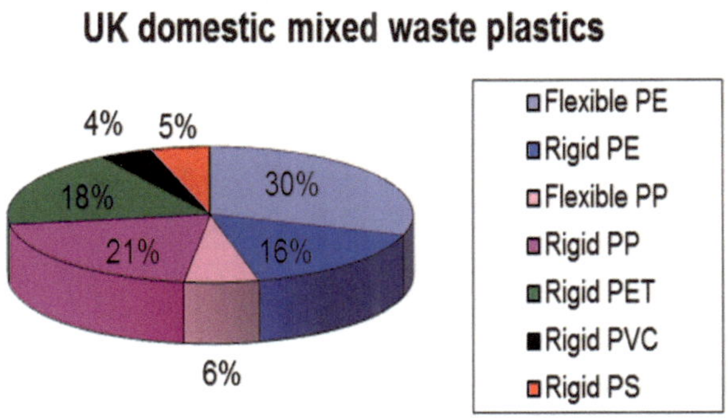

By blending specific polymers having known melt flow characteristics it is possible to utilise the attributes of different plastics in order to improve stiffness and flexibility whilst increasing creep resistance and impact strength. The use of blowing agents and other additives can prevent the creation of uneven gaseous voids within the product, creating a more uniform honeycomb, as shown in the photograph of a Hahn Plastics recycled plastic profile cut with a domestic circular saw.

In order to avoid the problems of domestic waste plastic and to ensure structural integrity of the finished product, extensive testing is used to arrive at proprietary blends of post-industrial waste plastic. Post-industrial waste includes end-of-life products such as wheelie bins and industrial packaging such as crates and drums, with the benefit that the source and melt flow of the polymer is usually known. The actual blends used are commercially sensitive; however, an indication of the properties of blended polymers can be seen in the following test results from four separate blends produced by I-plas, the forerunner to Hahn Plastics Ltd. in July 2010. Testing was undertaken by Independent Polymer Technology Ltd:

Average tensile test properties for four materials

	Blend 1	2	3	4
Emod (MPa)	831.42	1444.64	1143.40	886.08
Sec Mod (MPa)	413.14	756.06	527.22	354.12
Yield Stress (MPa)	12.42	15.26	12.60	8.50
Yield Strain (%)	5.06	2.58	2.85	2.40
Max Stress (MPa)	13.24	15.26	12.10	8.72
Strain at Max Stress (%)	7.94	2.64	2.62	2.70
Elongation at Break %	13.94	2.76	2.66	2.76

In the above results, modulus values were recorded as the gradient of a line between 0.05% strain and a 0.25% strain as specified in ISO527-1 and designated Emod. A second secant modulus Sec Mod was recorded between 0.5and 1%.

A second technical report produced for I-plas, by Hodzic (2010) examined the potential effect of UV radiation and microbiological attack in recycled plastic components. Four different components were tested, including two that had been exposed to outdoor environmental conditions for 8 years, with continuous contact with soil during that period in one case. The conclusion to the report noted "both strength and modulus, as the main parameters of the material's load bearing performance, have shown similar patterns in all systems and any reduction in properties was constrained in the outer layer of 3mm thickness. Also any UV degradation would only affect the surface and not to a significant degree. The recycled polymer components thus preserve their load bearing capacity under normal environmental conditions if designed according to the standards".

In a paper "Development of recycled polymer composites for structural applications", Hugo and others (2011) investigated the mechanical and structural properties of a proprietary blend of recycled polymers from the same source as the above tests, using a range of different fillers. The study concluded that the "addition of small quantities of mica to glass fibre reinforced blends exhibited a significant synergy in tensile strength and modulus".

Hahn Plastics has its foundations in Intruplas Ltd., which was incorporated in 2000 to demonstrate the technical and commercial viability for intrusion moulding of waste plastic. Amongst the products developed by Intruplas were revetments used by British Waterways during construction of the Millennium Ribble Link canal. This history of polymer testing and product development led to the creation of the Ecocrib retaining wall system.

21.5 Ecocrib Case Study

Crib walls have been in use since long before the invention of plastics, with Missouri University of Science and Technology noting that in 1919, precast concrete elements began being cast for use in crib walls in the Cleveland USA area. Prior to the development of an intrusion moulded recycled plastic offering, crib walls were made from concrete or wood.

In the UK, the only British Board of Agrément (BBA) accredited wooden crib wall is Permacrib (BBA, 2016), manufactured using Radiata Pine, which is impregnated with a high concentration of copper azole preservative. Radiata pine is a major plantation species in the southern hemisphere (New Zealand, Australia and Chile) and therefore requires marine transport for use in other parts of the world. A European Commission (2013) report noted that emissions from marine transport account for 3% of global greenhouse gases and that international shipping was responsible for around 800 m tonnes of CO2 emissions in 2010, with this figure expected to increase to 5% by 2050.

The alternative to wooden crib walls was previously concrete. According to a National Ready Mixed Concrete Association (NRMCA, 2012) report, an average of 927 kg of CO2 is emitted for every 1000 kg of Portland cement produced in the USA.

The Ecocrib retaining wall is manufactured using intrusion moulded recycled plastic. The product was granted a BBA certificate in 2012 (BBA, 2012), with an assessed Design Service Life of 120 years—twice that of the wooden crib wall.

Ecocrib header widths are manufactured from 500 mm to 3000 mm and without the need for steel reinforcement, the Ecocrib components are far lighter than their concrete equivalent, with the heaviest item weighing 16.5 kg. In comparison, a 1500 mm concrete header component from the New Zealand company Gibbons Contractors Ltd. (2016) weighs 70 kg.

In 2013, (Ecocrib, 2016) Ecocrib was used during construction of additional lanes for the A14, which was the first example of Ecocrib being used for a major highways project. The Ecocrib mass gravity wall now supports over 400 linear metres of the A14 east bound carriageway, having been designed to support an environmental noise fence; withstand heavy goods vehicle loadings of 20 kN/m^2 and be in accordance with BSEN 1997-1:2004 Eurocode 7.

Ecocrib has been used by major UK housebuilders and was selected for the Center Parcs village at Woburn Forest, where the cut and fill strategy of the structural engineer required the construction of significant height and length retaining walls to support the Plaza hotel car park and a retaining bridge. Site-won fill material (Woburn sand) was used between the horizontal layers of geogrid reinforcement, thus helping to further reduce waste in addition to the 75 tonnes of plastic waste that was diverted from landfill through the use of Ecocrib.

21.6 Other Intrusion Moulded Waste Plastic Products

The Ecocrib product is typical of the use that can be made of intrusion moulded waste plastic, in that the end product itself was not new but replaced wood or concrete with an alternative material that might have been the first choice, had plastic been in existence when the product was first designed.

Other similar products have been designed and tested with interesting results. For example, work undertaken by Professor Chris Gorse of Leeds Met (now Beckett) University (Gorse, 2013) on an intrusion moulded recycled plastic kerb indicated that the kerb had an average Pendulum Test Value (PTV) of 49, giving the kerb a low slip potential classification. On sanding the recycled plastic kerb smooth to simulate wear, the product became more slip resistant, with the University recording an average PTV result of 64 as compared with 0–24 for high slip potential.

As part of the same exercise, the University also undertook and filmed some simple tests to compare the results from dropping concrete and recycled plastic kerbs.

Concrete kerb dropped from 0.5 m

The report concluded “We dropped concrete kerbs and edgings from approximately 200 and 500mm high onto a concrete surface, as expected, the concrete kerbs all fractured. R3 recycled plastic kerb was dropped from 200mm, 500mm, 1m and 2ms high. No deformation was evident at 200, 500 and 1m and at the higher height the kerb was scuffed on one corner. The scuff on the corner could have been repaired by light sanding of the surface”.

Recycled plastic kerb dropped from 2 m

The brittle nature of concrete kerbs, which results in damage during transit and installation again raises the question as to whether concrete would have been used for kerbing had intrusion moulded recycled plastic been available at an earlier date.

Intrusion moulded recycled plastic components can be sawn, drilled and screwed in exactly the same manner as wood and unlike their concrete counterparts do not create harmful dust when cut on site. No waste is created during the manufacture or installation as all surplus material can be reprocessed to form new profiles. When an intrusion moulded product eventually reaches the end of its useful life, it can be recycled again.

As recycled plastic does not rot, is resistant to termite attack and requires no maintenance, the original plastic lumber applications are now being expanded into products installed on a worldwide basis, such as grain store floors, cable-troughs and biofilter raised flooring.

21.7 Conclusion

Wood and concrete remain important materials for construction projects, and there is no suggestion that recycled plastic will replace these items entirely. However as demonstrated above, the development of more sustainable, longer-lasting, lighter, stronger intrusion moulded products provides the opportunity to question whether traditional materials are the most cost-effective choice in every application, or whether recycled plastic alternatives would have been the first choice, had plastic been available at an earlier date.

References

BBA. (2016). Agrément Certificate Permacrib 95/3115 p7.

BBA. (2012). Agrément Certificate Ecocrib 12/H194 p5.

British Plastics Federation. (2016). *A history of plastics* [Online]. Retrieved April 2016, from http://www.bpf.co.uk/plastipedia/plastics_history/default.aspx

Concrete Network. (n.d.). *Timeline of concrete and cement history* [Online]. Retrieved May 2016, from http://www.concretenetwork.com/concrete-history/

Ecocrib. (2016). *Case studies* [Online]. Retrieved April 2016, from http://www.ecocribwall.co.uk/case_studies.html

Environment Protection Agency—Standard Volume-to-Weight Conversion Factors, Appendix B, p. 60. Personal communication with a representative from the American Plastics Council. November 2, 1995.

Environmental Association for Universities and Colleges. (2016). *Conversion factors for calculation of weight to volume* [Online]. Retrieved from http://www.eauc.org.uk/page.php?subsite=waste&page=conversion_factors_for_calculation_of_weight_to_vo

European Commission. (2013). *Time for international action on CO2 emissions from shipping*.

Gibbons Contractors Ltd. (2016). *Gibbons Crib Wall Components p5* [Online] Retrieved May 2016, from http://www.gibbonscontractors.co.nz/gibbons-crib-walls-xidc108107.html

Gorse, C., et al. (2013). *R3 products recycled road Kerbs skid testing and evaluation*. Leeds: Leeds Metropolitan University.

Hodzic, A. (2010). *Technical report on the potential effects of UV radiation and microbiological attack in recycled plastic components manufactured at Lynwood Products Ltd (i-plas)*. Sheffield: The University of Sheffield, Faculty of Engineering.

Hugo, A.-M., et al. (2011). Development of recycled polymer composites for structural applications. *Plastics Rubber and Composites*. August 2011.

Missouri University of Science and Technology. *Crib Walls* [Online] Retrieved from web.mst.edu/~rogersda/umrcourses/ge441/online_lectures/.../GE441-Lecture6-2.pdf

NRMCA. (2012). *Concrete CO2 Fact Sheet*. Silver Spring MD: National Ready Mixed Concrete Association NRMCA publication number 2PC02 February 2012 from a survey of Portland Cement Association members.

PlasticsEurope. (2015). *Plastics—The facts 2015. An analysis of European plastics production, demand and waste data*. Brussels: PlasticsEurope.

WRAP. (2008). *Domestic mixed plastics packaging waste management options*. Banbury: Waste & Resources Action Programme.

Index

M. Dastbaz et al. (eds.), *Building Information Modelling, Building Performance, Design and Smart Construction,* DOI 10.1007/978-3-319-50346-2

H

I

O

P

Q

R